Neural, Novel & Hybrid Algorithms for Time Series Prediction

Neural, Novel & Hybrid Algorithms for Time Series Prediction

Timothy Masters

JOHN WILEY & SONS, INC.
New York • Chichester • Brisbane • Toronto • Singapore

Publisher: Katherine Schowalter
Editor: Diane D. Cerra
Managing Editor: Micheline Frederick

Designations used by companies to distinguish their products are often claimed as trademarks. In all instances where John Wiley & Sons, Inc. is aware of a claim, the product names appear in initial capital or all capital letters. Readers, however, should contact the appropriate companies for more complete information regarding trademarks and registration.

This text is printed on acid-free paper.

Copyright © 1995 by John Wiley & Sons, Inc.

All rights reserved. Published simultaneously in Canada.

This publication is designed to provide accurate and authoritative information in regard to the subject matter covered. It is sold with the understanding that the publisher is not engaged in rendering legal, accounting, or other professional service. If legal advice or other expert assistance is required, the services of a competant professional person should be sought.

The algorithms and programs in this book have been prepared with care and tested extensively. The publisher and author make no warranty of any kind, expressed or implied, of their suitability for any particular purpose. In no event will the publisher or author be liable for any consequential, incidental, or indirect damages (including damages for loss of business profits, business interruption, loss of business information, and the like) arising from the use or inability to use the programs and algorithms in this book, even if the publisher or author has been advised of the possibility of such damages.

Reproduction or translation of any part of this work beyond that permitted by section 107 or 108 of the 1976 United States Copyright Act without the permission of the copyright owner is unlawful. Requests for permission or further information should be addressed to the Permissions Department, John Wiley & Sons, Inc.

Library of Congress Cataloging-in-Publication Data:

Masters, Timothy.
    Neural, novel & hybrid algorithms for time series prediction /
  Timothy Masters.
        p.   cm.
    Includes index.
    ISBN 0-471-13041-9 (paper : alk. paper)
    1. Neural networks (Computer science) 2. Algorithms. I. Title
  QA76.87.M369 1995   95-31203
  519.5'5--dc20   CIP

Printed in the United States of America
10 9 8 7 6 5 4 3 2 1

I saw that under the sun
The race is not won by the swift;
The battle is not won by the strong.
Bread is not awarded to the wise,
Nor is wealth given to the intelligent,
And fame does not come to those of skill.
But time and chance happen to them all.

Ecclesiastes 9:11

Preface

Everyone wants to see into the future. Marketing analysts need to predict future sales. Economists want to predict economic cycles. Process control specialists need to predict furnace temperatures. And everyone wants to predict the stock market. Since time-series prediction is such an important topic, and since so much money rides on correct decisions, this subject has been extensively studied for decades. Many excellent books have been written on the subject. However, virtually all of these books fall into one of two categories. Some are highly theoretical, presenting breakthroughs in a totally rigorous fashion. The remainder offer broad overviews or practical advice in specific situations, but they leave out most of the implementation details. This book fills the void between those extremes by taking a four-way approach. First, each subject is treated in an intuitive fashion, appealing to the reader's common sense and experience. Then, the mathematical foundations of the subject are rigorously summarized, with references to additional advanced sources supplied as needed. Complete C++ source code is given to illustrate the practical implementation details. Finally, for the more important or obscure techniques, real-world data is used in example applications.

Over the years, many rigorous and effective methods for predicting future values of time series have been developed. The world's libraries are filled with texts devoted to this subject. It would be wasteful to create yet another rehash of the same material. On the other hand, there are some reasons for including traditional techniques in this new text. Among these reasons are the following:

- Some techniques may be extremely important to achieving good performance, yet frequently ignored by those responsible for making predictions. Brief summaries of these techniques are worthwhile. For example, Chapter 1 begins with a basic preprocessing checklist that is mandatory for all applications.

- There are many common tools that traditionally have an excessively narrow range of use. One such set of tools is ordinary lowpass, highpass, and bandpass filters. These have been extensively used for smoothing and related operations. However, this text shows how digital filters can be used as potent information selectors for enhancing valuable components, eliminating random noise, and splitting complex information sources into bite-size pieces.

- Some tools are widely used in other application domains, but rarely make their way into time series prediction. Wavelets are heavily used in science and engineering, especially in signal and image processing. This text explores the use of wavelets as feature detectors which can provide valuable assistance in predicting future movements of a time series.

- Finally, many algorithms and models have not been brought to their full potential by the time-series prediction community. For example, this text takes Box and Jenkins' venerable old ARIMA workhorse and extends it into a very general multivariate prediction model. Then, neural networks are married to this model to produce a powerful hybrid.

Several largely original algorithms also appear in this text. The most important is a totally general method of computing robust confidence intervals for any prediction model. Another is the ability to include very general seasonal terms in ARIMA models, avoiding being restricted to unit roots in the model equations. In one sense, these are not completely new techniques, as they are straightforward applications of standard statistical methods. However, they are new in that almost nobody uses them or even knows about them.

Chapter 1, *Preprocessing*, begins by reviewing the most critical issues in preprocessing. These issues are so crucial to good performance that everyone should be aware of them. Transformations, scaling, and event representation are featured. Centering and detrending, two of the most potentially valuable, yet overused and abused techniques, are covered in some depth. Several methods for reducing the dimensionality of a dataset, including principal components and linear discriminant functions, are shown. The chapter ends with an introductory look at digital filters as preprocessing algorithms.

Chapter 2, *Subduing Seasonal Components*, is almost a continuation of Chapter 1. Preprocessing methods for handling series that contain strong seasonal components are compared and contrasted. The many dangers inherent in the use of moving averages are especially emphasized, and effective alternatives are presented. An example using real-life data completes the chapter.

Chapter 3 provides a foundation for frequency-domain techniques. The discrete Fourier transform (DFT) is introduced in an elementary fashion. This leads naturally to the subject of power spectra. The value of visually examining the power spectrum is discussed, and the Savitzky-Golay filter is shown to be effective at taming the random variation inherent in the DFT spectrum. The cumulative power spectrum is presented as a simple but effective method for assessing the randomness of a series. The maximum entropy spectrum is introduced as a useful alternative to the DFT spectrum, and its strengths and weaknesses are discussed. Many illustrations comparing and contrasting the various spectrum alternatives are provided.

Chapter 4 continues where Chapter 3 leaves off, moving on to advanced frequency-domain techniques. Lowpass, highpass, and bandpass filters are presented in a relatively rigorous fashion, and complete C++ code is given for a variety of useful filters. In-phase, in-quadrature, and quadrature-mirror (QM) filters are treated in some depth, and their utility in practical applications is emphasized.

Chapter 5 is a gentle introduction to wavelets. The relatively theoretical treatment of QM filters in Chapter 4 is relaxed to a more intuitive level. The concept of the *presence* of a feature versus the *state* of the feature is presented in the context of QM filters. Frequency-domain versus time-domain scaling issues are discussed, and this leads to a solid intuitive motivation for wavelets. A real-world example of these issues reinforces the motivation, then the concept of wavelet filter banks is introduced. Morlet wavelets are presented in some detail, and their utility relative to other wavelet families is stressed. Finally, an example is given to demonstrate how the NPREDICT program supplied with this text can be used to implement Morlet wavelet filter banks.

Chapter 6 starts by reviewing Box and Jenkins' famous ARMA model. This model is extended to a multivariate version that also allows completely general seasonal terms. An efficient training algorithm, including C++ source code, is supplied. A powerful method for hybridizing the ARMA model with neural networks is given. Many important issues relating to designing and validating multivariate ARMA models are discussed. Finally, the closely related topic of lagged correlation (including partial and crosscorrelation) receives a thorough treatment.

Chapter 7 is entirely devoted to the subject of differencing. This includes both adjacent-point and seasonal differencing. Most texts lump this topic in with the ARMA chapter, thereby developing the ARIMA model. However, differencing is too important and too broadly useful to be cheated out of its full due. A detailed example of time-series differencing closes out the chapter.

Chapter 8 is arguably the most unique chapter in this text. Predictions are useless if they are not accompanied by some measure of confidence. Yet the subject of confidence intervals is shamefully ignored by many people. The few treatments of the subject that do exist almost always make unrealistic assumptions about the statistical distribution of the data. Normality of the error terms is a common assumption. This chapter presents a general algorithm that can be used to find practically distribution-free confidence intervals for any prediction model. Applications of this method are shown for both the ARIMA model and neural-network predictions.

Chapters 9 and 10 are intended for programmers who wish to pull subroutines directly from the programs on the accompanying disk and use them in their own programs. Many routines of broad utility are available. These include C++ and Pentium-optimized assembler versions of the complex-general mixed-radix fast Fourier transform, as well as a wide variety of deterministic and stochastic optimization algorithms and several numerical and statistical routines.

Chapter 11 is a user's manual for the NPREDICT program supplied on the accompanying disk. DOS and Windows NT versions of this program (both source and executable) are provided. The NT version uses only the Win32s API subset, so it should be able to run under Windows 3.1 if the Win32s extensions are present. A fairly thorough validation suite is also supplied to assist users who recompile the code for their own purposes.

The program code in this book is in C++. An attempt has been made to use designs unique to C++ only when they are vital. Thus, much of the code can be easily adapted to C compilers. Vector and Matrix classes, as well as operator overloading, have been deliberately avoided. Such specialized constructs contribute little or nothing to program efficiency and complicate the lives of programmers who wish to use these algorithms in other languages. Strictly C++ techniques have been used only when they significantly improve program readability or efficiency. All of the C++ code in this book and on the accompanying disk has been tested for strict ANSI compatibility. The author uses Symantec C++ version 6.1 and Borland C++ version 4.5. It is anticipated that any ANSI C++ compiler should be able to compile the code correctly.

Contents

1. Preprocessing 1
    Things Everyone Should Know 3
        Transformations 3
        Scaling 6
        Status and Event Flags 8
    Centering and Detrending 9
        Centering 9
        Detrending 12
    Data Reduction and Orthogonalization 16
        Optimizing Factors Using Class Membership 22
    Filtering for Information Selection 29
        What Is a Filter? 29
        Why Filter a Time Series? 30
        Filter Parameters 33
        A Basic Filtering Example 35
        An Alternative Bandpass Filter 38
        Another Argument for Information Preservation 40
        Revealing Trends 41

2. Subduing Seasonal Components 43
    The Problem of Harmonics 44
        Moving Averages and Alternatives 45
    A Seasonality Example 48
        Summary of the Examples 50

3. Frequency-Domain Techniques I: Introduction 61
    Introduction to the Frequency Domain 63
    The Discrete Fourier Transform 66
        Computing the Discrete Fourier Transform 69

The Power Spectrum 70
    Frequency versus Period 72
    Aliasing and the Nyquist Frequency 74
    Data Windows 75
    Code for Computing the Power Spectrum 80
    Smoothing for Enhanced Visibility of Spectral Features ... 81
    The Cumulative Power Spectrum 83
The Maximum Entropy Power Spectrum 90
Power Spectrum Examples and Summary 92
    Summary of Power Spectrum Usage 99

4. Frequency-Domain Techniques II: Filters and Features 101
What Is a Digital Filter? 102
Filtering in the Frequency Domain 104
    Padding to Avoid Wraparound Effects 106
    Endpoint Flattening to Reduce End Effects 108
    Shaping the Filter 110
Bandpass Filters 112
Lowpass and Highpass Filters 114
    An Alternative Bandpass Filter 115
Implementation Details and Sample Code 116
    Summary of Basic Filters 120
In-quadrature Filters 121
    In-quadrature Filters for Change Detection 125
Quadrature-Mirror Filters 129

5. Wavelet and QMF Features 135
Feature Presence versus Feature State 136
    An Introductory Example 138
    Summary of Filter Outputs 139
The Width Dilemma 145
    One More Implication of Time-Domain Extent 148
    A More Realistic Example 150
    Summary of this Example 160

Wavelet Features	166
Arranging Filters According to Width	167
What Is a Wavelet?	170
The Morlet Wavelet	172
Code for the Morlet Wavelet	175
Implementing Wavelets with NPREDICT	177

6. Box-Jenkins ARMA Models 181

Overview of the ARMA Paradigm	182
Duality Between AR and MA Models	184
ARMA Models	185
Homogeneity, Periodicity, and Stationarity	185
Computing the Parameters	187
Univariate ARMA Prediction	191
Predicting with Predicted Shocks	193
Multivariate ARMA Models	200
Computing the Error and Predicting	203
Training the Multivariate ARMA Model	211
Computing Initial AR Parameter Estimates	211
Iterative Refinement of All Parameters	215
The Complete Training Algorithm	217
Designing and Testing ARMA Models	223
Examining the Power Spectrum	229
Multivariate Models	232
Computing Lagged Correlations	234
Autocorrelation	234
Crosscorrelation	235
Partial Autocorrelation	235
Partial Crosscorrelation	237
A Multivariate Example	237

7. Differencing 251

Stationarity	253
Identifying Nonstationary Behavior	254

	Removing Nonstationarity	259
	ARIMA Models	261
	Seasonal Differencing	261
	Spotting the Need for Seasonal Differencing	265
	Computational Considerations	268
	An Example of Differencing	268

8. Robust Confidence Intervals ... 273

   Overview ... 274
   Sampling the Prediction Errors ... 276
      Collecting Errors for Neural Network Models ... 277
      Collecting Errors for ARMA Models ... 284
   Compensating for Differencing and Transformations ... 292
   From Errors to Confidence Intervals ... 302
   Confidence in the Confidence ... 306
      Foundations ... 307
      Bounding the Confidence Bounds ... 310
      Code for Bounds on the Confidence Limits ... 318
   Multiplicative Confidence Intervals ... 323
   Confidence Intervals in Action ... 325

9. Numerical and Statistical Tools ... 333

   Random Numbers ... 334
      Uniform Random Numbers ... 334
      Gaussian (Normal) Random Numbers ... 336
      Multivariate Cauchy Deviates ... 336
      Other Random Number Generators ... 337
   Singular Value Decomposition ... 338
   Deterministic Optimization ... 339
      Bounding a Univariate Minimum ... 340
      Refining a Univariate Minimum ... 342
      Conjugate Gradient Minimization ... 344
      Levenberg-Marquardt Minimization ... 347
      Powell's Algorithm ... 350

Stochastic Optimization	351
Primitive Simulated Annealing	352
Generic Traditional Simulated Annealing	354
Stochastic Smoothing	355
Eigenvalues and Eigenvectors	357
Fourier Transforms	358
Data Reduction and Orthogonalization	362
Autoregression by Burg's Algorithm	364

10. Neural Network Tools ... 367
- Training and Test Sets ... 368
- Generic Network Parameters ... 369
- Generic Learning Parameters ... 370
- Generic Neural Networks ... 372
- Multiple-Layer Feedforward Networks ... 373
  - MLFN Training Considerations ... 374
- Probabilistic Neural Networks ... 374

11. Using the NPREDICT Program ... 377
- Using This Manual ... 378
- General Commands ... 379
  - Exiting the Program ... 379
  - The Audit Log File ... 379
  - Progress Indicators for Slow Operations ... 380
  - Command Control Files ... 381
- Working with Signals ... 381
  - Reading Signals from a Disk File ... 382
  - Saving a Signal to a Disk File ... 383
  - Generating a Signal ... 384
  - Modifying Signals ... 385
  - Copying and Subsetting a Signal ... 388
  - Displaying Signals ... 388
  - Displaying Correlations ... 390
  - Displaying Spectra ... 390

The Power Spectrum and its Relatives	390
Smoothing the Power Spectrum	392
The Maximum Entropy Spectrum	393
Data Reduction and Orthogonalization	393
An Example Ignoring Class Membership	400
An Example Using Class Membership	401
Filters	402
Lowpass, Highpass, and Bandpass Filters	403
Quadrature-Mirror Filters and Morlet Wavelets	404
Padding	404
Moving Average	405
Autocorrelation and Related Operations	405
Box-Jenkins ARMA/ARIMA Prediction	406
Specifying the Model and Signals	407
Training the Model	409
Predicting with the Model	411
Confidence Limits for ARMA Predictions	415
Example of ARMA Prediction Confidence	418
Predicting with Known Shocks	419
Saving and Restoring ARMA models	422
Neural Network Training and Test Sets	423
Specifying Inputs and Outputs	423
Class Names as Outputs	424
Cumulating the Training or Test Set	425
Clearing Information	426
Examples	428
Neural Network Models	429
The Probabilistic Neural Network Family	429
Kernel Functions	431
Multiple-Layer Feedforward Networks	432
General Training Considerations	433
Cross Validation Training	435
Specific PNN Training Parameters	436
PNN Progress Reports	437
Specific MLFN Training Parameters	438

MLFN Progress Reports	442
Testing a Trained Network	444
Saving a Trained Network	446
Neural Network Prediction	446
An Example of Neural Network Prediction	450
Confidence Intervals for Predictions	452
Example of Network Prediction Confidence	455
Alphabetical Glossary of Commands	456
Validation Suite	477
Appendix	**479**
Disclaimer	479
Disk Contents	479
Hardware and Software Requirements	482
Compiling and Linking the NPREDICT Programs	483
The DOS Version	484
The Windows NT Version	484
Data Files	485
Financial Data	485
Climate Data	486
Sunspot Data	488
Extra Data	488
Bibliography	**489**
Index	**509**

Neural, Novel & Hybrid Algorithms for Time Series Prediction

Neural, Novel & Hybrid Algorithms for Time Series Prediction

1

Preprocessing

- Transformations

- Scaling

- Flag and event variables

- Centering and detrending

- Data reduction and orthogonalization

- Filters for information selection

Preprocessing is the key to success. This is a bold opening to the chapter, but its truth is so profound that it must be emphasized. Even a primitive prediction model can perform well if the variables have been preprocessed in such a way as to clearly reveal the important information. And the best prediction model in the world is rendered helpless by variables that present the necessary information in complex and confusing ways. Therefore, this text begins with a discussion of preprocessing.

Designing and writing this chapter was frustrating because some of the decisions about which topics to include were very difficult. Many of the standard and traditional techniques are so useful that their consideration in practical problems is mandatory. On the other hand, since these methods are so common, they have also been extensively covered in the literature. This fact led to a painful choice: *Many of the most important preprocessing methods have been largely omitted from this text to avoid duplication.* The majority of the material in this chapter has been chosen because it is important *and* it has not received much coverage in other texts. With this in mind, the reader is directed to some other books that extensively discuss preprocessing in the context of neural networks and other nonlinear models:

- General aspects of variable coding and transformation are covered in Masters (1993). Methods for handling nominal, ordinal, interval, and ratio data are presented. Some important transformations and scaling issues are also discussed.

- Several highly specific coding, scaling, and transformation issues primarily aimed at signal and image processing are covered in Masters (1994). These include methods involving data in the complex domain, Fourier and Gabor transforms, wavelet features, and high-order moment variables.

- Many standard statistics texts treat transformation in depth. One of the best is Acton (1959). Readers should have no trouble finding good sources by just scanning a library shelf.

- Readers who have a solid background in mathematics and want to consult the ultimate authority on transformation should consult Kendall and Stuart (1969, 1973, 1976). Be warned that the level of mathematics in this text is extremely sophisticated. However, the treatment is superb if you can wade through it.

Things Everyone Should Know

In order to avoid totally ignoring the most important preprocessing techniques, this chapter begins with a cursory overview of the concepts everyone should consider in applications involving neural networks or other prediction models. Readers should consult the previously mentioned texts for further details. This is only a brief summary and is not as thorough as is needed for professional use.

Transformations

A cardinal rule in any data processing task is that the numbers that present information about the state of a process should take on values that are commensurate with the practical meaning of the current state. That's a real mouthful. Let's examine this statement more deeply and use some examples.

The first principle of transformation is that *decisions are necessarily arbitrary in many situations*. Occasionally we may have prior knowledge telling us that some particular transformation has a desirable mathematical property and therefore should be chosen. This doesn't happen very often. Usually, we have to examine a plot of the data and make an intuitive decision based on visual examination. Therefore, readers should not take the advice given here as Universal Truth. There is always room for discussion. Fortunately, it is almost always the case that a lot of latitude for error exists. As long as the transformation is roughly correct, that's good enough.

The concept of a variable reflecting the important information about a state is crucial. Here is a silly example, but one that is frighteningly close to situations that can inadvertently appear in applications. Suppose we are measuring the weight of a person every day, perhaps hoping to use this series to predict future weights. What if we decided that whenever the weight is below 150 pounds, we use pounds as our unit; but above that weight, we switch to kilograms to keep the numbers more manageable. Confusion would surely overcome any prediction model unless provisions for this crazy system were made. The numbers do not represent the state in a consistent manner.

Now look at an example that is superficially less absurd but nearly as dangerous in practice. Suppose we measure the amount of rainfall each day in the state of Florida. The majority of the observations come in at small numbers. But every so often, when a hurricane passes through, the rainfall for a day (or several) is gigantic. If one were to plot a histogram of this data, the majority of the observations would cluster at the low end of the range, with a few oddballs at the high end. The middle would be largely empty. Do those huge observations rate such extreme values in the measured variable? Is it worth abandoning such a large area of middle ground just to provide a dramatic dichotomy in the data? Probably not. Any prediction model faced with data like this would most likely see everything in black and white: hurricane or normal weather. The relatively tiny but important day-to-day variations would be swamped out. In situations like this, it is virtually always advisable to use a compressive mapping function to move the extreme observations down toward the majority. Naturally, they should still stand out. Hurricanes are important information. They should simply not stand out so much that all other information is obliterated in their wake.

An even more interesting situation arises when the data naturally falls into clusters, but the relative position of each case in the cluster is as important as the cluster to which the observation belongs. This frequently occurs when the value of the measured variable is determined by more than one underlying effect. It is not uncommon for effects that cause relatively little change in the variable to be as important as those that cause great change. For example, suppose we observe the waiting room of a walk-in health clinic served by three doctors. We measure the waiting time of each patient and record the average at the end of each day. When all three doctors are present, perhaps the waiting time is around ten minutes. But sometimes one of the doctors is called to the hospital for an emergency and is gone for half the day. Now the waiting time jumps to the vicinity of 30 minutes. And on rare occasions, two doctors are away for some reason. When this happens, waiting time can exceed an hour. This data may naturally cluster around three values. Despite this clustering, the difference between an average wait of 28 minutes on one day and 31 minutes on another day may be important to the prediction model. This small-scale variation can become lost in the dominant variation caused by changes in the number of available doctors. When this happens, exotic mapping transformations may be needed to equalize the distribution of information across the range of possible values. The most popular method is *histogram equalization*. This extremely powerful and effective transformation algorithm can be found in most image-processing texts.

In summary, the following items should be considered when designing variables that will be presented to a prediction model:

- Think about the relationship between the variable and the information it is designed to present. Are these quantities commensurate?

- If the data lies in clusters that obscure important local information, first try to analyze the reason for the clustering and separate the effects if possible. Otherwise, consider a remapping of the data, probably by histogram equalization, to more equitably distribute the observations.

- Many types of data exhibit multiplicative responses to underlying effects. This sort of data generally has the property that its local variation is proportional to its mean value. For example, the daily variation of a $100 stock is roughly on the order of ten times the daily variation of a $10 stock. Such data virtually always requires a log transformation.

- If the data occasionally contains values much larger than the central tendency, a compressing transformation is appropriate. The log is an extreme compressor, while the square root is relatively mild.

- If the data requires compression of extremes, but the values are not always positive, the cube root (or, rarely, a higher odd root) may be a good choice.

- If the data contains discontinuities, beware: They can spell disaster. Many times, a little creative thought can remove the discontinuity. For example, the angular position of an object (such as the state of a rotating shaft or the azimuth of a vehicle) has an obvious discontinuity when the angle passes from its maximum (360 degrees) to its minimum (0 degrees). Recoding this single variable as the *pair* of variables defined by the sine and the cosine of the angle removes the discontinuity perfectly.

- If certain locations in the range of a variable have special meanings, consider recoding the variable as a set of fuzzy membership functions. See any good text for more on this topic.

Scaling

Most linear models, such as discriminant analysis and general linear hypothesis methods, are intrinsically immune to scaling difficulties. Multiplicative rescalings of variables mathematically cancel, leading to carelessness when nonlinear models like neural networks are used. Unfortunately, scaling is crucially important to most nonlinear models. We will briefly consider the most important aspects of the problem and summarize the steps that need to be taken to ensure compliance with the most important rules.

The strongest and most direct way that scaling influences most nonlinear models is through the implied relative importance of the variables. When more than one variable is supplied to a model, as is nearly always the case, most nonlinear models implicitly or explicitly assume that variables having large variation are more important than variables having small variation. This occurs for both input and output. For example, suppose we are predicting two variables simultaneously. Perhaps one variable is the closing price of a $100 stock, and the other is the closing price of a $1 stock. Most training algorithms minimize an error criterion involving the mean or sum of (probably squared) errors across all outputs. Thoughtless use of such a criterion will cause the training algorithm to devote inordinate effort to minimizing the prediction error for the $100 stock, where the major gains lie, while largely ignoring the $1 stock. The possibility that 100 times as many shares of the $1 stock are at stake is not taken into account.

As a short aside, simultaneous prediction of several variables should be done only in those rare instances when it is absolutely necessary. If the reader always takes care to use separate models for each prediction, this problem will not occur for outputs! But sometimes, as with multivariate ARMA methods, multiple predictions are an intrinsic part of the model. When this is the case, the scaling of each variable must be consistent with its relative importance.

Relative importance plays a role for input variables, too. Probabilistic neural networks that use the same sigma weight for each variable automatically assign an importance to each variable in rough proportion to its variance. Also, many input preprocessing schemes (such as some versions of principal components) are strongly affected by the relative variance of each variable. The reduced set of variables generated by these methods will tend to favor information from high-variance inputs.

Scaling can impact models in indirect ways as well as in the direct ways just discussed. Most neural network training algorithms are able to adjust input weights over a reasonably wide range, but they have difficulty driving weights to extreme values. Training of multiple-layer feedforward networks is compromised when the input variables have widely disparate scalings that necessitate wide weight ranges. Offsets can also adversely impact training. If a variable is centered around a large value, the input bias will need to be large to compensate. This is difficult for most training algorithms to achieve. It can be done, but it is not easy. At best, training will be slowed.

How should variables be scaled for best results? There is no clear answer. Luckily, there is a lot of margin for error. As long as they are *approximately* centered and scaled to reasonable values, everything should be fine. The most important point is that all variables be scaled so that their variation is commensurate with their importance. In the absence of better information, equal scaling is usually the best bet. Some users scale variables to have unit variance. This is fine. A rule as simple as keeping numbers to single-digit quantities (−10 to 10) is also good. Do whatever is convenient, but do *something*. Here is an effective scaling procedure that should be applied separately to each variable (after transformations, if any, have been done):

1. Compute the mean across the training set. If there are any wild outliers, *which should not often be the case if an effective transformation was used*, use the median instead of the mean.

2. Subtract the mean (or median) from each observation in the training set to center the variable around zero.

3. Compute the standard deviation across the training set. Alternatively, compute the interquartile range. The latter is more effective if there are outliers (which there shouldn't be).

4. Divide each observation in the training set by the standard deviation (or interquartile range) to equalize scales.

5. Train the model with this centered and scaled data.

6. When the trained model is put to use, apply the *same* centering and scaling to the newly observed data.

Status and Event Flags

Our natural inclination is to think of variables as always having numeric values. But sometimes we want to define a variable in terms of a status or an event. A brokerage house may consider a stock to be a buy, sell, or hold. The Fourth of July may have special significance to a model. A patient for whom we want a prognosis report may have a tumor in a special class. How do we encode this sort of nonnumeric information? Here are some of the more important rules:

- Even when a variable is not explicitly numeric, we should still attempt to keep it near zero on average. Therefore, if a binary flag responds to a rare event (such as a date being a national holiday), use 0 to encode the default, and use 1 to encode the rare event. Conversely, if a binary event is nearly equiprobable, use 1 to encode one state and −1 to encode the other state.

- To encode class membership information in which the classes have an implied order relationship, such as grade A, B, or C indicating quality, there are two possibilities. We could use a single variable and choose its value according to the class. But a frequently better approach is to use one fewer variables than there are classes. For the lowest-rank class, set all variables equal to −1. For the next lowest class, set the first variable equal to 1 and leave the others at −1. For the next, set the first *two* variables to 1 and the others to −1. For each successive class, change one more variable to 1 until the highest class has all variables at 1. This is a lot of variables, but many nonlinear models respond well to this encoding.

- For classes that have no implied order relationship, such as gender or race, use one variable for each class. Set the variable corresponding to the case's class equal to 1, and set all other variables equal to 0. *Never* encode unordered classes by using different values of a single variable.

- Always remain open to deriving flag variables from logical relationships in other variables. For example, an old market saying asserts that nothing is more bullish than setting a new high. Market predictors might want to define a variable that is normally 0, but is 1 whenever a new record is set.

Centering and Detrending

The most basic preprocessing technique is centering a signal so that it has a mean of zero. A close runner-up is detrending so that the overall trend of the signal across its extent is flat. This section will briefly examine these two operations.

Centering

Data centering is a common procedure. It is easy to do and undo, and it is a fundamental early step in many standard techniques. However, a great number of people treat the subject of centering too lightly. In many applications, blatantly noncentered data is fed to prediction models that suffer accordingly. Conversely, many people automatically center every data stream, blissfully ignoring the possible negative consequences of unwarranted centering. This section explores a few of the relevant issues.

Some applications *seem* explicitly to require centering at some stage. For example, suppose we are modeling a physical process that is known *a priori* to hover around a fixed location. Suppose also that we difference the data so we are working with changes rather than raw values. Random errors may cause the mean of the differenced data to be nonzero, even though we know the physics of the process demands a zero mean. When this happens, we face a decision that is far more complex and important than most people realize. Do we, or do we not, center the series to force a mean of zero? For some prediction models, the nonzero data mean would be erroneously reflected in the model's predictions. This causes projected future values to trend away from their correct neighborhood when the series is summed to counteract the differencing. In these situations it may be advantageous to center the differenced signal before modeling it. Other models may not be fooled by the illegitimate offset, and their predictions are damaged by centering when none is theoretically required. The bottom line is this: Know your data, know your model, and use common sense. If you are sure that the variable has a true mean of zero, but the mean of your sample is significantly nonzero, investigate further. Try to determine whether the prediction model is sensitive to the sample's mean. If so, centering may be good. If not, keep the data intact. If in doubt, try it both ways. But by all means, expend some real effort. This is a potentially serious problem. The decision of whether to center can be crucial to performance.

Even when centering is not an explicit part of the model, it can be useful to help the model do a good job. For example, many neural network training algorithms have difficulty driving weights to extreme values. If one or more input variables have a mean that is significantly different from zero, their optimal bias weights will probably be so far out in left field that learning will be impaired. Even though the model has the theoretical capability of learning to handle the data, centering will almost invariably speed training and enhance the quality of the results.

On the other hand, some models may not need centering to counteract algorithmic weaknesses in their components. The probabilistic neural network has no such weaknesses. Even so, centering can still be beneficial. The reason has to do with the way arithmetic is performed in computers. When two numbers are subtracted to produce a difference that is small relative to the magnitude of the subtracted numbers, numerical precision suffers. Although different algorithms react differently in this regard, it is a good general rule that the more unbalanced the data, the more loss of precision occurs.

In summary, there appears to be a strange contradiction in this advice. When there is no reason to believe that the data is intrinsically centered due to the nature of the generating process, we should often go ahead and forcibly center it. On the other hand, when we know that the process produces data that has a true mean of zero, we must be wary of centering the data. What gives? The answer becomes clear when we ask the right question: Are we making an unnatural imposition? When the mean is not physically constrained by the process to a certain *known*, *fixed* value, we are almost always safe shifting the center any way we desire, including shifting it to zero in order to reap the numerical benefits of this operation. But when we know the true mean of the underlying population, we had better be cautious about shifting the data. The measured values probably have real meaning in terms of their deviations from the mean. Forcibly centering the data changes the observed deviations, which, in turn, changes the information content of the series. This can be deadly.

It's unfortunate that such an important decision has to be based on such flimsy evidence. Perhaps the series does have a true mean of zero. What difference does it make whether we know that fact? The truth of the matter is that sometimes it's just a judgment call. For this reason, we should try it both ways if there is any doubt. In the end, we are left with a tradeoff between the clear practical benefits of centering badly noncentered data versus the distortion induced by making the final values represent deviations from the *observed mean* as opposed to deviations from the *true population mean*. This is the key.

It should be made clear that this entire discussion is unnecessary if we take care to use the same fixed centering constant for all training samples and for the data that is eventually given to the trained prediction model. In other words, if we decide to automatically and individually center every separate data stream that is given to the model, we must pay strict attention to the preceding discussion. There is a crucial distinction between the modified data being relative to the fixed population mean (if any) versus being relative to the mean of that particular collection. Sometimes one is better, and sometimes the other is better. It depends on the application. But we can often avoid the issue entirely by choosing a single fixed value for approximate centering and sticking with it forever. Here is a common and effective technique when the central tendency of the data remains reasonably constant: Collect a sample of the data that is representative of what will be encountered in practice. Compute its mean, then decree that this mean, or any fixed number near this mean, will be subtracted from this variable from now on. The result is that individual data streams will not be exactly centered. But for most prediction models, that's fine. If the central tendency does not wander too much, the numerical and algorithmic benefits of centering will be largely reaped, while there will be no loss of potentially important information regarding the baseline for computing deviations. In the event that inferior results are obtained compared to individually centering each data stream, it is likely that the variable is (perhaps borderline) nonstationary. In this case, we are better off handling the problem with differencing or filtering as discussed elsewhere in this text. Precomputing and then universally employing a fixed constant for each variable adds to the complexity of the task, but it is almost always worthwhile.

Finally, in most situations, the most appropriate value to use to center a series is its mean. This quantity has a host of valuable statistical properties. However, if there are any significant wild outliers, the mean may not be a good representative of the series' central tendency. In this case, the median may be a better choice. But in this situation, a red flag must always be raised. *Why are there outliers?* Are they erroneous? Are we missing an important effect? Should we be applying a transformation? Outliers are more than a nuisance. They are potentially valuable flags.

Detrending

Differencing a series automatically removes a trend by converting the trend to an offset. This is often a welcome side effect of differencing. But sometimes we do not want to difference a series. A significant trend can be troublesome to many models, so it should usually be removed.

First, let us think about why a trend can cause problems. The answer is simple: confusion of the issues. Suppose a series consists of two prominent components: a steady increase in its central tendency, and predictable variation around that central tendency. Any model that is required to predict this series would have to deal with a variety of prediction rules. The important rules, those that apply near the end of the series, could not be deduced easily from data early in the series where the numbers are different. This is an extremely serious problem that should not be underestimated. But if we were to remove the trend before training the model, data from the entire extent of the series could contribute equally to learning the correct patterns. After the predictions are made, the trend could be reapplied to the data.

Some trends vary slowly across time, so they must be approached with smoothing or filtering. These subjects are discussed in Chapters 4 and 5. In this section, we consider only a single uniform linear trend that encompasses the entire extent of the series. There are many methods for detecting and removing a linear trend. One of the oldest is to look at the value of the last point in the series relative to the first point. Divide their difference by the length of the series to get an average trend. Despite the primitive nature of this method, it is still in use in some circles, and it has many staunch adherents. Its obvious weakness is its sensitivity to subversion by random noise in those two points. That's too much power to place in just two samples.

The NPREDICT program supplied with this text bases the trend on all points in the series. It computes the straight line that minimizes the mean squared difference between the line and the observed data. This is a common approach. Mathematics for this algorithm can be found in Masters (1993) and (1994).

There are situations in which detrending should not be used, even though the data exhibits a clear linear trend. The most obvious case is when the physics of the process dictates that there should not be a trend. It may be that random errors have conspired to induce a trend where none is expected. If so, we should not compound the problem by explicitly modeling the trend, even applying it to predictions! Most models, neural and otherwise, do not readily learn long-term trends unless special provisions are made. Therefore, an illegitimate trend will probably be ignored by the model, which will instead keep its predictions in the neighborhood of the overall central tendency. This is good.

A more subtle situation in which explicit trends, linear or slowly varying, should not be removed is when the value of the data affects the rules that govern its changes. Suppose consumer buying patterns depend on the price of a commodity. If the price is steadily increasing across time, it may be that the rules that reigned ages ago no longer apply. Thus, we would not want to imply that they do by explicitly removing the trend. Rather, we primarily want the recent data to influence the predictions. Or it may be that the price slowly rises and falls in possibly random cycles. We would want the predictions to be most strongly based on the historical data that is closest to the current values.

The problem of illegitimately removing a trend can be illustrated by an example contrived from a sample of a periodic waveform. Figure 1.1 is a sample of a cosine wave. Figure 1.2 shows the result of differencing the original series. Observe that this series starts out below zero and ends above zero. Automated detrending interprets this as an upward trend. Thus, the "detrended" series shown in Figure 1.3 exhibits an obvious downward trend even though it is numerically flat. When a neural network is trained on this series and used to predict future values, the result is shown in Figure 1.4. As is almost always the case, the neural network has failed to learn the trend. The prediction stabilized around the overall central tendency. When the predicted series is retrended to compensate for the original detrending, the known extent of the series is returned to its correct values, but the prediction is conspicuously wrong, as shown in Figure 1.5. This error is compounded in Figure 1.6 when the series is integrated to compensate for the original differencing. It's not a pretty sight.

How did a problem of this magnitude occur? It happened because we detrended a series that did not contain a legitimate trend. The apparent trend was the result of the particular sample we gathered. Although this is an extreme case, deliberately contrived to emphasize the problem, we should always be wary of detrending. It's a powerful tool, and one that is often necessary. But like all powerful tools, it is easily misused. Never detrend unless there is a clear reason for doing so.

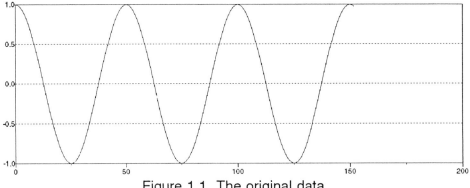
Figure 1.1 The original data.

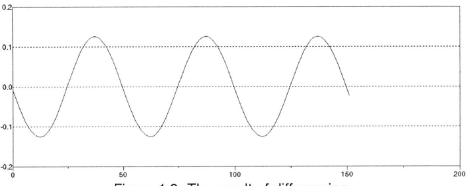
Figure 1.2 The result of differencing.

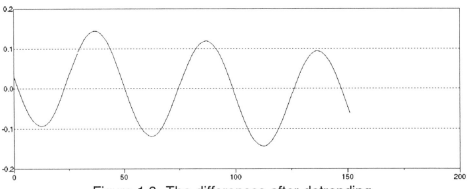
Figure 1.3 The differences after detrending.

Centering and Detrending 15

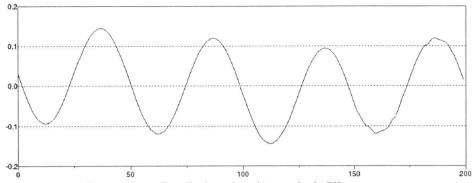

Figure 1.4 Predicting the detrended differences.

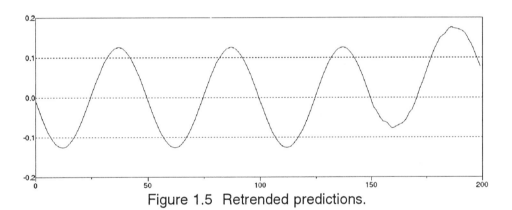

Figure 1.5 Retrended predictions.

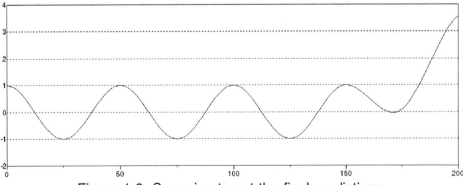

Figure 1.6 Summing to get the final predictions.

Data Reduction and Orthogonalization

In many situations, a multitude of variables are available for measurement, far more than we could possibly use in any prediction model. An ideal approach would be to try as many subsets of them as possible, perhaps being guided by an intelligent selection algorithm. Brute force selection of an effective subset is nice when we can do it. But when time, economics, or other practical issues preclude this approach, there is an alternative that is often useful. Several algorithms exist by which a large number of variables can be used to compute a smaller number of new variables that capture most of the information in the original set. The mathematics and philosophical justification of these algorithms are discussed in Masters (1995). This section considers some of their practical aspects, especially as related to time series prediction and classification.

One point should be made clear right up front: The methods of this section are potentially dangerous. The key to reducing the number of variables is that some information must be discarded. Unless there is perfect dependence in the dataset, it is mathematically impossible to represent the information in a set of continuous variables by means of a smaller set of variables. Something has to go. What we try to do is to throw away the information that is presumably least important. Many years of experience indicate that we have a very good chance of attaining this goal. But it is not guaranteed. We could easily discard the very information that we need and never even know we did so. It's unlikely, but possible. Be warned.

The data reduction methods of this section all use the same basic principle: Compute new variables (often called *factors*) as linear combinations of the old variables. The weight vectors are designed in such a way that some optimality criterion is maximized. Later, we will examine the three different criteria available here. For now, let us think about the concept of computing new variables as weighted sums of the old.

The classic illustration involves the height and weight of people. Measure these two variables across a large population. If one had to capture the essence of the data in a single variable, that variable would probably be a measure of size. A person's height plus his or her weight would furnish more information about the pair of measured variables than any other linear combination. The sum is not the total picture, though. Some people are short and big around, while others are tall

Data Reduction and Orthogonalization

and thin. The difference of height and weight, a measure of obesity (or lack thereof) also contains information about the original pair of measured variables. These two new variables, taken together, capture all of the original information. Which is more important? The answer to that question depends entirely on the application. If you discard the obesity factor, keeping only the measure of size, and your application involves something related to body fat, you are in trouble.

When there are only two variables, the relationship between their correlation and the computable factors can be easily seen. Look at Figure 1.7, which shows scatterplots of two different situations. This figure illustrates one type of factor, *principal components*. More will be said about this method of data reduction later. For now, we will concentrate on the general concept.

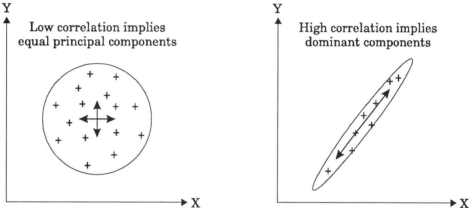

Figure 1.7 Principal components of pairs of variables.

The left half of the figure shows the situation in which the two measured variables are largely independent of each other. This might happen if we measure a person's weight and his or her hair color (on some continuous color scale). When the variables are this independent, it is useless to try to reduce their number. All of the original variables are needed to convey the total amount of information.

The right half of this figure shows a completely different situation. Perhaps the horizontal axis is a person's height and the vertical axis is the length of his or her legs. In this case, data reduction is much more of a possibility. The majority of the variation among the observed subjects lies along one axis. Tall people tend to have long legs. Thus, if we compute a single new factor, perhaps the sum of height plus leg length, this single variable will tell us nearly everything that the pair of original variables told us. It is probably safe to say that the remaining small bit of information, discrepancy between leg length and height, may be safely ignored in most applications.

It's time to be a little more specific. Readers who want full mathematical details should see Masters (1995). Here we will be satisfied with an intuitive discussion of principal components.

Computation of principal components starts with the goal of finding a single factor (linear combination of original variables) that accounts for the majority of the variation across the training set. There is to be no other linear combination that has higher variance. The rationale for this choice is that more variation presumably means more relevance. This thesis may be questionable in some applications, but it is nonetheless our working hypothesis. If we can capture the majority of the training set's variation in a single variable, we have (hopefully) captured the most important information in the training set.

Once a single dominant factor has been found, the next goal is to find a second factor that captures the majority of the information remaining in the original variables *that is not explained by the first factor*. After this second factor is found, we look for a third that captures more of the remaining information than any other possible factor. It can be seen that at every step we have a set of variables that accounts for the most information of any possible set of the same number of variables. This definition of an optimal reduced set of variables has been found to be generally excellent in most applications. There is a strong (but not universal) tendency for the important information to be concentrated in the first few principal components, while the system noise falls mostly in the later components that are discarded.

A nice byproduct of the principal component algorithm just described is that the computed factors are independent of each other. There is no redundancy in the information they convey. In mathematical terms, the principal components are *orthogonal*. Geometrically, they are all at right angles to one another. The implication for us is that we can use fewer model inputs when we ignore redundant information.

There is one especially important decision to be made when computing principal components. Suppose we measure a person's height in millimeters, and we measure their hair color with a variable that ranges from zero to one. The variation of the height variable will tremendously exceed that of the hair color variable. Despite the fact that these two variables are essentially independent, we will find that a single "new" variable defined as the height (plus an insignificant multiple of the color) accounts for the vast majority of the total information of the two variables. In any automated system, this single variable will be deemed sufficient, and hair color information will be almost totally lost. This is obviously an incorrect operation. The problem results from the disparate scales of the two measured variables. It is usually in our best interest to standardize the original measured variables before computing principal components. The only exception to this rule is when the disparate scaling is meaningful, in that variables having higher variance can be deemed to be more important. Even then, prescaling of some sort should be seriously considered. The NPREDICT program supplied with this text includes the option to standardize inputs automatically. Unless there is a strong reason to do otherwise, this option should always be specified when principal components of raw data are computed.

Before leaving principal components, let us look at one example from time series analysis. Suppose we have a single series, such as the one shown in Figure 1.8. Consider the information contained in the current and two lagged values of the series. This is a total of three variables. It is safe to assume that their variances are identical, since they are lagged values of the same series. Thus, forced standardization is not needed. We can compute the principal components with the following NPREDICT rules:

```
INPUT = x 0-2              ; Current and two lags
CLASS = dummy              ; Any name will do
CUMULATE TRAINING SET      ; This builds the training set
ORTHOGONALIZATION TYPE = PRINCIPAL COMPONENTS
ORTHOGONALIZATION STANDARDIZE = NO
DEFINE ORTHOGONALIZATION = princo
```

Note that we need to specify a class name even though it is ignored. This is explained in Chapter 11. The default maximum of three principal components are computed. The information printed in the audit log file is as shown below:

```
Percent      ----->    87.6583     9.4962     2.8455

Variable  1 ( 0)        0.1677     0.6458     0.7057
Variable  2 ( 1)        0.1801     0.0025    -1.3257
Variable  3 ( 2)        0.1701    -0.6394     0.7076
```

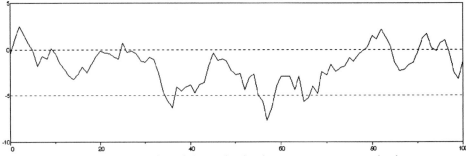
Figure 1.8 A signal for principal components analysis.

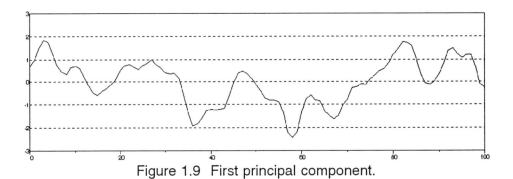
Figure 1.9 First principal component.

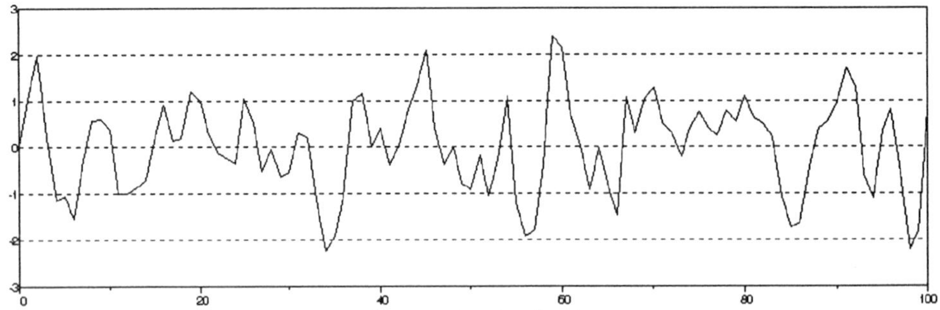

Figure 1.10 Second principal component.

The first thing to notice is that almost 88 percent of the variance of the three input variables can be explained by just one principal component. This factor is defined as essentially the mean of the three adjacent values. (The weights for the three lags are almost equal.) If we were forced to condense the information in the three original variables into one variable, their mean would capture the maximum information relative to any other linear combination of the variables. *Whether this is the information that our application would find most beneficial is another question entirely.* But when the criterion is sheer information content, the first principal component is the winner. Figure 1.9 graphs this factor. Observe that it is essentially a lowpass filtered version of the original signal. By the way, this and the other two principal components are computed in NPREDICT by means of the following commands:

```
NAME = factor1 , factor2 , factor3
APPLY ORTHOGONALIZATION = princo
```

If we allow ourselves a second new variable, the second principal component provides an additional 9.5 percent of the total information. These two new variables taken together account for more than 97 percent of the variation in the original variables. How is this factor being computed? Look again at the audit log output on page 19. This factor is the difference between the current value and the value at a lag of two. The lag-one value is ignored. This factor is graphed in Figure 1.10

Although the last principal component contains almost no information (at least in terms of training-set variation), it is interesting to examine its weights. This factor is defined as the signal value at a lag of one relative to the mean of the two adjacent values. Sometimes a careful pondering of the factor weights will yield valuable insights into an application.

The preceding example is unrealistic in one sense. Only rarely will we try to reduce as few as three variables into just one or two. More likely, we will be attempting to reduce dozens of variables into a dozen or so. In this situation there is generally less likelihood of discarding valuable information. It is interesting to examine the principal components of the previous signal when we use a total of ten lags. The first five factors are shown in the following table:

```
Pct -->  59.6906    21.1669    8.7689    4.4606    1.9470

(0)       0.0487   -0.1349    0.2197   -0.3248    0.5479
(1)       0.0570   -0.1391    0.1688   -0.0713   -0.2472
(2)       0.0644   -0.1197    0.0394    0.2005   -0.4943
(3)       0.0687   -0.0819   -0.1130    0.3148   -0.0573
(4)       0.0708   -0.0315   -0.2202    0.1533    0.3515
(5)       0.0711    0.0216   -0.2196   -0.1619    0.3065
(6)       0.0702    0.0749   -0.1145   -0.2990   -0.1411
(7)       0.0678    0.1169    0.0342   -0.2063   -0.4344
(8)       0.0621    0.1358    0.1638    0.0548   -0.1234
(9)       0.0529    0.1307    0.2199    0.3325    0.4456
```

Once again, the first principal component is essentially a moving average, and this single factor accounts for the majority of the information contained in the series. The second principal component pits recent values of the series against more distant values. The third pits central values against early and late values. Then interpretation becomes more difficult. The important point is that by using only half of the original variables, we can account for 96 percent of the total variation. Although we should always remain open to the possibility that the five principal components we discarded may contain some vital information, the fact of the matter is that this situation is not common. In most practical applications, experience indicates that the most important information appears right up front, and the noise is concentrated in the later components. This is not a sure bet, but it's usually a good one.

Optimizing Factors Using Class Membership

When the application at hand directly involves classification, or when the task can be cast as classification, we can usually do better than principal components. Even econometric prediction may fall into this category: Think about buy/sell decisions. The likelihood of discarding valuable information is decreased by taking class membership into account when the optimal factor weights are computed. If we force the model to focus on class membership, we can tilt the odds in our favor.

As an example, suppose we have two classes. Look at Figure 1.11, which illustrates a possible scatterplot of the data. Class separation this well defined is something we only dream of, but it makes for good examples.

Data Reduction and Orthogonalization

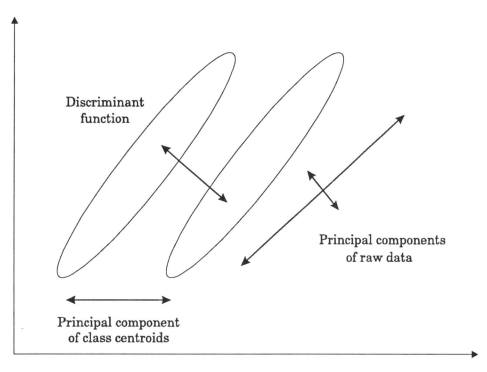

Figure 1.11 Factors when there are several classes.

The principal components of the raw data, which totally ignore class membership, would resemble the two crossed vectors shown at the far right side of this figure. Observe that the single variable defined by the first principal component would be virtually useless at discriminating between the two classes. The second component would be very good, but in practice we have no way of knowing this without explicitly testing it, an expensive proposition.

A naive but sometimes effective method of data reduction that enhances class separation is to compute the principal component(s) of the class centroids. When there are only two classes, there can be only one principal component, because two points determine a line. The principal component of the class centroids in Figure 1.11 is shown at the bottom of the figure. Note that due to the significant overlap of the classes along this dimension, the utility of this single variable would be fair to poor.

If we are willing to assume that the shape of the distribution within each class is similar for all classes, there is an algorithm for computing new variables that optimally separate the classes. These new variables are called *linear discriminant functions*. The single one that would be defined by the data in Figure 1.11 is shown near the center of that figure. Note that class separation in this direction is fabulous.

Once again, the reader is reminded that this discussion of data reduction is necessarily brief, focusing on practical aspects of the technique. Detailed mathematics and philosophical justifications can be found in Masters (1995). That said, let us end this section with a look at linear discriminant functions in a time-series context.

Suppose we measure two variables, x and y. We also have two classes. For the sake of this demonstration, assume that x is worthless for distinguishing between the classes; its distribution is the same in both classes. The x series used here, as sampled from the first class, is graphed in Figure 1.12. The x series sampled from the second class is in Figure 1.13.

So that we have something meaningful to demonstrate, the y series are designed to be different for the two classes. Their basic structures are similar for both classes, but the mean in the second class is slightly greater than in the first class. The y signals, as sampled from the first and second classes, are shown in Figures 1.14 and 1.15, respectively.

Let us arbitrarily choose to use the current and lag-one values of each series. This gives a total of four raw variables. The NPREDICT commands for computing the optimal linear discriminant function for this data are shown below. More information concerning these commands can be found in Chapter 11.

```
INPUT = x1 0-1
INPUT = y1 0-1
CLASS = class1              ; These signals represent first class
CUMULATE TRAINING SET       ; Start building the training set

CLEAR INPUT LIST            ; Start a new input list
INPUT = x2 0-1
INPUT = y2 0-1
CLASS = class2              ; These represent second class
CUMULATE TRAINING SET       ; Append to the training set

ORTHOGONALIZATION TYPE = DISCRIMINANT
DEFINE ORTHOGONALIZATION = discrim
```

Data Reduction and Orthogonalization

The maximum possible number of linear discriminant functions is one less than the number of classes. This follows from simple linear geometry, as two points define a line, three points define a plane, and so forth. Thus, the audit log record for this operation consists of just one column for the single discriminant function. This record looks like the following:

```
Percent      ----->   100.0000
Variable   1 ( 0)      -0.0436
Variable   2 ( 1)       0.0016
Variable   3 ( 0)       0.5063
Variable   4 ( 1)       0.5316
```

It should come as no surprise that the first two weights, corresponding to signal x, are nearly zero. The two weights for signal y are significant and practically equal. This equality is also to be expected, since the relative values of adjacent points have no importance. Only the series mean is offset.

In a real application, we would want to compute the values of this factor (discriminant function) using the samples from each raw series. The new series would form part or all of the training set for some model, neural or otherwise. The following NPREDICT commands compute the two new signals, one for each class:

```
CLEAR INPUT LIST
INPUT = x1 0-1
INPUT = y1 0-1
NAME = dfact1
APPLY ORTHOGONALIZATION = discrim

CLEAR INPUT LIST
INPUT = x2 0-1
INPUT = y2 0-1
NAME = dfact2
APPLY ORTHOGONALIZATION = discrim
```

A plot of the new factor for the first class is shown in Figure 1.16, and that for the second class is in Figure 1.17. Note that the former is almost always negative, while the latter is almost always positive. If we were to use these signals as inputs to neural or other models, it should be apparent that we would be handing the model its data on a silver platter. The classification task can be performed with a great deal of accuracy based on this single variable alone. In fact, if our model contained no other inputs and based its classification on the trivial rule of choosing the class via the sign of the discriminant function, it would be wrong only for those points having a positive value in Figure 1.16 (and a negative value in Figure 1.17 for the other class). We have nearly performed a miracle. Not only have we greatly simplified the task of the classifier, but we have also reduced the number of input variables from four to one! Not bad.

One final warning, along with a little advice, is in order. The power of this technique is tremendous. It allows us to compute quickly and easily relatively few variables that can be shown to be optimal classifiers under many conditions. Computation of linear discriminant functions as new inputs for neural networks or other models is highly recommended. However, these new variables *should not be relied on alone*. Remember that they are strictly linear. Much information that could be of great use to nonlinear classifiers like neural networks has been discarded. Cautious researchers will use linear discriminant functions *in addition to* other carefully selected variables, perhaps even including some or all of the variables that went into the new factors. Consider them a free (or very cheap) ride. They are quick and easy to compute, we will never be burdened with many of them since there can be, at most, one less than the number of classes, and the information they provide generally ranges from good to excellent. It's an offer you shouldn't refuse.

Some alert readers may be confused over the fact that the plotted values of the discriminant functions are positive for one class and negative for the other class. Straightforward multiplication of the original variables by the weights listed in the audit log file does not produce these results. What happened? The answer is that the original variables are automatically (and invisibly to the user) centered before the weights are applied. This subtraction of the grand mean vector induces some nice mathematical properties, something that is surely hinted at by the apparent balance between positive and negative values observed in this example. Readers who are interested in the complete details should see Masters (1995) for a rigorous mathematical explanation of the entire process.

Finally, the file ORTHOG.CON in the \EXAMPLES directory of the accompanying disk contains the full text of several orthogonalization techniques very similar to the examples appearing in this chapter.

Data Reduction and Orthogonalization 27

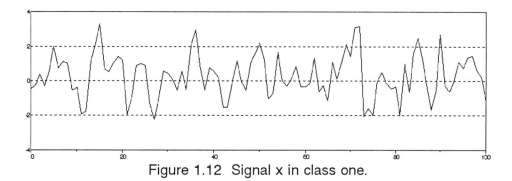

Figure 1.12 Signal x in class one.

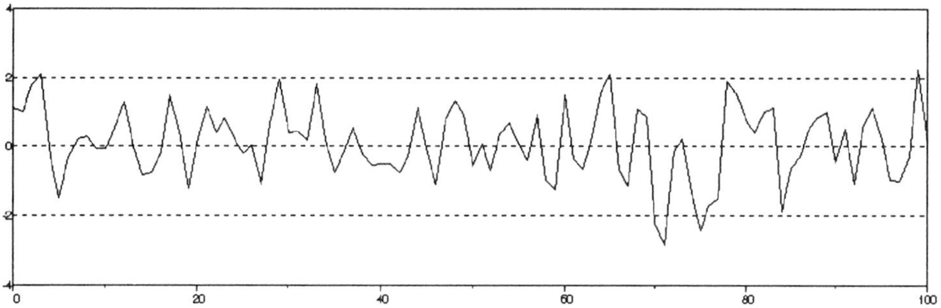

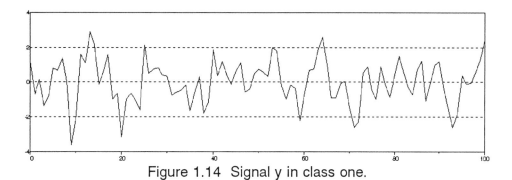

Figure 1.13 Signal x in class two.

Figure 1.14 Signal y in class one.

Preprocessing

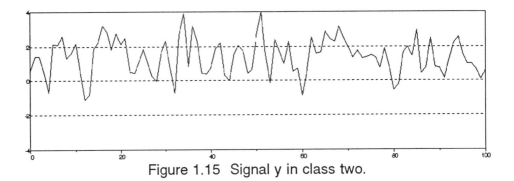

Figure 1.15 Signal y in class two.

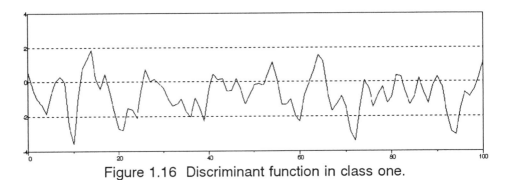

Figure 1.16 Discriminant function in class one.

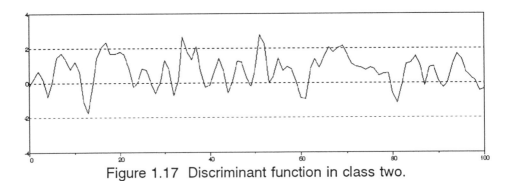

Figure 1.17 Discriminant function in class two.

Filtering for Information Selection

This section contains motivational and instructional material concerning the use of lowpass, highpass, and bandpass filters for selecting important information while suppressing undesirable information in a time series. Closely related information can be found in two other areas of this text. Chapter 4 contains technical details on the mathematical and computational aspects of filters. Some readers might want to skip ahead temporarily and read Chapters 3 and 4 before proceeding in this chapter. However, this is not necessary. Also, Chapter 5 contains examples of several real-world problems in which filtering plays a major role. That chapter also illustrates use of the NPREDICT program (included on the accompanying disk) to filter time series. The material in that chapter is an excellent supplement to the relatively elementary and introductory material here.

What Is a Filter?

Chapter 4 will go into considerable detail about the nature and process of filtering a time series. In this section, we will be satisfied with an intuitive approach. Everyone who listens to music is aware that the sounds emanating from the source contain a variety of frequencies. The low pedal notes of a pipe organ, or the thump of a bass drum, may be so low in frequency that we can actually feel the vibrations in our bodies. The shrill sound of a fife may pierce our skull with its extremely high-frequency notes. A similarly large range of frequencies may be present in the daily closing price history of a commodity. There may be a deep, slow cycle that rises and falls over decades. If the commodity is seasonal, an intermediate-frequency periodic component may repeat once every year. And there are always the rapid daily fluctuations that keep the spice in a commodity trader's life.

Whether our time series is a sound sampled thousands of times per second, or the daily closing price of pork bellies, or the temperature inside an industrial reaction chamber measured every five seconds, the various periodic components in the time series each contain different sorts of information. Some of this information may be useful for answering certain questions, while other bits of information may be superfluous random noise. The series that we observe and record is an amalgam of the many individual pieces of information.

A filter is a means of separating the various periodic components of a time series into individual series. There are three basic filter types. A *lowpass* filter passes only those components whose frequency is lower than a specified threshold. This filter is mainly used for smoothing a time series. After lowpass filtering, a noisy series will have the rapid jitter removed, leaving only the slower variations. A *highpass* filter does just the opposite: Relatively slow variation is removed, leaving only the rapid changes. Finally, a *bandpass* filter passes only those components lying within a specified range of frequencies. Although this type of filter is sometimes used to emphasize a particular component, its most common use is to *remove* a troublesome source of deterministic noise. This will be discussed later.

Why Filter a Time Series?

There are three primary reasons why we might want to filter a time series before invoking a neural network or other model for making predictions. These are:

- Troublesome sources of noise can be reduced.

- Important sources of information can be emphasized.

- A single series that is an unwieldy conglomeration of important sources of information can be split into two or more bite-size pieces. Our neural network will be grateful.

Examples of each of these uses will appear soon. However, it is good to discuss them in general terms before delving into specifics. Let's start with the first item. When might we want to use a filter to remove troublesome noise? A classic example is smoothing. Suppose we are tracking a measured variable in an industrial manufacturing process. This variable (and probably some other information) is used by a neural network to predict when the reaction will be complete. But what if the measurement is subject to significant jitter? Every time we take a sample, the observed value is randomly too high or too low, with a set of consecutive measurements only tending to stay near the unknown correct value. This does not help the neural network to do its job. A good solution is to pass the raw data continuously through a lowpass filter. The noise that masks meaningful information will be removed, with only the clean data getting through to the neural network.

The second point is similar to the first. Emphasizing important information is closely related to removing troublesome noise. The difference is largely philosophical: Do we know more about the noise or about the information? In the example just presented, we apparently did not know a lot about the nature of the information, but we knew there was meaningless high-frequency noise to be suppressed. Now let's look at a slightly different example. Suppose we are following the price of a publicly traded stock. Perhaps our model assumes that short-term price fluctuations are vital indicators, while the overall level of the stock price and slower variations are irrelevant to our task. We want to emphasize the short-term movements. A highpass filter is the answer. Only rapid local variation will be presented to the model, all slower variation having been suppressed.

The third point is probably the most important, yet it is commonly overlooked. More often than not, the information in a time series arises from different processes, each of which presents itself as an individual component that varies at a characteristic rate or range of rates. When these separate information sources are combined into a single series, their individual contributions become lost in the sum. At some time slot, one component might be unusually high, conveying a vital piece of information, while another component might be unusually low, providing another vital signal. But the net effect is a sum near zero, which will probably not produce any excitement at all. This problem is far more common and serious than most people realize.

This effect is illustrated in Figure 1.18. The original series, shown at the top of the figure, is not particularly interesting. But the lowpass-filtered version of the series clearly reveals a slow underlying motion whose value at any given time might well be of use. And the highpass-filtered version contains much potentially useful information about local variation around the general height. In particular, the sharp downward spike about two-thirds of the way across does not stand out in the original series because it comes at a time when the slow component is unusually high. By providing both sets of information to the prediction model, its job is greatly simplified. Even though there are now more inputs for the model to handle, the information content of the inputs is far better defined than would be the case with the original measured series. This is virtually always a tradeoff well worth making, especially for neural networks.

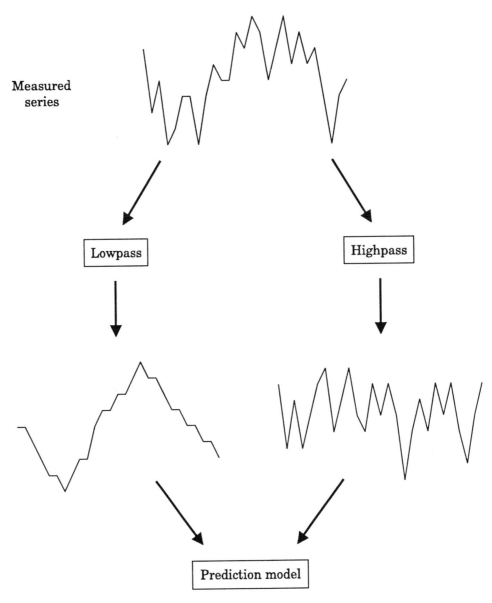

Figure 1.18 Splitting into separate bands reveals hidden details.

Filter Parameters

There is a vast sea of filter designs from which one can make a choice. This text employs only one basic type, with all filters used here belonging to this one family. Chapter 4 provides details concerning the exact design and construction of the members of this filter family. For now, be aware that these filters, whether lowpass, highpass, or bandpass, are characterized by two parameters: frequency and width. For lowpass filters, the frequency is the cutoff below which periodic components are passed and above which periodic components are obstructed. The width is the transition range over which the response of the filter goes from one extreme (unimpeded passage) to the other extreme (total cutoff). As will be seen in Chapter 4, it is impossible to design a practical filter that has a sharp transition, fully passing all periodic components below the threshold and fully stopping all components above it. There is always a gray zone of frequencies that are partially passed, and the size of this gray zone is a crucial determinant of the filter's performance.

Highpass filters have the same two parameters, but their operation is obviously the reverse of lowpass filters. Periodic components whose frequency is above the specified cutoff frequency are passed, while those below are cut. There is also a gray area of partial passage.

Bandpass filters of the type used in this text have the property that there is one single frequency, called the *center frequency*, that they fully pass. Frequencies on both sides of this center frequency are cut, with the degree of passage decreasing as the distance from the center frequency increases. The width parameter determines the width of the frequency range over which passage is relatively unimpeded.

The frequency parameter is always in the range 0.0 to 0.5. A thorough discussion of the meaning of this parameter can be found in Chapter 3. The important point is that *the reciprocal of the frequency parameter is the period of the periodic component*. For example, suppose we measure a variable once a day, and we want to keep only those components whose period exceeds 30 days. We would employ a lowpass filter whose frequency is 1/30 = 0.0333. If we want to keep only those components whose period is less than 14 days, we would use a highpass filter whose frequency is 1/14 = 0.071. Finally, suppose we measure a variable once a month, and we want to focus on annual variation, with slower trends and more rapid movement eliminated. We would use a bandpass filter having a frequency of 1/12 = 0.0833.

The width parameter is trickier to specify. There is no simple calculation to provide the correct value. It is an arbitrary choice. Unfortunately, a real tradeoff is involved. Temporarily turn to page 113 and look at Figure 4.5. This shows the effect of three different width parameters on the width of a bandpass filter. For a lowpass or highpass filter, exactly the same shape holds in the transition region. (See, for example, Figure 4.6 on page 114.) The smallest width, 0.01, produces a very sharp transition. If this width were used for a lowpass or highpass filter, the gray area of partial passage would be very small. The transition from full passage to full attenuation would be almost instantaneous as far as frequency is concerned. The largest width in that figure, 0.1, produces quite a wide transition region. The taper from full passage to complete cutting is slow and gradual. There is an extensive range of frequencies in the gray area of partial passage.

Our initial inclination, especially for lowpass and highpass filters, would be to make the width tiny. After all, a rapid transition from full passage of the components we want, to full blockage of the components we don't want, seems desirable. Unfortunately, there is a high price to pay for a rapid transition. In order to filter a time series, contiguous groups of observations are examined. To compute, say, the filtered value for the 150th time slot, the filter may utilize the original values of the time series for time slots 100 through 200. This is an unavoidable fact of filtering. Nevertheless, it is disagreeable. There is something unsettling about an observed value at time 100 affecting the filtered output at time 150. If the observations were made monthly, an observed value for some month would have an impact on filtered values 50 months (over four years) before and after that observation. A little more locality would be nice.

Therein lies the tradeoff. Sharper transitions imply less locality. In fact, there is a simple rule-of-thumb that applies to the filters used in this text. Divide 0.8 by the width parameter to get the most distant observation that affects each filtered output. Naturally, there is no magic cutoff. The observations nearest the output time slot affect that time slot most strongly, with the effect gradually tapering to zero as the time distance increases. The heuristic limit computed with this formula is the time distance at which the impact of observations has dropped to insignificantly small values. For example, the sharpest filter of Figure 4.5, having a width of 0.01, would have a time extent of 0.8/0.01 = 80 observations. In other words, each filtered output is affected by original series values up to 80 observations earlier and later. The widest filter has a time extent of 0.8/0.1 = 8 observations. This is extremely reasonable. By the way, readers who would like more details on this phenomenon, which is essentially the Heisenberg Uncertainty Principle, should see Masters (1994) for a thorough treatment.

In practice, choice of a width is usually not too difficult. Figure 4.5 on page 113 is always a good place to start. This figure can be used to estimate, for each of the three displayed widths, the degree of attenuation at any distance from the specified frequency. Users who need more accuracy can always use Equation (4.2) on page 113 to compute attenuation for any frequency and width. However, going to that extreme should almost never be necessary. A width of 0.05, having an associated time extent of 0.8/0.05 = 16, is a good all-purpose value that should suffice for many applications. If the frequency was chosen arbitrarily and its exact value is not too critical, move up to a wider width to reap the benefits of shorter influence distances. If, for some reason, a sharp dropoff at the specified frequency is critical, use a smaller width, but recognize that the price is less localization in time.

A Basic Filtering Example

It's time for a filtering example. The three major filter types, lowpass, bandpass, and highpass, are demonstrated. Figure 1.19 plots monthly averages (auction results) of three-month T-Bill interest rates from 1931 through 1994. The variation is obviously a combination of slow trends and rapid fluctuations.

Filters can be used to separate these components. Figure 1.20 is the output of a lowpass filter. The extensive smoothing allows this new dataset to convey an excellent indication of the overall level of interest rates without the data being contaminated by distracting fluctuations.

At the other extreme, Figure 1.22 is the output of a highpass filter. Large-scale trends are totally absent. Instead, this new dataset captures the very short-term changes that cause the interest rates to deviate rapidly from the slow trends. This series obviously presents an entirely different perspective on interest rates than the lowpass series.

Finally, Figure 1.21 shows what's left in the middle. These are the intermediate-term changes that reflect neither long-term trends nor extremely rapid changes. Although this series bears a superficial resemblance to the highpass series, close examination reveals many interesting differences.

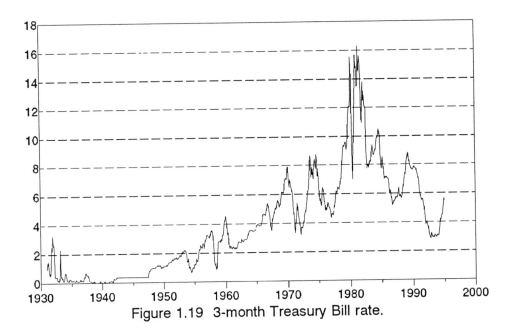
Figure 1.19 3-month Treasury Bill rate.

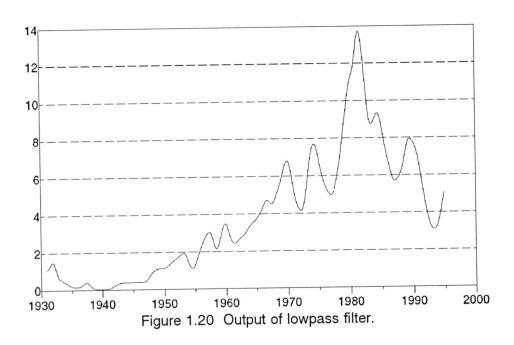
Figure 1.20 Output of lowpass filter.

Filtering for Information Selection

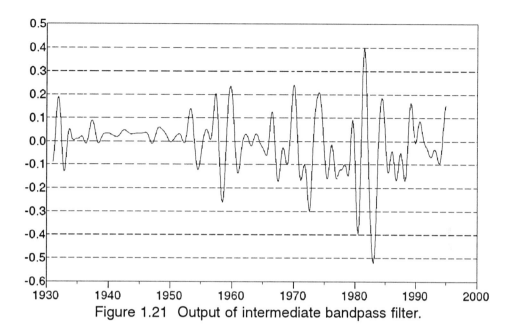

Figure 1.21 Output of intermediate bandpass filter.

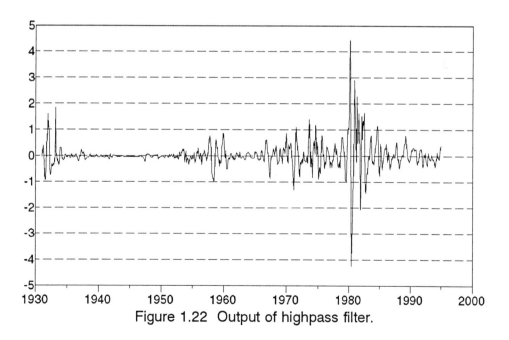

Figure 1.22 Output of highpass filter.

An Alternative Bandpass Filter

The discussion so far implies that the only way to apply a bandpass filter to a series is to filter the series directly, using a single bandpass filter having a specified center frequency and width. This is a common and easy approach. However, there is an alternative that is often more appropriate. In fact, the series just shown in Figure 1.21 was produced with this alternative method.

The inspiration for the method is this: In many situations it is reasonable to assume that *all* of the information in the original series is important. We do not want to discard any of it. But naive (or perhaps even sophisticated) application of a lowpass filter, a highpass filter, and one or more intermediate bandpass filters may discard some information. At best, the total information content of the filter outputs will inequitably represent the original information, emphasizing some frequencies and downplaying others. This may be acceptable for many applications, but it is not necessarily desirable.

The solution relies on the fact that the filters in this text (and the NPREDICT program supplied with the text) are linear. In particular, a complex filter consisting of several individual simple filters can be applied to a series by separately applying each simple filter to the series and summing their outputs. So suppose we have applied a lowpass filter and a highpass filter to a series. To ensure that all of the information in the original series is equitably captured, we want to apply a bandpass filter that is the exact complement of the first two filters. Direct computation and application of this filter would be relatively difficult from a logistical standpoint (though quite easy mathematically). But linearity provides the user with a simple method for accomplishing this goal: Subtract the lowpass and highpass filtered outputs from the original series. The net result is exactly the same as applying a complementary bandpass filter to the original series. Not only has the information in the series been split into three separate bands, but the sum of these filtered outputs is equal to the original series. The information split is perfect. As a general rule, this is the preferred approach. Direct bandpass filtering is appropriate primarily in situations in which it is known in advance that some (usually narrow) range of frequencies is of special interest. Alternatively, we may know that a band is undesirable (60-cycle hum, for example) and must be eliminated. This is done by subtracting the bandpass output from the original series. But unless there is reason to do otherwise, bandpass filters should usually be implemented by subtraction.

More information on bandpass filtering by subtraction can be found elsewhere in this text. Chapter 4 discusses the specific methods used in the NPREDICT program. Also, Chapter 5 contains examples of the technique in the context of solving real-world problems with real-world data. Finally, some readers may be curious about the parameters used to generate Figures 1.20 through 1.22. The lowpass filter has a cutoff frequency of 0.0 (a common choice that results in attenuation starting immediately) and a width of 0.04. By referring to Figure 4.5 on page 113 (or Equation (4.2), which produced the figure), it is seen that this filter has dropped its response significantly by a frequency of 0.04 cycles per sample. This is equivalent to a period of $1/0.04 = 25$ months. The lowpass filter passes practically no components having a frequency of twice the width, corresponding to a period of $1/0.08 = 12.5$ months. Monthly variation within a year is nearly totally suppressed.

The highpass filter has a frequency of 0.08 and a width of 0.04. Given the fact that the legal range of filter frequencies is 0.0 to 0.5, a cutoff frequency of just 0.08, corresponding to a period of $1/0.08 = 12.5$ months, seems tiny. However, this is a common choice. Because of the reciprocal relationship between the frequency and the period, most of the interesting information in a series is usually concentrated in the lower end of the legal frequency range. The entire remainder of the legal range, from 0.08 to 0.5, concerns only components having a period of from two months to 12 months. Although in terms of frequency, 0.08 is very near the bottom of the range, in terms of period it is just about in the middle (in some sense!). By using a cutoff of 0.08 for the highpass filter, we capture all of the variation within each year.

What about the bandpass filter output obtained by subtracting the lowpass and highpass outputs from the original series? There isn't much left for it, but what is there is worthwhile. The lowpass filter started at zero, dropped substantially by 0.04, and cut off almost completely by 0.08. The highpass filter started at 0.08, dropped substantially by 0.04, and cut off by zero. All that's left is a frequency range centered at 0.04 (a period of about two years) that contains the information in the partial gap caused by the two primary filters having reduced response there. But this is often a useful band. It contains neither the slow multiyear trends nor the rapid monthly fluctuations. Sometimes this information is of little use. But even so, there is always the fact that, if nothing else, it provides the missing link connecting the lowpass and highpass outputs with the original series. Why take a chance by discarding this information?

Another Argument for Information Preservation

The preceding discussion encouraged the reader to preserve all of the information in a time series if at all possible. The bandpass filter example demonstrated an obvious choice. However, information can be inadvertently discarded quite easily. Let's look at a general scenario. A popular sequence of events for predicting a time series consists of three steps:

1. Remove slow trends.
2. Use a model to predict the remaining variation.
3. Reinsert the trend, extending it in some way.

This is a potentially dangerous approach. The slow trend that is removed may provide vital information about the underlying physical process. For example, suppose we are following the price of a publicly traded stock. We may use a filter to remove large-scale variation, being concerned only with variation that spans a period of several months at most. But by doing so, we are eliminating the possibility that the absolute price of the stock influences the nature of the local variation. What if one pattern of daily or monthly variation applies when the price of the stock is relatively high, and a totally different model applies when the price is low? It is *crucial* that absolute level information be supplied to the prediction model.

Another way in which valuable information can be accidentally discarded is by differencing a series. This is discussed in more detail in Chapter 7. However, since we are on the subject of information preservation, and since differencing is really not much more than a potent highpass filter, a point should be made. Many people do not try to predict the true value of a series. They predict the *change* of the series relative to the previous observation. There is nothing wrong with this procedure. The problem is in how it is implemented. If the raw series is differenced, and only the differences are given to the model, trouble is likely. *Both* the raw data and the differences should usually be supplied to the prediction model.

The bottom line is this: Never blithely throw away information. You never know what might be important. If a lowpass filter is applied to a series, and the output is subtracted from the original series to remove slow trends, give the lowpass output to the model, too. The same applies to noise elimination with a highpass filter. Those high frequencies might be more than just noise!

Revealing Trends

Sometimes we do not care about predicting exact values of a series. Instead, we may be most interested in relatively long-term trends. Economists, social scientists, and marketing executives often fall into this camp, while industrial control engineers probably do not. Trends are easily computed by first removing high-frequency noise from a series via a lowpass filter, then differencing the output. A second lowpass filter applied to the differences may be beneficial. Figure 1.23 is the result of differencing, then lowpass filtering (cutoff=0, width=0.04) the series shown in Figure 1.20 on page 36. Note how clearly the long-term interest rate trends are revealed. Some experts contend that the most profitable stock market trading strategies are based on predicting this sort of trend series rather than predicting actual prices or price changes.

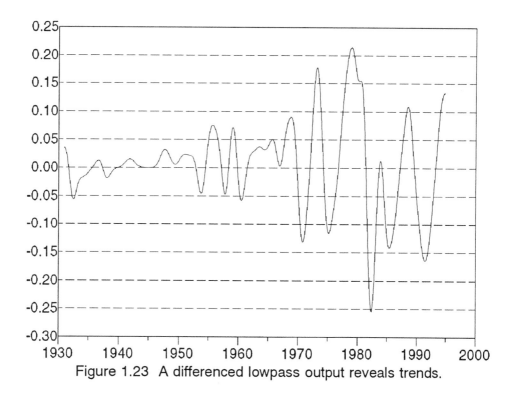

Figure 1.23 A differenced lowpass output reveals trends.

2

Subduing Seasonal Components

- Moving average filters

- Seasonal differences

- Splitting harmonics

This short chapter is a continuation of the previous chapter in that it involves preprocessing. However, the topic of dominant periodic (or *seasonal*) components contains so many diverse and important subtopics that it deserves a chapter of its own.

Chapter 1 discussed the use of filters to separate different groups of periodic components so that each band can be handled in relative isolation. It was also stated that bandpass filters are capable of emphasizing or cutting specific (usually narrow) bands. This may lead to a dangerous temptation. Many time series, especially those found in econometrics or in studies of natural phenomena, are dominated by a strong periodic component. Sales figures of a popular Christmas gift have tremendous annual cycles. Sunspots have a well-known periodicity. Very often, a cyclic component so overwhelms all other aspects of the data that it absolutely *must* be handled separately in order to effectively process all of the information in the series. It is tempting to apply a narrow bandpass filter to the series to isolate the periodic component, then subtract the filtered output from the original series to remove the seasonality. *This is almost always a poor solution.* This chapter briefly examines why this is so and offers several effective alternatives.

The Problem of Harmonics

The reason that a single bandpass filter is not usually effective at removing seasonal components is that most repetitive phenomena contain additional components that repeat at multiples of the basic rate. A pure periodic component, at least from the viewpoint of a bandpass filter, is a smooth sine wave similar to that shown in Figure 3.1 on page 64. The seasonal components found in practice have more rapid cycles within the outermost cycle. Consider monthly measurements that (approximately) repeat yearly. This is typical of sunscreen sales, outdoor temperatures, employment rates, and sundry other series. There is an obvious cycle that repeats once per year. But within that year are smaller ups and downs caused by components that repeat twice a year, three times a year, four times a year, and so forth. These variations at integer multiples of the basic component are called *harmonics*.

Usually, we are not directly concerned with harmonics. The important point is that harmonics are the reason a bandpass filter optimized for the period in question is not effective. Even if a bandpass filter eliminates the yearly component, the harmonics remain and cause problems by continuing to mask subtle indicators. However, we will soon see an example demonstrating possible interest in and use for the harmonics. In this (relatively rare) situation, a bank of bandpass filters is required, one for each harmonic. This can quickly get out of hand.

Moving Averages and Alternatives

One of the most popular methods of removing a seasonal component is by means of a *moving average filter*. When the period to be removed is of odd length, this filter averages one period of contiguous points to compute the filtered output for the time slot of the central point in the contiguous group. When the period is even, it uses one additional point, but it weighs the outermost two points half as much as the inner points. For example, suppose we are dealing with monthly data, and we want to remove annual seasonality. To compute the filtered output for June, compute the sum of the original value for June, plus the previous five months (January through May), plus *half* the previous December, plus the next five months (July through November), plus *half* of next December. Divide by 12 to get the filtered value for June. Repeat this procedure for every month. It can be shown that this process removes all vestiges of annual seasonality from the series, including the harmonics.

There is a lot to recommend a moving average filter. It certainly is easy to compute. And it does a thorough job of removing seasonality. However, it has significant drawbacks. Its effect on the various other frequency components in the series, those not related to the seasonal cycle, is irregular. It is a *kind* of a lowpass filter, but not a really good one. A great deal of high-frequency energy gets through a moving average filter. At the same time, this high-frequency energy is not totally unimpeded. It gets partially cut, and the degree of cutting is variable. To summarize, the effect of a moving average filter on a time series is not straightforward. There are more advanced and improved versions of the filter that repeatedly apply filters of varying length. However, in the end, it's still a primitive filter with numerous problems. It does not deserve its popularity.

Even though filter response functions are not rigorously discussed until Chapter 4, an intuitive examination of the filters presented in this chapter is beneficial. Look at Figures 2.1 through 2.3.

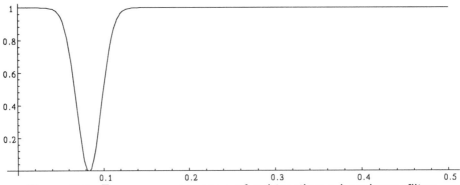
Figure 2.1 Frequency response of subtracting a bandpass filter.

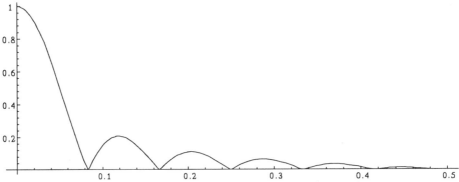
Figure 2.2 Frequency response of a 12-month moving average.

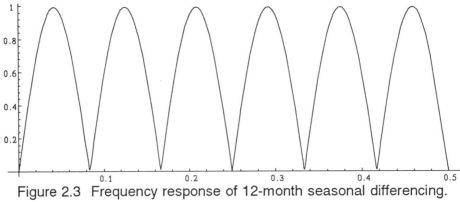
Figure 2.3 Frequency response of 12-month seasonal differencing.

If a narrow bandpass filter centered at a frequency of 1/12 = 0.083 is applied to a time series, and the filtered output is subtracted from the original series, the effective filter that results has a response function similar to that shown in Figure 2.1. The fundamental frequency of the seasonal component (annual cycles) is eliminated. But the harmonics that appear at multiples of the fundamental frequency are fully passed. This approach does little or no good.

Figure 2.2 is the frequency response of a 12-month moving average filter. Not only is the annual component eliminated, but the cycles that repeat 2, 3, 4, 5, and 6 times per year are eliminated as well. This thoroughness is necessary in order for the filter to do a good job of removing the confusing effects of seasonality. However, it is not perfect. Observe how much high-frequency energy still gets through to a small but significant degree. This inconsistency is annoying at best.

There are alternatives. Multidimensional moving average filters exist. They repeatedly apply filters of varying lengths. Although this approach is *sometimes* an improvement over the basic model, it is a kludge and is not pretty. No more will be said on this issue.

Another technique is to apply a lowpass filter that cuts off slightly below the fundamental frequency. The output of this filter contains no remnants of the seasonal component, since the harmonics are all above the fundamental frequency. This output can be subtracted from the original series to recover the seasonal information plus all other information. Often, these two series alone are sufficient. Optionally, the residual difference series can be split using one or more bandpass filters. The exact nature of the splitting depends on the application and is left to the reader's discretion.

The final alternative is called *seasonal differencing*. This is an excellent choice in many applications. It is performed by subtracting from each observation the value exactly one period earlier. In other words, to compute the annually differenced series value for the month of January, subtract the previous year's January value from this year's value. Do this for every month. Figure 2.3 is the frequency response of this operation (which really *is* a filter). The fundamental and all harmonics are fully eliminated, just like a moving average filter. But the other frequencies are preserved in a much more consistent manner. The slow trends are also strongly cut, so if this information is important (and it almost always is), a lowpass filter must be used to supply slow variations to the prediction model. Nothing more about seasonal differencing will be said now, as a large part of Chapter 7 is devoted to this topic. Simply be advised that it is an excellent technique.

A Seasonality Example

Let's examine these seasonality options in the context of a real dataset. Figure 2.4 is the M1 money supply (not seasonally adjusted) for the years 1960 through 1994. The obvious multiplicative trend is easily removed by taking the log and differencing once. (The motivation for this is made clear in the beginning of Chapter 1.) Figure 2.5 is the result of this pair of operations. Note the obvious annual cycle.

The structure of the seasonal component is not clear with such a large-scale display. Figure 2.6 is a magnified view of the beginning of the series. Observe that the annual variation is not at all like the smooth curve shown in Figure 3.1 on page 64. A lot of rapid variation within each year is apparently present. This is confirmed in Figure 2.7, which is the power spectrum of the differenced log series. The spectrum will be thoroughly discussed in Chapter 3. For now, note the narrow peaks at the annual fundamental (very weak!) and at each harmonic. For clarity, the horizontal axis of this graph is labeled in cycles per year instead of the 0.0–0.5 that has been used for filter response functions.

Figure 2.8 is the result of seasonally differencing the series. All traces of regular repetition have been removed. Moreover, this operation has revealed several interesting features that were completely obscured in the raw series (the differenced log). The sharp spikes around 1981 and 1987 are invisible in the raw data but obvious after seasonal differencing. It is likely that a wealth of other information not so plain to the naked eye has also been uncovered. It should be emphasized again that seasonal differencing is an extremely powerful tool when dealing with series having a strong seasonal component.

The elimination of the annual fundamental and its harmonics is apparent in Figure 2.9, the power spectrum of the seasonally differenced series. The fact that the spectrum does not completely touch zero at these points is due to the smoothing nature of the maximum entropy method used to compute this spectrum. This method is discussed in Chapter 3.

Figure 2.10 is the series produced with a 12-month moving average filter. It is a combination of the slow trends plus a very significant quantity of rapid local variation. This ugly mixture is typical of moving average filters and is a major argument against their use. Although the high-frequency variation included in this filtered output probably contains some useful information, its value is compromised by the fact that the selection process is not straightforward (as evidenced by Figure 2.2). A further complication ensues when the filtered output is subtracted from the original series to procure the seasonal information. This is illustrated in Figure 2.11. The problem is that the nature of the residual series is not well defined. Since many nonseasonal components are present in the filtered series at different levels, they will be removed by subtraction in unusual ways. Interpretation of the residual series is not easy.

Figure 2.12 is the result of applying a lowpass filter to the differenced log series. Its cutoff is 0.0 with a width of 0.03. This causes the filter to reject almost totally all frequencies above twice the width, 0.06. This is comfortably below the fundamental frequency of 1/12 = 0.0833. Compare this figure to the moving average output in Figure 2.10. All of the slow trends in the original series are preserved in the lowpass output, making this filtered series an excellent indicator of overall level. The moving average contaminates the trends with troublesome jitter. This jitter is quite possibly important, and it should probably be supplied to the prediction model. But mixing it with the trend series is not the right approach! The combination of the two sources of information is potentially confusing. The lowpass output is clearly a superior vehicle for conveying level information.

In order to retain the seasonal and other high-frequency information, the lowpass output should be subtracted from the unfiltered series. This difference is shown in Figure 2.13. In many applications, this residual is a good source of information to supply to the prediction model. It contains both the seasonal component (with all of its harmonics) and other rapid variation. However, it may sometimes be appropriate to split the content even more. If there is no strong seasonal component, the best approach is usually to apply a highpass filter to extract the high-frequency variation, then subtract the highpass output from the lowpass residual to isolate the middle band. This was discussed at the end of Chapter 1. Our concern here is with seasonal data. The best approach to further splitting is to use separate bandpass filters at each harmonic. Be warned that this generates a *lot* of new data. The tradeoff should be carefully considered. But for the sake of argument, let us proceed in this direction.

Figure 2.14 is the result of applying a narrow bandpass filter to the lowpass residual (the unfiltered series minus the lowpass output). This filter is centered at the fundamental (0.0833) and has a width of 0.02. Figures 2.15 through 2.19 are similar bandpass outputs (all with a width of 0.02) at 2, 3, 4, 5, and 6 cycles per year, respectively. There is obviously independent information present in each of these series. The degree of importance of this information depends on the particular application.

Perhaps the most interesting series of all is the grand residual. Figure 2.20 was produced by subtracting all of the bandpass filters from the lowpass residual. This series is the remainder after the slow trends and all seasonal information has been removed from the original series. One way of looking at the grand residual is to see it as the information that has nothing to do with regular variation. This is the irregular variation contained in the original series. It is not unusual for this series to be more useful than any of the harmonics.

It is interesting to compare the grand residual series with the seasonally differenced series that appeared in Figure 2.8. They are very similar. There is a 12-month offset due to the fact that the first year in Figure 2.8 is essentially discarded: The differences do not become defined until a full season has passed. But the features in the two series are substantially similar. This is not surprising, since the frequency response of 12-month differencing shown in Figure 2.3 on page 46 is similar to that produced by subtracting lowpass and narrow bandpass outputs from a series. This similarity is driven home even more by comparing the power spectrum of the grand residual, Figure 2.21, with the power spectrum of the seasonal differences, Figure 2.9.

Summary of the Examples

This preceding set of examples illustrated some methods for handling a time series having a strong seasonal component. It was shown that the common traditional method of applying a moving average filter has a serious problem: Its effect on different frequency bands is irregular. A series filtered by means of a moving average is a confusing combination of smoothed trends and rapid local variation. Seasonal differencing was seen to be extremely effective, but full discussion was postponed until Chapter 7.

It was pointed out that a good approach is to split the series using a lowpass filter to isolate the trends, then subtract that series from the original to isolate the residual. Additional splitting of the residual using narrow bandpass filters for each harmonic may be effective, but excessive splitting adds tremendously to the number of variables to be handled. Finally, if such splitting is done, the grand residual may be the most important piece of information of all.

A Seasonality Example

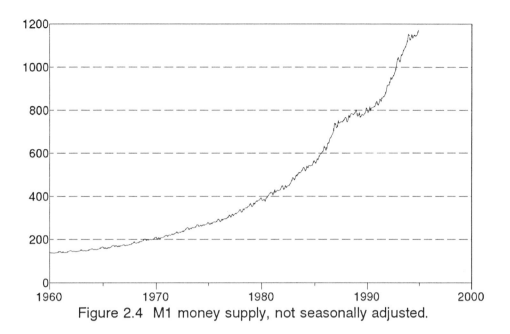

Figure 2.4 M1 money supply, not seasonally adjusted.

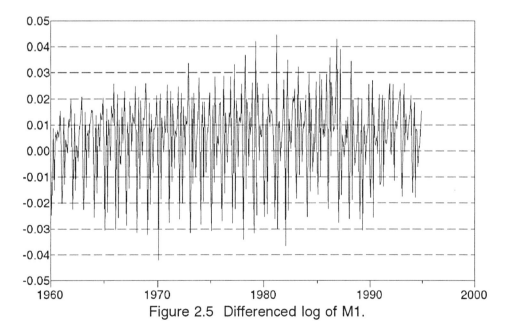

Figure 2.5 Differenced log of M1.

52   *Subduing Seasonal Components*

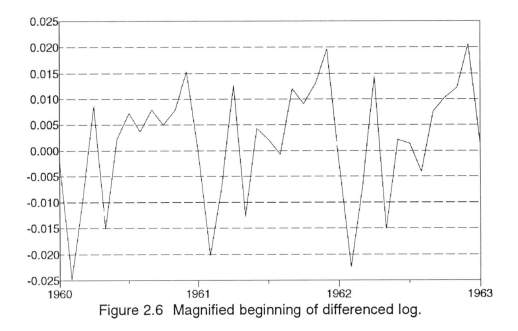

Figure 2.6   Magnified beginning of differenced log.

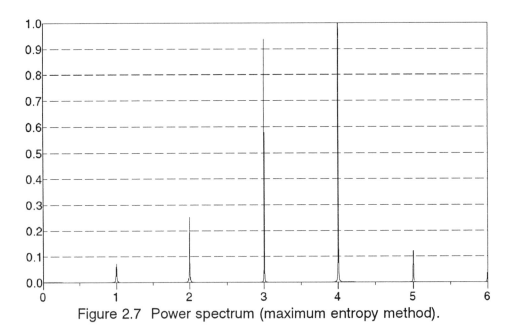

Figure 2.7   Power spectrum (maximum entropy method).

A Seasonality Example

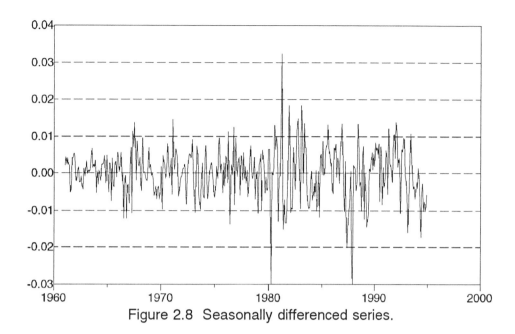

Figure 2.8 Seasonally differenced series.

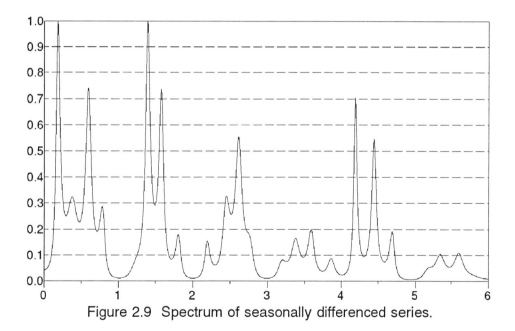

Figure 2.9 Spectrum of seasonally differenced series.

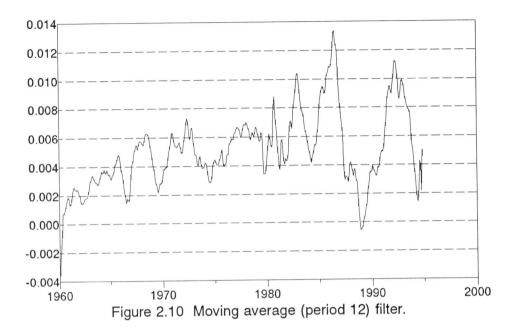

Figure 2.10 Moving average (period 12) filter.

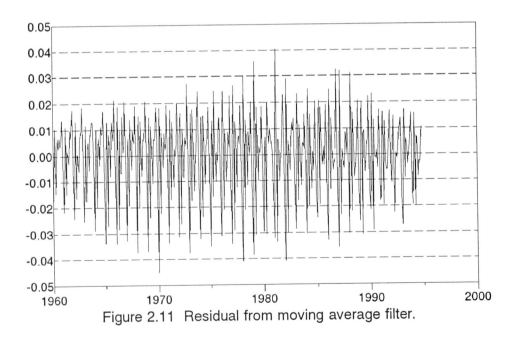

Figure 2.11 Residual from moving average filter.

A Seasonality Example

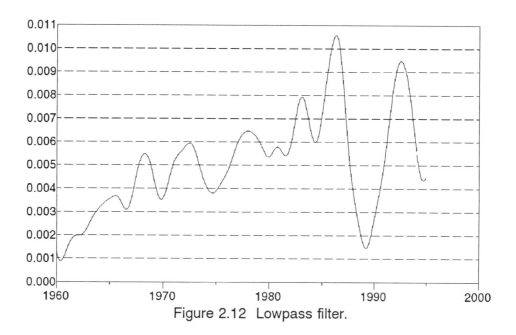
Figure 2.12 Lowpass filter.

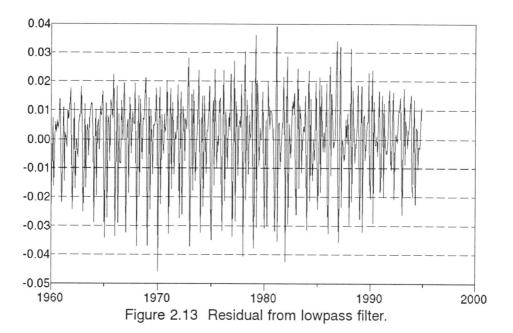
Figure 2.13 Residual from lowpass filter.

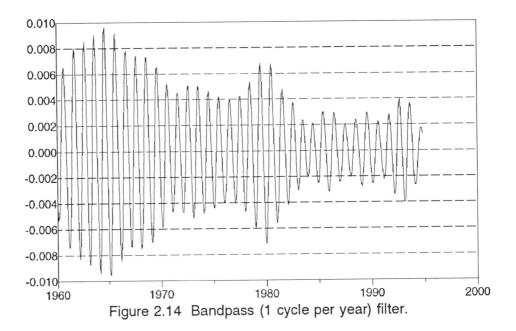

Figure 2.14 Bandpass (1 cycle per year) filter.

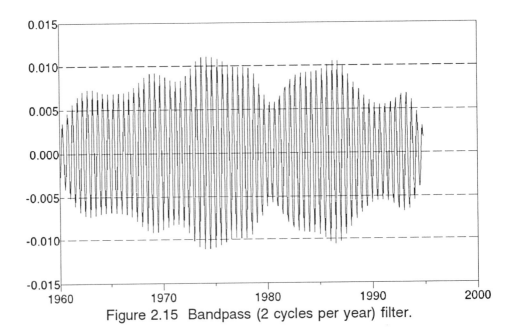

Figure 2.15 Bandpass (2 cycles per year) filter.

A Seasonality Example

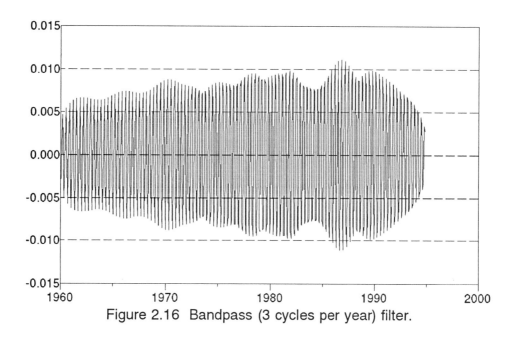

Figure 2.16 Bandpass (3 cycles per year) filter.

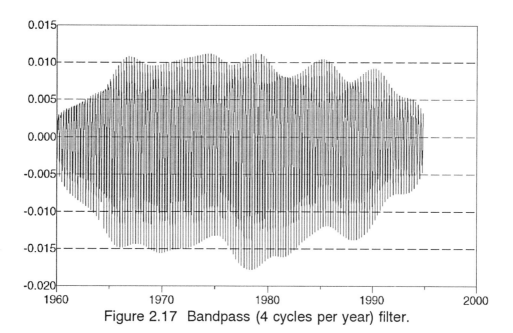

Figure 2.17 Bandpass (4 cycles per year) filter.

58   *Subduing Seasonal Components*

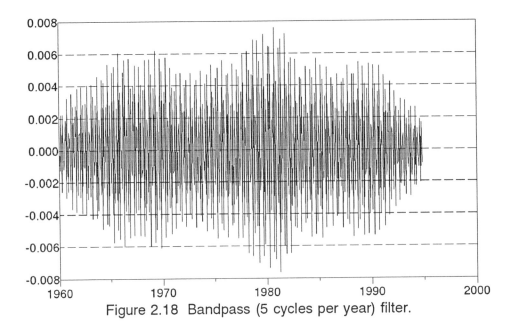

Figure 2.18 Bandpass (5 cycles per year) filter.

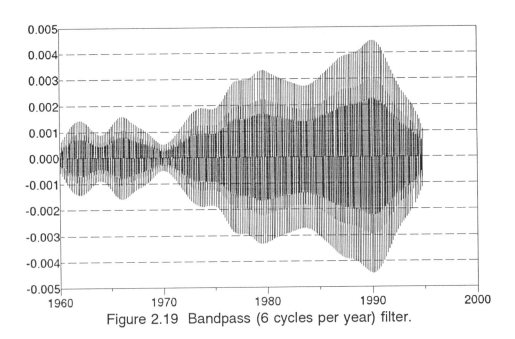

Figure 2.19 Bandpass (6 cycles per year) filter.

A Seasonality Example

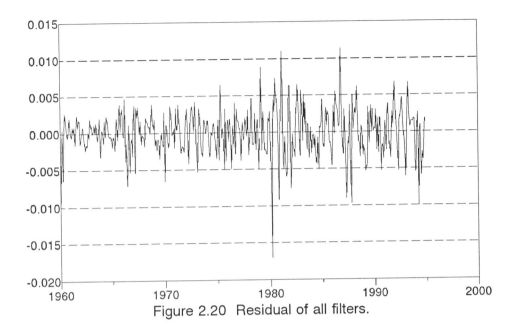

Figure 2.20 Residual of all filters.

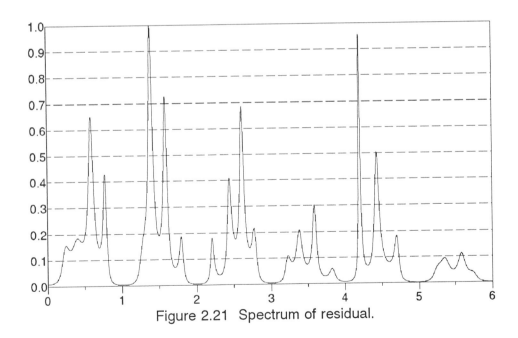

Figure 2.21 Spectrum of residual.

3

Frequency-Domain Techniques I: Introduction

- The discrete Fourier transform

- Aliasing and the Nyquist limit

- Data windows

- The discrete and cumulative power spectrum

- Smoothing with Savitzky-Golay filters

- The maximum entropy spectrum

Material on frequency-domain techniques is scattered throughout this text. Keeping all of that material together in one location is not as helpful to the reader as separating it according to use. Therefore, this chapter begins with the following directory showing where associated material may be found:

- The use of lowpass, highpass, and bandpass filters for selecting and rejecting certain types of input information is discussed in Chapter 1, *Preprocessing*. Examples in that chapter feature datasets specially selected to illustrate specific points.

- Wavelets as feature detectors are presented in Chapter 5, *Wavelet and QMF Features*. Again, some of the examples in that chapter are built around carefully contrived datasets.

- Power spectrum information can aid the user in selecting ARMA models (Chapter 6) and differencing options (Chapter 7).

- All of these uses for frequency-domain techniques are illustrated by applying the NPREDICT program to real-life data. These actual applications appear in several chapters.

So what is left for this chapter and its sequel? These two chapters cover the mathematical and computational details that are shared by all of the techniques. For example, several other chapters refer to the *Nyquist frequency*. You will find its definition in this chapter. Other chapters may instruct the reader to apply a lowpass filter to a series. Chapter 4 tells precisely how to do so. Readers who are only interested in using NPREDICT or another program to apply frequency-domain techniques may safely skim this chapter. The material here is primarily aimed at two groups of readers: those who want to fully understand the principles underlying the techniques, and those who want to write computer programs to execute the techniques.

This chapter and its sequel are tightly knit and should be read in order. The discrete Fourier transform, which is the foundation of nearly all frequency-domain techniques, is covered here. The power spectrum, while not strictly necessary for understanding the filters described in Chapter 4, provides a solid intuitive concept that helps the reader follow filter developments. Finally, this chapter contains many general ideas, like aliasing, that universally apply to all frequency domain techniques.

Introduction to the Frequency Domain

A time series is a collection of samples of a variable measured at different times. Unless otherwise indicated, it is assumed that the measurements take place at equally spaced intervals. Since the data is spread out across time, the series is said to be in the *time domain*. But this is not the only way of expressing the information contained in the data. There are an infinite number of other domains that could be used to express exactly the same information. For example, in Chapter 7, we see that it may be useful to compute the change of each sample relative to the previous sample. As long as the first point in the original series is available along with the changes, this new series in what may be loosely termed a *differenced domain* conveys exactly the same information as the original series. But even though the two series are equivalent in that they hold the same information, they do not necessarily have the same utility. Turn to Figure 6.20 on page 233. That raw series and the series computed by differencing it plainly highlight dramatically different aspects of the information content. The perfect redundancy of a raw and differenced series is mathematical only. In most practical applications, little or no redundancy would be found.

The same principle applies to other domains in which the information contained in a time series may be expressed. One of the most useful of these is the *frequency domain*. To acquire an intuitive feel for frequency-domain representations of time series, look at Figures 3.1 through 3.6. The first five of these figures are pure waveforms: They are sine or cosine waves that repeat an integer number of times across their timespan. Figure 3.6 is the sum of these five pure waveforms. Observe that even though it is constructed from only five simple waves, it is relatively complex and is reminiscent of many series that might be encountered in actual applications.

These figures depict a situation in which a complicated series may be represented by a set of simple series. Is it *always* possible to break down a complicated series into the sum of simple series? Under what conditions can it be done? Is there a practical way to do it? Are some collections of simple series better than other collections? How many component series do we need? These questions and many others will be answered in the next few pages.

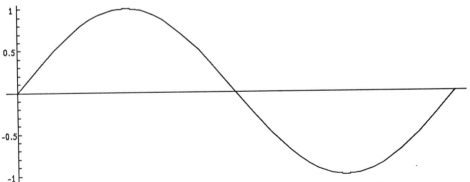
Figure 3.1 Sine; amplitude = 1.0, frequency = 1.0.

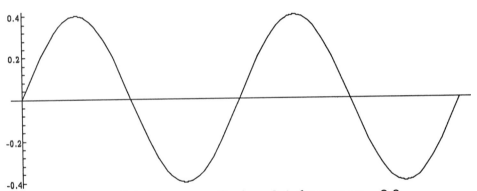
Figure 3.2 Sine; amplitude = 0.4, frequency = 2.0.

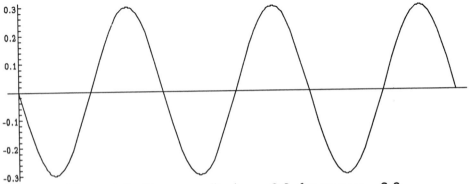
Figure 3.3 Sine; amplitude = −0.3, frequency = 3.0.

Introduction to the Frequency Domain

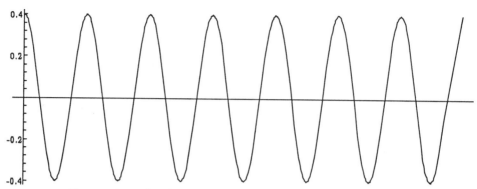

Figure 3.4 Cosine; amplitude = 0.4, frequency = 7.0.

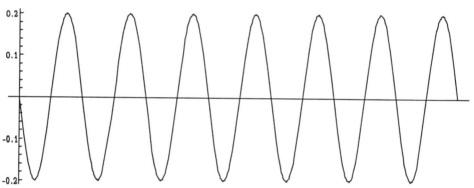

Figure 3.5 Sine; amplitude = −0.2, frequency = 7.0.

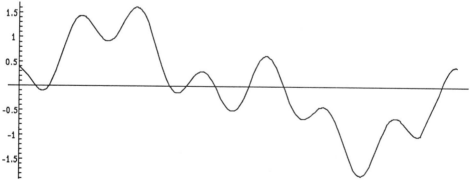

Figure 3.6 Sum of previous five waveforms.

An issue to be briefly examined and then summarily swept under the rug is the issue of *continuous* versus *discrete* time series. The majority of results presented in signal-processing texts concern signals that are defined on a continuous time domain. Values of the signal are known at all times in the extent being studied. The six signals we just studied are continuous. Under very reasonable conditions, any continuous signal can be represented as the sum of pure sine and cosine waves. Despite the great beauty of this fact, it does us no good when we are working with discretely sampled data. For the remainder of this text, we will assume that the time series is discrete, having been sampled at equally spaced time intervals.

The Discrete Fourier Transform

A small but significant issue to be resolved is the precise nature of the frequency-domain representation that we want to use. In (somewhat) mathematical terms, we need to decide upon a *basis* for the representation. A basis is (for our purposes) a set of simple functions that can be combined to produce whatever series we want to represent. In the six figures that were just studied, the basis was a set of sine and cosine waves that repeated an integer number of times during the total extent of the series. There is an infinite number of choices for a frequency-domain basis. Many of them are useful. This text will deal exclusively with one commonly used basis: the one implied by the *discrete Fourier transform* (*DFT*). Like the example just studied, this basis uses pure sine and cosine waves at integer multiples of a fundamental frequency that covers exactly one cycle during the extent of the measured series. This representation has several advantages: It lends itself to simple intuitive interpretability. It is parsimonious in that the DFT representation contains exactly as many numbers as were in the original series in the time domain. It is unique in that any given time series has exactly one representation in this basis, and any given series in this basis has exactly one time-domain series associated with it. Last but not least, it is quick and easy to move back and forth between the time domain and the DFT domain.

The only small disadvantage for some people is that the DFT representation of a time series is complex (in the mathematical sense, as in imaginary numbers). Readers who squirm at the thought of imaginary numbers should not worry, though. Everything will be kept nice and simple. Those who would like a gentle but more detailed review of complex numbers in the context of signal processing should see Masters (1994). This chapter will avoid all unnecessary details.

It's notation time. Let $x_0, x_1, ..., x_{n-1}$ be a time series consisting of n equally spaced samples of a variable. Although, in general, it is perfectly legal for this variable to be complex, we will assume that it is strictly real, as this is the usual situation in time series prediction tasks. Its discrete Fourier transform is a series of n complex numbers that we shall call $w_0, w_1, ..., w_{n-1}$. This alternative representation of the original series may be computed with the formula shown in Equation (3.1).

$$w_j = \sum_{k=0}^{n-1} \left[x_k \cos\left(\frac{2\pi jk}{n}\right) + x_k \sin\left(\frac{2\pi jk}{n}\right) i \right] \quad (3.1)$$

This formula looks fierce, but it is really quite simple. It consists of two nearly identical terms. The first term is the dot product of the time-domain series with a series defined by the cosine function. The second term is the dot product of the series with a sine series. This second term is also multiplied by the imaginary i. For readers uncomfortable with imaginary numbers, the best approach is to think of this new w series as two separate series. One of them, which we shall call the *real* part, is computed using cosines. The other series, which we call the *imaginary* part, is computed with sines. With this mode of thinking, no actual use of complex numbers is needed. Treat the discrete Fourier transform as having provided two ordinary real series as opposed to a single complex series. Mathematically speaking, there is really no great sin in this approach. Feel free to indulge.

It is worthwhile to examine Equation (3.1) a little more closely. What cosine (and sine) series are being dotted with the original series? Suppose $j = 0$, so we are computing the first term of the DFT series. The cosine terms are all equal to one, so the real part of w_0 is the sum of the time-domain series. The sine terms are all zero, so the imaginary part of w_0 is guaranteed to be zero. (This fact will be important later. Remember it.) Suppose $j = 1$. As the summation index k sweeps across its range, the angle in the cosine and sine terms sweeps from zero to just under two pi. In other words, the trigonometric functions sweep out exactly one period across the time extent. In general, these two functions will cover j complete periods when computing w_j.

Frequency-Domain Techniques I: Introduction

Alert readers may now be crying foul. Several paragraphs ago it was asserted that the DFT is parsimonious in that it produced exactly as many numbers in the frequency-domain representation as were in the time-domain representation. Yet we just saw that a time series containing n samples produces a DFT containing n *pairs* of numbers. This is twice what is expected. What gives? The best answer is found by looking at two examples. The following small table shows a 6-point time series in the first column, the real part of its DFT in the second column, and the imaginary part in the third column. Notice anything unusual?

Point	Raw data	Real part	Imaginary part
0	3.0	10.0	0.0
1	1.0	0.5	2.6
2	5.0	-3.5	0.9
3	-2.0	14.0	0.0
4	4.0	-3.5	-0.9
5	-1.0	0.5	-2.6

There is a symmetry around point three. The real parts above and below this center point ($n/2 = 3$) are equal, and the imaginary parts are equal in magnitude but opposite in sign. This is not limited to series having an even number of points. Look at the next table, which shows the effect of adding one more point to the series, bringing n up to seven. The symmetry still exists, although now it is around a fictitious center point at $n/2 = 3.5$.

Point	Raw data	Real part	Imaginary part
0	3.0	11.0	0.0
1	1.0	1.6	3.3
2	5.0	0.2	2.1
3	-2.0	3.2	-10.5
4	4.0	3.2	10.5
5	-1.0	0.2	-2.1
6	1.0	1.6	-3.3

Count how many *unique* numbers there are in the DFTs shown in these two tables. Do not count the imaginary part of the first DFT term, as it is always guaranteed to be zero. Also, do not count the imaginary part of the middle term when n is even. Although we will not prove it here, this number is also guaranteed to be zero. There are six unique numbers in the first table and seven in the second table.

This is equal to the number of points in the time-domain series. These examples illustrate the following important general rule:

- Only the terms from zero through $n/2$ (truncating the fraction if n is odd) of the DFT of a real time series are unique. The imaginary part of term zero is always zero. The imaginary part of term $n/2$ (if n is even) is always zero. Thus, there are exactly n unique numbers defined by the DFT.

Most practical time series applications do not directly use the DFT of a series as a source of features. However, readers who wish to do so should always keep this rule in mind. It would be a terrible error to blindly throw redundant data at a neural network or other model. Take care to use only the unique components of the DFT.

The concept of symmetric redundancy around the center point has many profound consequences far beyond accidental inclusion of superfluous input data. Later, we will see how redundancy can be used to speed computation of the DFT. Also, it will have implications to filtering and convolutions that are vital to wavelet applications. This text is not an appropriate forum for advanced details on DFT theory. Inquisitive readers are encouraged to consult Masters (1994) for a much more detailed treatment of the subject in the context of feature extraction for neural networks.

Computing the Discrete Fourier Transform

Look back at Equation (3.1) on page 67. What needs to be done to compute a term in the DFT? We must multiply each of the n points in the time series by a pair of trigonometric functions, then sum. This must be done for each of the terms in the DFT. The total number of arithmetic operations for a complete DFT computed this way is on the order of n^2. When the series is long, this is a lot of arithmetic. Even fast modern processors can get bogged down in the sheer mass of computation. Thankfully, various trigonometric identities can be used to drastically reduce the workload.

An algorithm called the *fast Fourier transform* (*FFT*) was invented several decades ago. It works by recursively splitting the series in half, transforming each half separately in a quarter of the time that would be taken to transform the entire series, then quickly merging the results. The number of operations in this algorithm is proportional to $n \log n$, a vast improvement over n^2. Unfortunately, the original algorithm requires that n be a power of two. This restriction is, at best, an annoyance. Modern *mixed-radix* algorithms avoid this problem by factoring n into (hopefully) small prime numbers and using an extremely sophisticated method of splitting the series this way. Probably the best explanation of this marvelous algorithm is in Elliot (1987). No details will appear here. Suffice it to say that this text supplies on the accompanying code disk both C++ and Pentium optimized assembler versions of the mixed-radix FFT. Methods for using this set of subprograms are described on page 358 of this text.

One small warning about the mixed-radix FFT is in order. It is tempting to consider this algorithm almost magical since we can give it a series of any length. Remember that its performance rests entirely on its ability to factor n into small prime factors. If you give it a series whose length happens to be prime, it reverts to brute-force implementation of Equation (3.1). You may wait a long time for your results.

The Power Spectrum

We have already seen (from the explanation after Equation (3.1)) that each DFT term w_i is related to a periodic component that repeats exactly i times across the measured extent of the time series. But treating each term as a real part and an imaginary part is tedious and confusing. When what we really want to know about the time series is how much of its energy is due to different frequency components of interest, there is a better way. There are two common ways of expressing complex numbers. One way is to use their real and imaginary parts, which is what we have been doing. The alternative is to express them as a magnitude and a phase angle. This is illustrated in Figure 3.7.

A complex number may be plotted on a Cartesian coordinate plane by letting its real part determine its horizontal position and its imaginary part determine its vertical position. The location of this point may be expressed either as its (x, y) position (where x is the real part and y is the imaginary part), or in polar coordinates as a length and an angle. The Pythagorean theorem tells us that the length (usually called the *absolute value*) of a complex number is the square root of the sum of the squares of the real and imaginary parts. The phase angle is defined by the arctangent of the ratio of the parts.

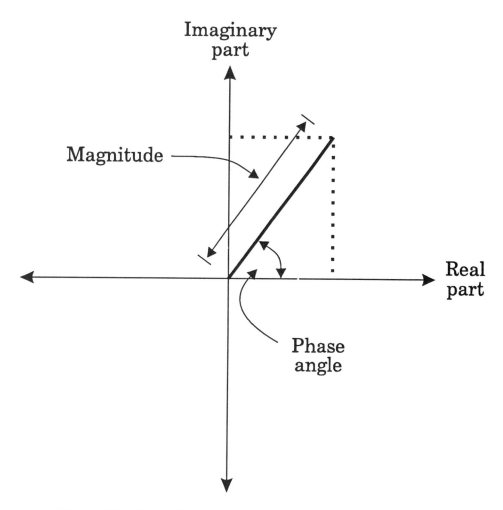

Figure 3.7 Alternative representations of a complex number.

We state without proof that the energy due to a single component of the DFT is equal to the square of its absolute value. This should make sense to anyone with a background in physics or electrical engineering. The mean absolute value of a component is its absolute value divided by n, so the power of a component is computed by dividing its squared absolute value by n^2. Finally, all DFT terms except w_0 and $w_{n/2}$ (if n is even) occur twice due to symmetry. Therefore, the power in these frequency bands should be doubled. This heuristic development culminates in the following definition of the discrete power spectrum (or *periodogram*) of a time series:

$$P_0 = \frac{|w_0|^2}{n^2}$$

$$P_i = 2\frac{|w_i|^2}{n^2} \qquad 0 < i < n/2 \qquad (3.2)$$

$$P_{n/2} = \frac{|w_{n/2}|^2}{n^2}$$

Equation (3.2) is extremely useful. It tells us how to use the DFT to estimate the power contained in each discrete frequency band. The need for this information is expressed several times in this text. On page 229 we see that the power spectrum helps determine the form of an ARMA model. And on page 265, the spectrum is used to assess the need for seasonal differencing. Finally, examination of the power spectrum of the residuals of any model is always advisable. This is a very powerful tool that should be available to everyone who wishes to do time series prediction.

Frequency versus Period

Up until now we have been content with stating that w_i represents a periodic component that repeats exactly i times across the time period defined by the series x_0 through x_{n-1}. It doesn't often happen that the precise length of the time interval over which measurements were taken has any useful meaning. Chances are the interval was almost randomly selected, or perhaps it was given to us by default. In any case, it is rarely useful to speak of a periodic component that "repeats 17 times over the time period from March 12, 1972 to September 28, 1993." It is generally far more useful to speak of a component that, for example, repeats itself every 365 days.

This is easily done. If a periodic component repeats itself i times over n samples, the period of repetition is obviously n/i. We already saw that the highest frequency component in the DFT before the numbers fold back and repeat is at $n/2$. This very special frequency is discussed in the next section. For now, be aware that this tells us that the shortest period resolved by the DFT is two samples. And the longest period is n samples: one repetition across the entire interval.

The Power Spectrum

Sometimes we want to use specific units when the period or frequency is specified. Let the Greek letter *delta* (Δ) be the time interval separating successive samples as measured in the desired time units. Its reciprocal ($1/\Delta$) is sometimes called the *sample rate*, since it is the number of samples per unit time. Then, the highest frequency in the DFT (corresponding to $w_{n/2}$) is $1/(2\Delta)$ cycles per time unit, and its period is 2Δ time units per cycle. The lowest frequency (corresponding to w_1) is $1/(n\Delta)$ cycles per time unit, and its period is $n\Delta$ time units per cycle. The frequency corresponding to w_i is $i/(n\Delta)$ cycles per time unit, and its period is $n\Delta/i$ time units per cycle.

An example may clarify these details. Suppose we measure a quantity four times a year. If we want the basic time unit to be a year, we have $\Delta = 0.25$ years per sample, or a sample rate of four samples per year. Assume we have ten years of data, so $n = 40$ samples. If we compute the DFT of this series, we get 40 complex numbers. But only the first 21 of these (0 through 20) are unique. The other half of the transform folds back in symmetric redundancy. The lowest frequency resolved (w_1) is 0.1 cycles per year, with a period of ten years. The highest detectable frequency (w_{20}) is two cycles per year, with a period of one-half year per cycle. Each w_i corresponds to a frequency of $i/10$ cycles per year (a period of $10/i$ years). We could just as well let the basic time unit be a month, in which case $\Delta = 3$ months per sample. Now, w_i corresponds to a frequency of $i/120$ cycles per month or a period of $120/i$ months.

What about w_0? This corresponds to a frequency of zero cycles per time unit. Electrical engineers call this the DC (direct current) component. In terms of the power spectrum, P_0 is the squared mean of the time-domain series. Again, electrical engineers will be totally comfortable with this fact. Also, statisticians recognize this as the variance component due to a constant offset from zero. For everyone else, just be aware that w_0 has nothing to do with periodic components. It is only the effect of the time-domain series not being centered at zero, and it should almost always be ignored.

There are three principal facts that the reader should remember from this section. These are the following:

- The total duration of measurement determines the lowest frequency that can be detected (one cycle per that time interval). A longer time span provides the ability to detect lower frequencies. The total time interval has no effect on the highest detectable frequency.

- The sample rate (number of samples per unit time = $1/\Delta$) determines the highest detectable frequency ($1/(2\Delta)$ cycles per unit time). This rate has no direct effect on either the number of bands resolved or the lowest detectable frequency.

- The frequency range from zero through $1/(2\Delta)$ cycles per unit time is resolved into $n/2$ bands. Thus, the total number of samples determines the resolution, with more samples providing finer resolution.

Aliasing and the Nyquist Frequency

We have already seen that the highest frequency component that can be detected has a period of two samples. This makes intuitive sense: This component goes high on one sample and low on the next. Nevertheless, it is instructive to consider what happens when a higher frequency periodic component is present. Look at Figure 3.8. The cross-hatches along the time line depict the sample times. The solid curve is a periodic component whose period is less than two samples. The dotted line is the periodic component that will *appear* to be present as a result of sampling too slowly. Note that this effect has nothing to do with the DFT or any other aspect of our technique. It is a fundamental physical limitation. There is no way around it.

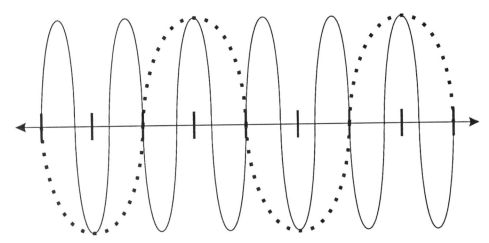

Figure 3.8 Aliasing in action.

We have been speaking of the *highest detectable frequency*. Actually, this term is dangerously misleading in that it implies that higher frequencies may be present, but we just can't detect them because they don't show up in our analysis. After examining Figure 3.8, it should be apparent that the situation is considerably worse. They *do* show up. Unfortunately, they show up wrong. They appear as an *alias* at a lower frequency. Sometimes aliasing is unavoidable. But whenever possible, we should consider the possibility and try to sample at a sufficiently high rate.

This magic frequency limit corresponding to $n/2$ has a name: the *Nyquist frequency*. There is a very famous theorem called the *Nyquist sampling theorem*. It (roughly) states that if we have a continuous waveform that does not contain any frequencies higher than some limit f, we can capture all of the information in the waveform by sampling at a rate of at least $2f$ samples per unit time. This is an extremely profound statement. The waveform is continuous: It has well-defined values at all times. Yet we can sample it at a finite number of discrete points and thereby know to mathematical perfection the waveform's values at all of the infinitude of intervening times. CD technology makes use of this fact by sampling the recorded musical waveform about 44 thousand times per second. This implies the ability to capture perfectly all of the information up to 22 thousand cycles per second. Since the limit of human hearing is about 20 thousand cycles per second, this is sufficient.

Data Windows

Suppose we measure a variable once a month for 40 months. If we compute the DFT of this series, we will resolve frequency bands of 1, 2, 3, 4, ..., 20 cycles in that 40-month span. These correspond to periods of 40, 20, 13.3, 10, ..., 2 months. But what if the series contains a significant and interesting component that doesn't happen to fall exactly on one of these discrete frequencies? Our hope is that it would be picked up by the nearest band slot. For example, suppose there is a strong component with a period of 12 months (a frequency of 3.33 cycles per the 40-month span). It would be nice if this component caused P_3 to be large, since this is the nearest integer frequency slot. We might even be willing to live with a little bleeding into P_4. Alas, this does not exactly happen. There is a strong tendency toward this end, but there is also a lot of bleeding into other slots, some of them disturbingly distant. Figure 3.9 shows the distribution of bleeding when a 40-point series is directly transformed with the DFT.

The horizontal axis is the difference between the frequency of the true component and the discrete frequency whose content is in question. The vertical axis is the relative magnitude of the DFT term in that slot due to the true periodic component. In the example just given, the periodic component has a frequency of 3.33. Consider DFT bin 3. The difference is 0.33, so we see that w_3 will have considerable magnitude. This is good. Bin 4, with a difference of 0.67, will also have moderate magnitude. This is acceptable. However, consider w_5, which has a frequency difference of 1.67. The magnitude here is almost 0.2, quite large for being almost two frequency slots away from the true frequency. Worst of all is the fact that the leakage lobes decrease terribly slowly. The graph shown here suddenly stops, but that is only due to running out of space on the printed page. The lobes really extend to infinity. Observe that frequency slots over four bins away are still picking up around 10 percent of the contribution of the component. Make no mistake: This is an intolerable situation. We absolutely, positively, cannot live with so much bleeding into distant frequency bins. We must do something about it, no matter what the price.

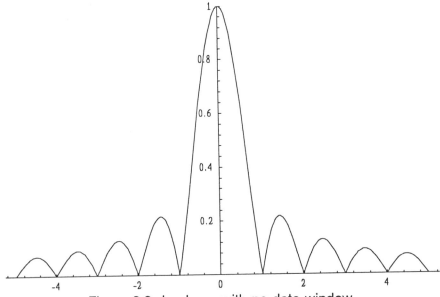

Figure 3.9 Leakage with no data window.

Luckily, there is a way of taming leakage, and the price is not extraordinarily high. What we do is gradually taper the ends of the time-domain series to zero before transforming with the DFT. The official terminology is that we apply a *data window*. A data window is a roughly bell-shaped curve that multiplies the time series. Near the center of the series, the multiplier is near 1.0, so the center is unaffected. At the beginning and end, the multiplier is near zero, so that the series starts and ends at very tiny numbers. The multiplier tapers smoothly across its extent. The most popular data window is the *Welch* data window. Its values are given by Equation (3.3). To apply this window to a time series, multiply each raw value x_i by the corresponding window multiplier m_i.

$$m_i = 1 - \left(\frac{i - 0.5(n-1)}{0.5(n+1)}\right)^2, \quad i = 0, \ldots, n-1 \qquad (3.3)$$

To optionally counteract the shrinking effect of a data window, divide the DFT terms by the RMS value of the window. This is not required in most applications, because the main interest is usually in the relative spectral values. However, the compensation shown in Equation (3.4) largely removes the downward bias attributable to the window. Note that since the DFT is linear, this compensation can equivalently be done to the raw data.

$$w'_i = \frac{w_i}{\sqrt{\frac{1}{n}\sum_{i=0}^{n-1} m_i^2}} \qquad (3.4)$$

Nearly all inexperienced people utterly abhor applying a data window. They see it as throwing away data. In a way, it is throwing away some information. However, the actual amount of information discarded is not as much as might be thought. (Statisticians will be interested in knowing that we end up with about 72 percent of the original degrees of freedom—a *very* reasonable price.) And windowing *must* be done. Look at Figure 3.10, which shows the leakage associated with the Welch data window. The near lobes are considerably smaller than the near lobes associated with using no data window. But an even more important effect is that the amplitude of the Welch lobes drops

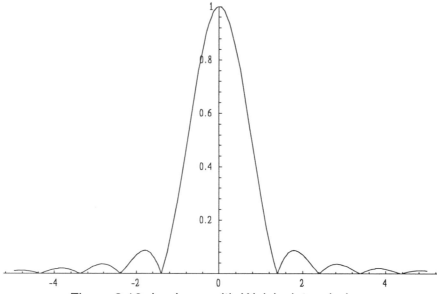
Figure 3.10 Leakage with Welch data window.

rapidly as the distance from the true frequency increases. This implies that leakage to distant bins is almost nonexistent, which is crucial to virtually all applications.

There is a temptation that must be avoided. Every so often, someone comes up with the seemingly wonderful idea of only tapering the ends of the series. The idea is to start about 10 to 20 percent of the way inside each end of the series and taper to zero from there, leaving the interior untouched. The feeling is that the best of both worlds is obtained: The ends taper to zero, yet little data is discarded. The problem is that the leakage of this method is not much different from that due to using no window at all. To put it bluntly, this idea doesn't work. The improvement is so slight that it's just not worth it. Bite the bullet and go all the way. The gain is well worth the price.

One last picky detail remains. It usually happens that the time series being transformed is fairly well centered. But if it is strongly offset from zero, the tapering operation introduces a bizarre effect. Figure 3.11 illustrates the result of applying a Welch data window to a stationary high-frequency series originally centered at 5.0. The effect of this distortion is the introduction of spurious components at very low frequencies. In most practical applications, this is no real problem. If the user is seriously interested in low-frequency components, he or she almost certainly will have already centered the series to make the interesting components stand out. Most of the time, the areas of interest are well above the areas that suffer from this effect. Nevertheless, it is good practice to address this problem in advance. The solution is simple: Compute a weighted mean of the time series using the window function as the weight. Then subtract the weighted mean from each point in the series before applying the window. This mean is given by Equation (3.5).

$$\bar{x} = \frac{\sum_{i=0}^{n-1} m_i x_i}{\sum_{i=0}^{n-1} m_i} \tag{3.5}$$

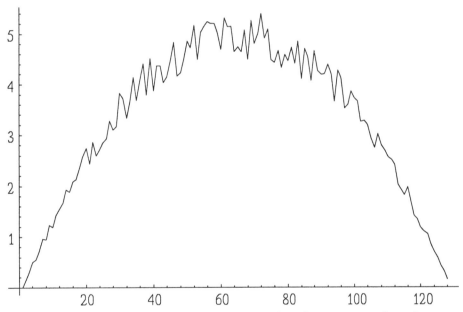

Figure 3.11 Windowing an uncentered series causes distortion.

Code for Computing the Power Spectrum

Power-spectrum computations are quite straightforward, so no great amount of space will be devoted to the topic. Here is a heavily edited code fragment from the module SPECTRUM.CPP on the accompanying disk. This illustrates the following points: applying a Welch window, packing the series for the efficient DFT described on page 358, calling the appropriate DFT routine, and computing the power and phase. This code assumes the length of the series is even. See the original module for details on handling an odd-length series. See page 358 for an explanation of the pairwise packing of the series in the real and imaginary vectors.

```
fft = new FFT ( n / 2 , 1 , 1 ) ;                   // Use the compact method

wsum = dsum = 0.0 ;                                 // Compute weighted mean
for (i=0 ; i<n ; i++) {
   win = (i - 0.5 * (n - 1)) / (0.5 * (n + 1)) ;    // Welch window
   win = 1.0 - win * win ;                          // Equation (3.3)
   wsum += win ;
   dsum += win * x[i] ;                             // Equation (3.5)
   wsq += win * win ;                               // For Equation (3.4)
   }
dsum /= wsum ;
wsq = 1.0 / sqrt ( n * wsq ) ;                      // See Equations (3.4) and (3.2)

for (i=0 ; i<n/2 ; i++) {
   win = (2*i - 0.5 * (n - 1)) / (0.5 * (n + 1)) ;
   win = 1.0 - win * win ;                          // Welch window
   win2 = (2*i+1 - 0.5 * (n - 1)) / (0.5 * (n + 1)) ;
   win2 = 1.0 - win2 * win2 ;
   win *= wsq ;                                     // See Equations (3.4) and (3.2)
   win2 *= wsq ;
   real[i] = win * (x[2*i] - dsum) ;                // Careful users have an even length
   imag[i] = win2 * (x[2*i+1] - dsum)               // So we can alternate the data
   }

fft->rv ( real , imag ) ;                           // And use the efficient DFT method
real[n/2] = imag[0] ;                               // The rv routine put real Nyquist in
imag[0] = 0.0 ;                                     // Imaginary zero, which is truly 0.
imag[n/2] = 0.0 ;                                   // This is also truly zero
```

```
for (i=0 ; i<=n/2 ; i++) {
  power[i] = real[i] * real[i]  +  imag[i] * imag[i] ;
  if (i  &&  (i < n/2))
    power[i] *= 2.0 ;                          // Equation (3.2)
}

for (i=0 ; i<=n/2 ; i++) {
  if ((fabs(real[i]) > 1.e-40)  ||  (fabs(imag[i]) > 1.e-40))
    phase[i] = atan2 ( imag[i] , real[i] ) ;
  else
    phase[i] = 0.0 ;
}
```

Smoothing for Enhanced Visibility of Spectral Features

The individual P_i terms in the discrete Fourier transform spectrum have an annoying statistical property: Their standard deviation is equal to their expected value. That's a *lot* of random variation! As a consequence, visual examination of a DFT power spectrum is likely to be intimidating. The individual components wildly dance up and down. Only extremely obvious spikes and broad tendencies stand out amid the chaos. Therefore, it is often in our best interest to apply some sort of smoothing operator to the power spectrum before displaying it for examination. Many of the lowpass filters described elsewhere in this text will do an acceptable (or even excellent) job of removing the local variation that can mask important larger features. However, one type of filter is often particularly appropriate for smoothing a power spectrum. This is the *Savitzky-Golay (SG)* filter. Its operation is illustrated in Figure 3.12.

This figure shows the power spectrum points in a section of the total transform extent. A peak obviously exists. Most lowpass filters smooth a series by computing a positively weighted average across an interval. As a result, peaks are attenuated when their lower neighbors are averaged in. A typical lowpass filter applied to this spectrum would produce the curve shown as a dotted line. Not only can we no longer observe the true height of the spectrum, but we may even doubt the significance of this peak.

The Savitzky-Golay filter works on an entirely different principle. In order to compute the filtered value at a given point, it examines the neighbors on both sides of the point. The number of

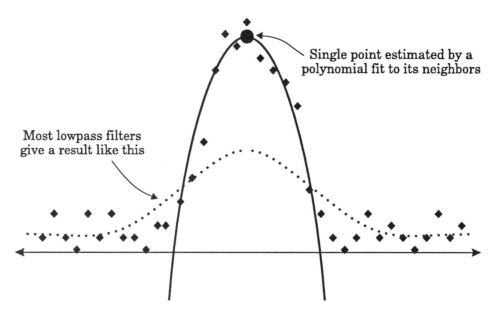

Figure 3.12 Savitzky-Golay versus lowpass filter.

neighbors examined on each side is called the *half-length* of the filter. It fits a polynomial, typically of degree two to six, to the point and its neighbors. The fit is done so as to minimize the mean squared error of the polynomial relative to the true spectrum values. The polynomial is evaluated at the position of the center point, and the value of the polynomial there defines the value of the filtered signal. Figure 3.12 shows a least-squares second-degree polynomial fitted to a small collection of points. The value of this polynomial at the center point is shown as a filled circle. Note how well the filtered output matches the apparent location of the peak. This is the advantage of the Savitzky-Golay filter.

This filter is not perfect for all situations. Since it operates by fitting polynomials to short sections of data, the data must reasonably support the assumption of polynomial behavior. If the raw spectrum does not approximate a polynomial across the extent (two half-lengths) of the filter's domain, results will not be excellent. Shorter filters are more likely to satisfy this assumption, but the price is less ability to smooth. The ideal situation is when you are lucky enough to know in advance that the spectrum is expected to vary smoothly and at a relatively slow rate. In this case, the filter can be made wide to gather as much information as possible, and the degree can be made high enough to follow the variations in that interval. If this is not a safe assumption, the width and degree must both be low. The bottom line is this: *The Savitzky-Golay filter assumes that the neighbors of a point contribute meaningful information by virtue of the slow polynomial-like behavior of the spectrum.* Look at how the neighbors of the central point in Figure 3.12 point the way to the peak. If this is not a characteristic of the spectrum, the utility of the Savitzky-Golay filter is questionable.

The mathematics of this filter is considerably beyond the scope of this text. A good explanation can be found in Press *et al.* (1992). The module SAVGOL.CPP on the accompanying disk contains complete source code for the filter.

The Cumulative Power Spectrum

Most of the applications in this text make use of the power spectrum in one of two ways. The spectrum may be used to identify local concentrations of energy, indicating lagged correlations or seasonal components. This use is best served by examination of the (optionally filtered) power spectrum. The other common reason for computing a power spectrum is to identify broad shifts in energy distribution. A series that has short-term positive autocorrelation has most of its spectral energy concentrated at low frequencies. Short-term negative autocorrelation shifts the spectral energy to the high end. These shifts may not be clearly visible in the spectrum. There is a simple technique for exposing broad spectral power shifts: Use the *cumulative power spectrum*.

A time series that is perfect white noise has a power spectrum that is perfectly flat: The P_i values are all equal. We already saw that the error variance of individual components is so high that assessing departures from equality by examining individual components is nearly hopeless. There are some statistical methods for computing significance by cumulating sums of neighbors, but these methods are fraught with numerous difficulties and are not generally recommended outside of certain special applications. But if we cumulate the spectrum along its *full extent*, we can effectively deal with the problem of high individual variance. Moreover, the cumulative spectrum can be fairly rigorously analyzed with elementary statistical techniques. It is not perfect at detecting all departures from equality, but it is a valuable tool for many applications.

The traditional approach to computing the cumulative power spectrum is to start with P_0 and work all the way up to the Nyquist frequency. This has the disadvantage that P_0 is the squared mean of the time-domain series, so offsets from zero strongly impact the cumulative spectrum, which is not usually desired. This section espouses an algorithm that discards P_0, starting the summation with P_1. With this in mind, here are a few definitions. Let S_i be the cumulative power spectrum from P_1 through P_i. It is defined in Equation (3.6).

$$S_i = \sum_{k=1}^{i} P_k \qquad (3.6)$$

Let $S = S_{n/2}$ be the total power in the spectrum (excluding the discarded DC component). Consider the fraction of the total power that is accounted for by the first i spectral terms. Denote this by s_i as shown in Equation (3.7).

$$s_i = S_i / S \qquad (3.7)$$

Let's think about the quantity defined in Equation (3.7). Except in the rarest pathological cases, s_1 will be a very small number, almost zero, because P_1 is generally a small fraction of the total power. As i increases, s_i will obviously never decrease, and will, in real life, steadily increase. The final value, $s_{n/2}$, is 1.0. If one were to plot s_i as a function of i, one would expect to see a line sloping upward from nearly zero to one. But what else can we say about this line? Think about the individual P_i terms in the sum. If the time series is truly white noise, all of the terms will be (theoretically) equal. Therefore, we expect the line to be (theoretically) straight. In particular, we expect that the fraction of total power accounted for up through term i should be equal to the fraction of the total number of terms cumulated so far. This is expressed in Equation (3.8).

$$s_i \approx \frac{i}{n/2} = \frac{2i}{n} \qquad (3.8)$$

Equation (3.8) is the theoretical expectation of s_i when the time series is white noise. But what if it's not white noise? If the spectrum happens to be shifted to the left (emphasizing low frequencies), the cumulative spectrum computed with Equation (3.7) will rapidly outpace the theoretical quantity expressed in Equation (3.8), not converging until much later. Conversely, if the spectrum is shifted to the right, the observed cumulative spectrum will be less than the theoretical quantity, taking a long time to catch up.

An interesting series can be computed as the deviation between the observed and the theoretical cumulative spectra. This series, called the *cumulative spectrum deviation*, is defined in Equation (3.9).

$$d_i = s_i - \frac{2i}{n} \tag{3.9}$$

If the spectral power is shifted to the left, d_i will take right off from its initial value near zero, then slowly drop back to zero much later. Conversely, if the spectral energy is shifted to the right, d_i will immediately plunge to negative numbers, then slowly work its way back up to zero. If the time series is white noise, d_i will tend to stay fairly near zero across its entire extent.

One of the nicest features of this method for assessing departures from spectral equality is that it lends itself to very easy statistical analysis (as long as we ignore the trivial distortion introduced by ignoring P_0). Let D be the maximum absolute value of all deviations d_i, as shown in Equation (3.10).

$$D = \underset{1 \leq i \leq n/2}{MAX} |d_i| \tag{3.10}$$

The distribution of the statistic D under the hypothesis of white noise is well known. It is called the *Kolmogorov-Smirnov distribution*. Tables showing the probability (α) of D exceeding fixed values (D_α) are provided in many standard statistics texts. Since most of our applications involve series containing at least several dozen points, we can safely use the asymptotic formula given by Equation (3.11). This formula is always conservative.

$$D_\alpha = \frac{\sqrt{-0.5 \ln(\alpha/2)}}{\sqrt{q}} \tag{3.11}$$

The quantity q is called the *degrees of freedom*. If the DFT is applied to the raw time series, $q = n/2-1$. (Recall that n is the number of points in the time series.) If a Welch data window was applied before the DFT, q must be multiplied by 0.72 to compensate for the loss of information inherent in windowing.

One use for Equation (3.11) is to test directly the hypothesis that the time series is white noise. Why might we want to test this hypothesis? Many of the techniques that we will see later in this text make use of two facts:

- We can be reasonably sure that our prediction model is correct if the residual error is white noise. If the residual is not white noise, this implies the existence of predictable relationships not accounted for by the model. It needs improvement.

- White noise has a flat power spectrum. The converse need not be true (although, in practice, it usually is true).

Thus, if the power spectrum of the residual fails the white-noise test, we ought to examine our model. Remember that passing the white-noise test is not a sure indicator that our model is excellent. But it is certainly comforting. And failing this test is a sure sign of trouble.

As an example, suppose the time series contains 200 points, and we want a significance level of 5 percent (a 5 percent chance of falsely rejecting the white-noise hypothesis). Assuming that we were intelligent and applied a Welch data window, $q = 0.72 * (200/2-1) = 71.3$. Plugging that fact and $\alpha = 0.05$ into Equation (3.11) gives $D_\alpha = 0.16$. To test the white-noise hypothesis, compute the maximum difference found by Equation (3.9). If it exceeds 0.16, we reject the hypothesis of white noise.

It is also possible to invert Equation (3.11) so that an observed D can be converted to a probability of observing a value so large if the time series were truly white noise. This is done in Equation (3.12).

$$\alpha = 2 \exp\left[-2 q D_\alpha^2\right] \qquad (3.12)$$

A more interesting use for this technique is to compute and display confidence intervals for the cumulative spectrum deviation. This way, a plot of d_i versus i is much more interpretable. When we see the deviation rise or fall away from zero, the presence of a pair of lines marking a confidence interval allows us to judge the severity of the deviation. The NPREDICT program supplied with this text displays $\alpha = 0.1$ confidence lines around a cumulative spectrum deviation plot. The meaning of these lines is that, assuming the time series is white noise, there is only a 10 percent probability that the deviation will hit a confidence line (either one) one or more times across the extent of the frequency domain. If the observed deviation hits a confidence line, we can be fairly safe in asserting that the times series is not white noise.

The NPREDICT program writes two of these statistics to the audit log file when a cumulative spectrum is computed. The maximum deviation given by Equation (3.10) is written, followed by its associated probability as expressed in Equation (3.12).

We conclude this section with a pair of examples. Figure 3.13 is the power spectrum of a time series that is primarily white noise but that has a tendency for each observation to correlate slightly with the previous observation. Careful study may indicate that there is a modest concentration of spectral energy in the low frequencies, as theory would dictate. But the deviation of the cumulative spectrum from its expected values, shown in Figure 3.14, brings out the unbalance. The deviation rises quickly and breaks through the confidence line (dotted) about a quarter of the way across. It doesn't start dropping back to zero until halfway to the Nyquist frequency.

The cumulative spectrum is primarily sensitive to general shifts of spectral energy. Local peaks, even strong peaks, are better detected with the (perhaps smoothed) spectrum. (In fact, the maximum entropy method, discussed in the next section, can be the best method for detecting and locating strong narrow peaks.) Figure 3.15 is the power spectrum of a mixture of white noise and a pure sine wave having a period of 4 points. Now look at the cumulative spectrum deviation in Figure 3.16. Despite the presence of an extremely strong periodic component, the confidence lines for white noise are not reached. The peak is too narrow in the frequency domain to significantly impact the cumulative spectrum.

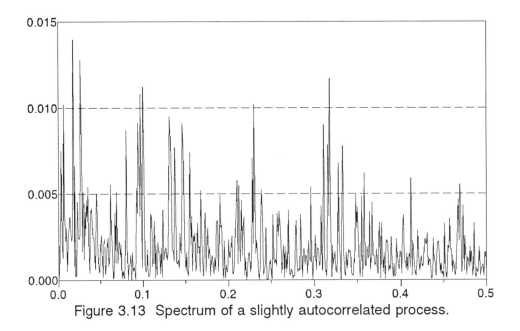
Figure 3.13 Spectrum of a slightly autocorrelated process.

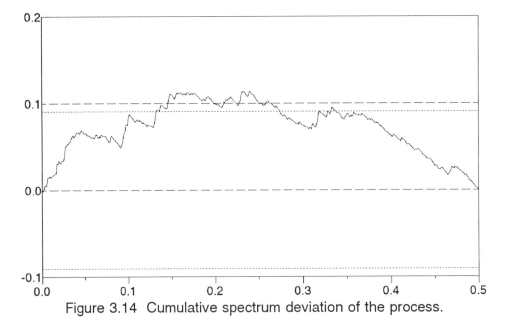
Figure 3.14 Cumulative spectrum deviation of the process.

The Power Spectrum

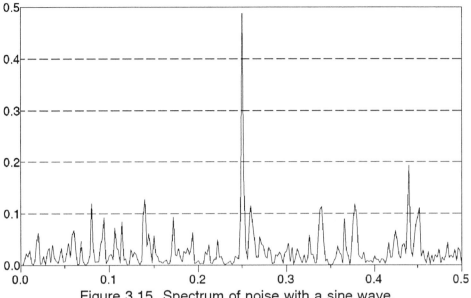

Figure 3.15 Spectrum of noise with a sine wave.

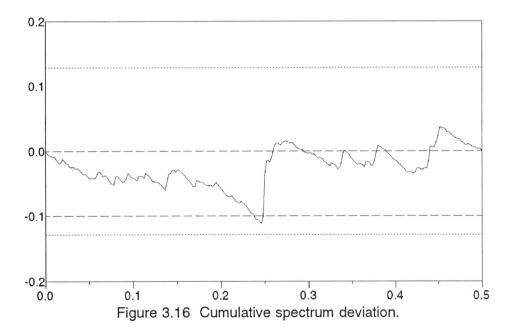

Figure 3.16 Cumulative spectrum deviation.

The Maximum Entropy Power Spectrum

Thus far, we have focused on use of the discrete Fourier transform to estimate the power spectrum of a time series. This method is fast, straightforward, and generally comfortable to use. Its mathematical properties are well known, and it has a highly deserved reputation for being safe and reliable. Nonetheless, it's not the only game in town. A competitor that is sometimes quite superior to the DFT is called the *maximum entropy* (*ME*) method. The philosophy and mathematics of the ME spectrum are extremely complicated and far beyond the scope of this text. Press *et al.* (1992) provides a brief overview, but the references cited there are needed for a full treatment. We will content ourselves with an examination of the advantages and disadvantages of the method.

Sometimes we can assume that the expected power spectrum of a time series is mostly flat, containing one or a few prominent narrow spikes. When this is true, the discrete Fourier transform may not be the best approach to computing a power spectrum. The tremendous error variance of each spectral estimate may cause narrow peaks of interest to appear deceptively small, while random noise may generate high peaks that distract our attention. If the true peaks are very narrow, lowpass filtering will do little good. Even the Savitzky-Golay filter cannot follow sharp spikes. This is the domain of the maximum entropy spectrum.

The ME spectrum has a remarkable ability to zoom in on extremely narrow features. At the same time, a properly computed ME spectrum will dramatically smooth areas of low spectral energy. The result is a best-of-all-worlds spectral display. The vast plains of low spectral power lie low and flat, while the important spikes jut out prominently. The spikes often have such sharp peaks that their precise location can be found to a resolution far higher than that obtained by the discrete Fourier transform. When everything works out well, an ME spectrum is a beauty to behold.

But everything doesn't always work out well. The principal problem is that the ME method has an absolute *craving* to find spikes of high spectral energy. If any are there, the ME method will certainly find them. Unfortunately, even if the true spectrum is essentially flat, the ME method will in all probability *still* find one or more spikes! It just needs to find something, *anything*. For this reason, the ME method should never be used alone. The DFT spectrum should always be computed, optionally filtered, and displayed. This serves as a reality check on the ME spectrum. The purpose of the ME spectrum is to alert us to the existence of subtle spikes and to locate their position precisely. The purpose of the DFT spectrum is to confirm that there is at least a good possibility of a mass of spectral energy at the indicated location. One does not need to demand a corresponding peak in the DFT spectrum. This would negate the utility of the ME spectrum. But one must expect at least a modest indication of spectral energy there.

The ME method requires the user to specify two parameters. One parameter is the resolution. This one is easy in that it defines the number of points in the frequency domain (from zero to the Nyquist frequency of 0.5 cycles per sample) at which the spectrum is to be computed. Recall that the resolution of the DFT spectrum is fixed at half the number of sample points (plus one if the useless DC term is counted). The same is not true of the ME spectrum. Any number of points can be resolved. In most cases the user will want to resolve several times as many points as exist in the time series. This takes full advantage of the ME method's ability to resolve narrow spikes tightly.

The other parameter is a little trickier to choose. This is the *order* of the spectral estimator. There is a vitally important tradeoff to consider. As a rough rule of thumb, the order should be several times the number of peaks that are to be resolved. However, larger orders have more of a craving to find potentially spurious peaks. They also smooth to a lesser degree. Therefore, *the order should be as small as possible, yet at least two or three times the number of peaks expected.* For many routine applications in which there is no reason to expect a large number of important peaks, a good order is about 20. There is virtually never any reason to use an order beyond 150, and an order beyond 50 would be unusual.

The advice in the previous paragraph applies quite broadly to most time series applications, especially those involving econometric data. On the other hand, there is a situation, usually limited to engineering applications, in which an order of around 150 may be desirable. If it is known in advance that one or more distinct peaks do exist, and great accuracy is desired in locating these peaks, low orders exert too much smoothing. A high order is needed to produce narrow peaks. In this situation, a relatively high order is required, despite the danger of finding false peaks. Remember that the ME method's craving to find peaks is generally satisfied by true peaks. As long as it can lock in on *something*, it will probably not get too inventive.

The computer code for computing the ME spectrum is surprisingly simple. However, since the supporting mathematics is extremely complex, there is no point in presenting any code here. The complete subroutine can be found in the module MAXENT.CPP on the accompanying disk.

Power Spectrum Examples and Summary

This section provides a few illustrations of the principles discussed so far. All except the last few of these examples are based on the same data model. This is a time series that is white noise modified by inclusion of a small autocorrelation at a lag of ten samples. In other words, it is an autoregressive series having a single moderate positive weight at a lag of ten. It's not hard to imagine physical or econometric series having the property that their current value is slightly impacted by the value at a fixed historical lag. The power spectrum of such a series is characterized by a peak at a period of ten (the lag), which corresponds to a frequency of 0.1 cycles per sample. Additional peaks at integer multiples (0.2, 0.3, 0.4 and 0.5) also appear. Finally, slow wandering of the series contributes to a concentration of energy at very low frequencies.

Figure 3.17 shows the 2000-point time series used in most of these tests. Since it is impossible to see any details, Figure 3.18 shows the beginning of the series. Observe that the lag-10 autocorrelation is too small to be visible. But it definitely shows up in the power spectrum. Figure 3.19 is the raw DFT power spectrum of the time series. Note the obvious concentration of spectral energy at integer multiples of 0.1.

This spectrum illustrates a serious problem of raw DFT-based spectral estimates. The huge error variance of each component makes interpretation difficult. Figure 3.20 is the result of smoothing the spectrum with a Savitzky-Golay filter having a half-length of ten points and using a second-degree polynomial. Note the distinct improvement. Figure 3.21 increases the half-length to 40 points, providing dramatically increased smoothing. Increasing the degree to four to allow greater versatility in the polynomial fit produces Figure 3.22. This would be beneficial for following sharper peaks, but it seems to do more harm than good in this case.

Power Spectrum Examples and Summary

Now we switch to maximum entropy spectra. Figure 3.23 is the order-20 spectrum of the time series. It's pretty tough to beat. The peaks are prominent, and the smoothing is phenomenal. Increasing the order sharpens the peaks slightly and brings in some superfluous peaks due to the random noise. Figures 3.24 and 3.25 are order-50 and order-150 spectra, respectively. Their quality depends entirely on whether you want to consider the extra peaks to be important information or undesirable noise effects.

The almost uncanny ability of the ME spectrum to locate important peaks while smoothing noise leads to investigation of its performance under more adverse conditions. A new time series having a lag-ten correlation is generated. But in this series, the correlation is very small. The raw power spectrum shown in Figure 3.26 barely shows the spectral energy at multiples of 0.1. The smoothed spectrum in Figure 3.27 doesn't help much. But the ME spectrum in Figure 3.28 still does a fairly respectable job of exposing the peaks.

We conclude this set of spectrum examples with an investigation of the situation in which a pure narrow-band periodic component exists. The spectral peak due to lagged correlation, which we have been studying, is relatively wide. For this last example, we add white noise to a sine wave having a period of 4.5 samples. The raw DFT power spectrum of this series, graphed in Figure 3.29, clearly exposes the narrow-band component. There is little or nothing to be gained by filtering a spectrum like this one, as can be seen in Figure 3.30. Even using such a narrow filter as this, with a half-length of just four points, substantially blurs the peak. Now look at Figure 3.31. This is the order-150 maximum entropy spectrum of the series. A smaller order would also locate the peak, but it would be relatively broad. An order this high is needed to achieve the narrow focus appropriate to the situation. This is obviously stiff competition for the DFT spectrum.

94   *Frequency-Domain Techniques I: Introduction*

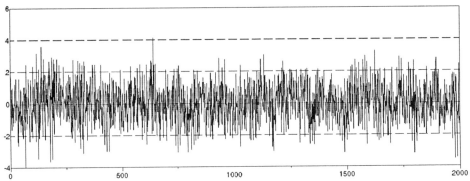

Figure 3.17 An AR series with moderate lag-10 correlation.

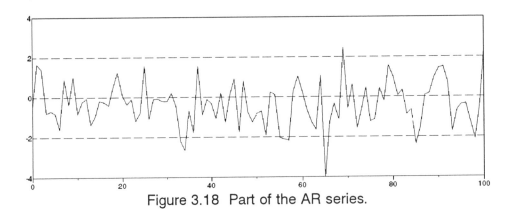

Figure 3.18 Part of the AR series.

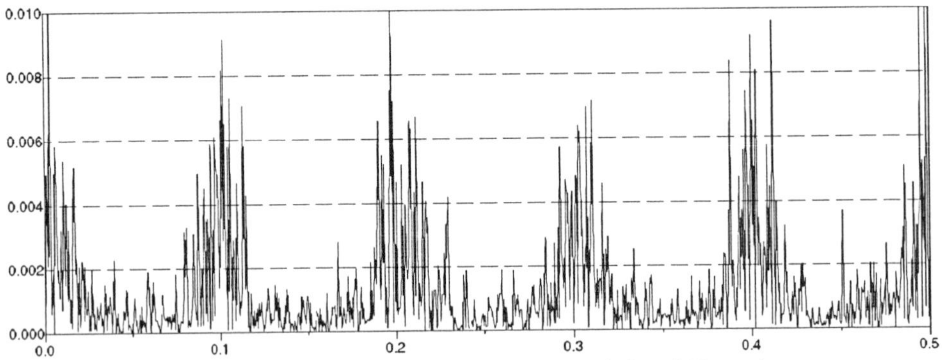

Figure 3.19 DFT power spectrum of the AR series.

Power Spectrum Examples and Summary

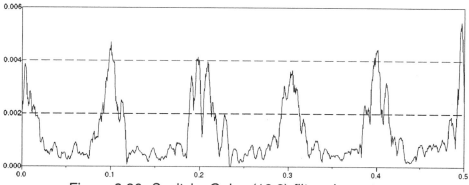
Figure 3.20 Savitzky-Golay (10,2) filtered spectrum.

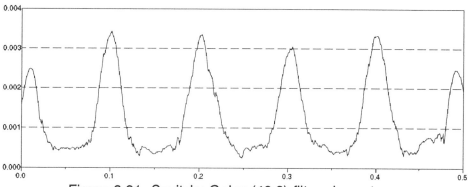
Figure 3.21 Savitzky-Golay (40,2) filtered spectrum.

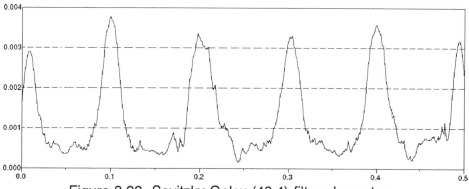
Figure 3.22 Savitzky-Golay (40,4) filtered spectrum.

96   *Frequency-Domain Techniques I: Introduction*

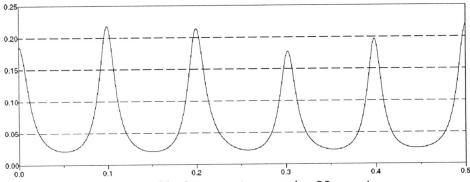

Figure 3.23   Maximum entropy order-20 spectrum.

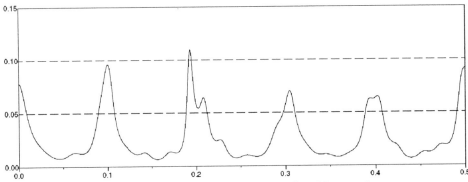

Figure 3.24   Maximum entropy order-50 spectrum.

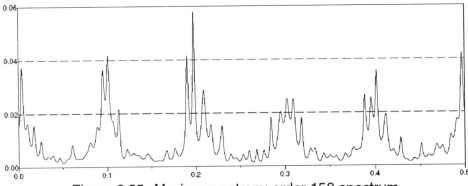

Figure 3.25   Maximum entropy order-150 spectrum.

Power Spectrum Examples and Summary

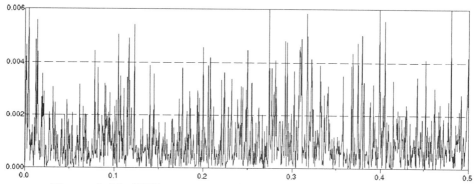

Figure 3.26 DFT spectrum of a very low correlation series.

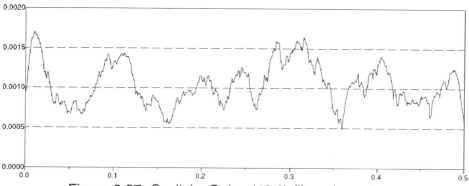

Figure 3.27 Savitzky-Golay (40,2) filtered spectrum.

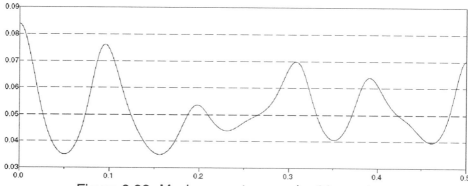

Figure 3.28 Maximum entropy order-20 spectrum.

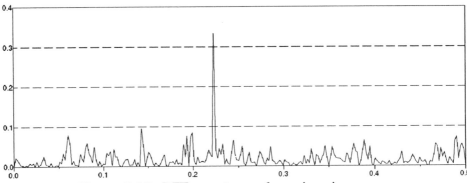
Figure 3.29 DFT spectrum of a noisy sine wave.

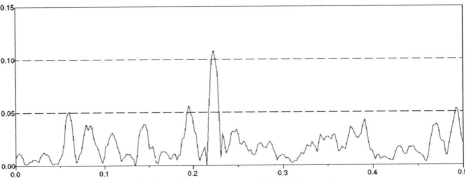
Figure 3.30 Savitzky-Golay (4,2) filtered spectrum.

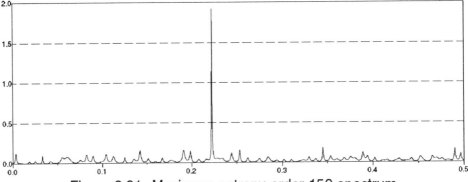
Figure 3.31 Maximum entropy order-150 spectrum.

Summary of Power Spectrum Usage

The previous section has presented practical details for computing and using the power spectrum of a time series. Let us briefly review why one might want to do so. All of these points are discussed more fully in other sections of this text.

- The spectrum generally enlightens us as to the nature of the data. The appearance of obvious periodic components should always be of interest. Sometimes the lack of an expected component may also be interesting. The DFT spectrum, possibly smoothed with a Savitzky-Golay filter, should always be the first item examined.

- Broad departures from a horizontal layout reveal underlying serial relationships. A heavy concentration of spectral energy at the low-frequency end indicates that the series is positively influenced by recent values. If the vast majority of the energy is at low frequencies, the series is probably nonstationary. Conversely, a high energy concentration at high frequencies indicates a negative correlation with recent values, perhaps due to excessive negative feedback in the control system. These departures are best detected with the cumulative spectrum deviation.

- An obvious narrow peak in the middle of the spectrum indicates a seasonal component that probably needs special treatment. A well-defined narrow trough might be due to a high negative correlation with a lagged value. This is often the result of excessive negative feedback with a significant time lag. Preprocessing with an ARMA model having a single lagged AR or MA term may be advisable. Narrow peaks are best detected with the maximum entropy spectrum.

- Other broad-based anomalies generally indicate some structure within the data. If such a spectrum belongs to the residual noise of a model, the model is less than excellent and needs work.

4

Frequency-Domain Techniques II: Filters and Features

- Overview of digital filters

- Filtering in the frequency domain

- Padding to avoid wraparound effects

- Lowpass, highpass, and bandpass filters

- In-quadrature filters

- Quadrature-mirror filters

In Chapters 1, 2, and 5 we see three valuable uses for digital filters in time series prediction. These are the following:

- In-phase filters (such as ordinary bandpass filters) allow us to emphasize important information, to reject irrelevant information, or to help a model by splitting a massive information source into separate, more digestible components.

- An in-quadrature filter lets us detect regions of change in a time series. Specialized versions can be optimized for particular tasks.

- Quadrature-mirror filters can be effective at locating and analyzing repetitive features that appear and disappear randomly in a time series.

Chapter 1 dealt only with basic applications of some of these filters. This chapter digs more deeply into their theory and provides details for readers who would like to program practical digital filters. This is really a sequel to Chapter 3, so it is important for the reader to be familiar with the material in that chapter, especially the introduction to the discrete Fourier transform at the beginning of the chapter and the *Power Spectrum* section starting on page 70.

What Is a Digital Filter?

We already know that a discretely sampled time series can be exactly represented as a sum of pure sine and cosine waves. Equation (3.1) on page 67 tells us how to compute the DFT coefficients for that representation. Equation (3.2) on page 72 tells us the contributions from each frequency band. In order to make the presentation flow smoothly, we sidestepped the topic of reconstructing the original signal ($x_0 \ldots x_{n-1}$) from the DFT ($w_0 \ldots w_{n-1}$). Reconstruction will appear shortly. For now, the important point is that the breakdown of a time series into pure periodic components is always possible. This leads to a natural explanation of the process of digital filtering: We simply emphasize those periodic components that we deem important, and we de-emphasize those that we deem superfluous. The remainder are left alone. How do we change the level of chosen frequency bands? There are at least two methods. Let us start by briefly examining one method.

Primitive digital filtering is common in traditional time series processing. Economists make heavy use of *moving-average* filters. For example, monthly sales figures for many goods exhibit prominent patterns that repeat on a yearly basis. By adding together each set of 12 successive observations and dividing the sum by 12 to get a mean, the period-12 component is totally eliminated. A new series computed this way may be more meaningful as a result of having strong local variation removed. Larger patterns of movement that may exist become more obvious.

A moving-average filter can be viewed as a weighted sum of contiguous observations, where the weights are defined by the nature of the filter. For example, the 12-month moving average just discussed uses a weight vector consisting of 12 equal components: (1/12, 1/12, ..., 1/12). There is no need for all of the components to be equal. An alternative filter sometimes used for removing yearly variation from monthly data has 13 components: (1/24, 1/12, 1/12, ..., 1/12, 1/24). Whatever the filter vector looks like, the operation is the same. The filter is dotted with the first section of the time series to produce the first output point. Then the filter is slid one point to the right and dotted with the next section of the time series to produce the second output point. This is repeated along the entire length of the time series. The name of this operation is *convolution*, and it is illustrated in Figure 4.1.

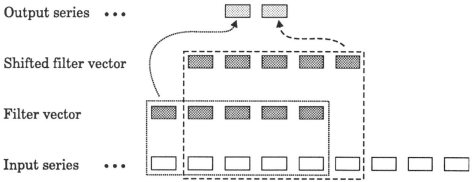

Figure 4.1 Filtering by time-domain convolution.

This figure employs a filter having five points—unusually short as filters go. Computation of two output points is shown. By arbitrary decision, each output point is placed so as to coincide with the center of the filter. This is not mandatory, but it is common practice.

A problem is immediately obvious. What happens at the beginning and end of the input series? Is it legitimate for the filter to hang over the ends before and after the known extent of the data? As can probably be guessed, there is no easy answer to this problem. The data simply isn't there. No matter what is done, it must be acknowledged that output points corresponding to partial sums are of dubious quality. There are some exotic formulas to compensate optimally for the lack of data, many of which are themselves of dubious quality. Many people ignore the overhang, acting as if the unknown input series points are all zero. A good approach is to use the mean of the series to fill out the overhang. This method works as well as many others, and it has a few optimality qualities of its own.

Another problem is not so obvious to the uninitiated, but it is extremely severe. Time-domain convolution is *slow*. Horrendously many multiplies and adds are required. Luckily, there is a fast alternative. Furthermore, this alternative has many other nice features, not the least of which is intuitive simplicity. Time-domain convolution is herewith abandoned in favor of a much better method.

Filtering in the Frequency Domain

We already know that the DFT can be used to deconstruct a time series into individual frequency components. Our goal is to emphasize or cut certain bands. So why not just transform the time series into the frequency domain, work our will on the new series, then reconstruct the times series using the modified frequency-domain series? It turns out that this is not only the easy way to filter a series, but it is also the fast way. It's easy because there is no need to laboriously compute the time-domain filter coefficients to be convolved with the input series. Since we almost certainly have defined the filter in terms of frequencies, it only makes sense to apply the filter directly to frequency terms. And it's fast because we just do a fast Fourier transform to get to the frequency domain, scale the individual terms in whatever way the filter dictates, then use the FFT once more to get back to the time domain. Two FFTs almost always beat one convolution.

Filtering in the Frequency Domain

Equation (3.1) on page 67 expresses the discrete Fourier transform of a time series in terms of the time-domain data. The original time series may be reconstructed from the DFT by means of Equation (4.1). There are only two differences between these equations: The sign of the *sin* terms are opposite, and the inverse DFT contains a factor of $1/n$.

$$x_j = \frac{1}{n} \sum_{k=0}^{n-1} \left[w_k \cos\left(\frac{2\pi jk}{n}\right) - w_k \sin\left(\frac{2\pi jk}{n}\right) i \right] \quad (4.1)$$

Even though the algorithm is straightforward, the digital filtering method used in this text (and many other texts) should be explicitly stated to avoid confusion. It is as follows:

1. Extend the time-domain series by appending additional constant values (such as its mean) to its end. This *padding* operation is described in the next section.

2. Use the fast Fourier transform to compute the DFT of the padded series as defined by Equation (3.1).

3. Multiply each w_i term in the DFT by whatever values are desired. This often means setting some of them equal to zero, leaving others unchanged, and smoothly tapering the DFT in between these extremes.

4. Use the fast Fourier transform to return to the time domain as defined by Equation (4.1).

5. Discard the extra values at the end of the series that are due to padding the series in the first step.

The algorithm just presented will be used for every digital filter in this text. It is fast, reliable, and easy to implement. The only aspect of the algorithm that distinguishes between the many types of filters to be discussed is step 3. The choice of the multipliers for the DFT terms is what defines a filter, so that will be the focus of the remainder of this chapter. However, there is one small but vital detail to deal with first.

Padding to Avoid Wraparound Effects

Review the discussion on page 104 which introduced the problem of what to do at the ends of the series where the filter overhangs the known data. It is tempting to assume that when we do not explicitly filter in the time domain, this overhang problem somehow magically disappears. It does not. Worse yet, it is cloaked in such a way that the extent of its effect is not obvious. The fact of the matter is that even when the filtering is performed in the frequency domain, *the effect is exactly equivalent to time-domain convolution with a filter vector*. The implicit time-domain filter overhangs both ends of the series, just as it would if the convolution shown in Figure 4.1 were used directly. The catch is that when we filter in the frequency domain, *we don't automatically know how long the time-domain filter is*. Later, we will learn how a good conservative estimate of its length may be computed. But for now, we need to realize that the problem is there, and it is serious.

Now that we know that DFT filtering automatically implies a time-domain filter that extends beyond the ends of the time series, the obvious question is this: What data does the filter use out there? Is it zero, which is the result when time-domain convolution simply truncates the filter at the ends? Or does some other set of numbers get included in the dot product with the filter vector? The answer may surprise many people. The DFT method wraps the time-domain series around in a circle, endlessly repeating copies to infinity. The effect is that when the filter extends beyond the far end of the time series, as happens when computing the last few output points, the data at the beginning of the series is included in the filter results. And when the filter extends past the beginning of the series, as happens when the first few output points are computed, the data at the end of the series is used. This effect, illustrated in the top half of Figure 4.2, is probably not what the user had in mind!

The wraparound implicit in DFT filtering is unavoidable. However, we can take action to lessen its impact. The solution is to append extra data to the end of the series. This data should be innocuous in some sense. It should have the property that, when it is included in the output computations, its effect is as benign as possible. When the last few output points are computed, the padding that appears in the right half of the implicit time-domain filter will be relatively innocuous. And when the initial output points are computed, the padding will wrap around, having relatively little impact on the left half of the filter. This is illustrated in the bottom of Figure 4.2.

Filtering in the Frequency Domain

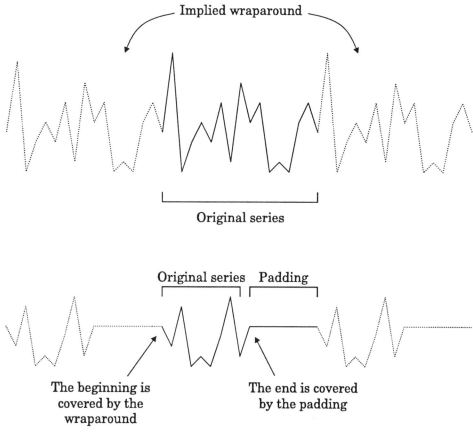

Figure 4.2 Padding ameliorates wraparound problems.

What do we use for the padding? Many people simulate truncation of the time-domain filter by padding with zeroes. This is often fine. However, if the time series is strongly offset from zero, the sudden jump to zero at both ends can have an adverse impact on many filters. A frequently superior approach is to compute the mean of the time series and use it for padding. This lessens the shock of the sudden transition from known to unknown data. Keep in mind that this may not always be the best approach. For example, we may know in advance that the expected mean of the series is zero. In this case, deviations of the observed mean from zero are random sampling error. Using the computed mean would be inferior to using the known value. But if we have no *a priori* knowledge of the true mean, the observed mean is a generally excellent choice for the padding constant.

Endpoint Flattening to Reduce End Effects

Padding is *always* necessary, and padding with zero (if that is the known underlying mean) or with the observed mean is generally a good approach. However, there is one situation in which padding alone may be insufficient to avoid serious anomalies at the ends of the series. If the series has a significant deterministic trend, or if it is nonstationary, padding the raw series with *any* fixed value will often introduce a dramatic jump at each end, where the series moves from its terminal value to the padding value. The effect on lowpass filters is to redirect the filtered output away from the input's (probably correct) terminal value and toward the padding value. The effect on bandpass and highpass filters is for them to view the transition as an extremely sharp transient that is dutifully included in the filtered output. This output transient can easily exceed all other transients in the series. Although these effects distort only the ends of the filtered output, the degree of distortion can be unbearable if the original series has wandered far from its mean. In order to be properly prepared for this possibility, one additional step is needed.

The underlying cause of this difficulty is the large instantaneous transition from the terminal value of the input series to the padding value. Anything that can be done to reduce or eliminate the transition would help. An easy and effective solution is to compute the trend line connecting the first point to the last, subtract this trend line from the input series, and then pad the series with whatever value appears at the ends after detrending. This totally eliminates the transition. The NPREDICT program included with this text moves both endpoints to zero and pads with zero. There is nothing magical about zero, but it is convenient. Endpoint flattening is illustrated in Figure 4.3.

This detrending operation is fairly benign, but it is not totally innocuous. There are three related effects that we must keep in mind. They are all caused by the fact that detrending introduces a large, primarily low-frequency, artifact into the Fourier transform of the series. This artifact usually dramatically reduces the magnitude of w_1, the first Fourier term (although in pathological situations it can actually increase it). This causes a (usually welcome) reduction in low frequency spectral energy. Unfortunately, the modifications are not totally limited to the low end of the spectrum. Some leakage makes its way into higher-frequency terms, all the way to the Nyquist frequency at the top! This may slightly add to or subtract from the spectral energy along the way, with the result being strongly data dependent.

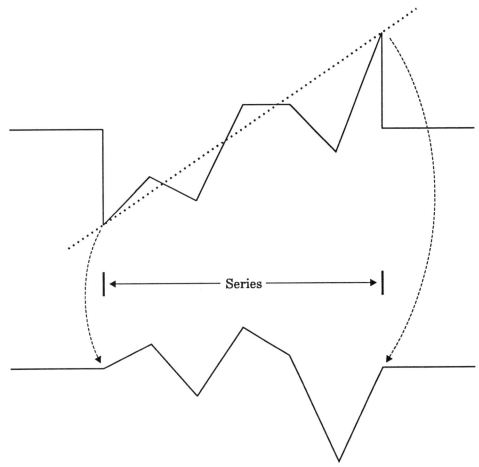

Figure 4.3 Endpoint flattening reduces endpoint distortion.

The practical effect is that some unwelcome distortion may be introduced into filtered outputs at *all* times and for *all* frequencies. But the degree of distortion is almost always negligibly small. This is a price that is well worth paying if the alternative is severe endpoint distortion. The only possible exception is for very-low-frequency filters, where the output distortion may be severe. A commonly effective solution is to reapply the trend line to the filtered output. This is discussed in more detail in the *Implementation Details and Sample Code* section starting on page 116. Low-frequency filters must be judged on a case-by-case basis. However, in practice, reapplying the trend line to the filtered output either works very well or is totally inappropriate. The decision is almost never difficult.

Shaping the Filter

Before studying particular types of filters, we need a generic study of filter shapes. Recall that the defining step of DFT filtering is multiplication of the frequency-domain series by a filter function. Frequencies corresponding to large values of the filter function are emphasized, while frequencies corresponding to small values are cut. There are few absolute restraints on the shape of the filter function. Its domain extends from 0.0 through 0.5, the complete range of the frequency-domain representation of the data. The filter function should never be negative, as that implies nasty effects that we won't even think about here. In general, its values range from zero to one, although this is only a tradition that may be ignored safely. Otherwise, there are no utterly firm constraints. However, there is one good habit that is very beneficial: The filter should not contain any sharp corners. Let us examine this aspect of the filter shape.

Suppose we want to filter a signal so as to retain only those periodic components that lie within a certain range of frequencies. All components lower than some minimum frequency or higher than a maximum frequency are to be rejected. We want to leave all components inside this frequency range unchanged. In other words, multiply them by one. Everything else is multiplied by zero. This filter function is illustrated in Figure 4.4a.

There are two problems with this approach. One problem is a phenomenon called *ringing*. It is reminiscent of the problem of sidelobe leakage introduced on page 75. Any periodic components that happen to have frequencies exactly corresponding to the frequencies implied by the DFT will be impacted exactly as shown in Figure 4.4a. However, there are undoubtedly other frequencies present—probably an infinite number of them. What about these components? Figure 4.4b shows the answer. These noninteger multiples of the fundamental frequency are enhanced and attenuated in varying degrees. Of particular importance, note that it is impossible to totally stop all components outside the passband. Some leakage is always present.

This ringing is not usually a serious problem, though. The real problem with a sharp filter corner is that *the implied time-domain filter vector is extremely long*. In other words, it extends across the time domain series, and ultimately over the ends, for an excessive length. When we compute the filtered output value at some time slot, the computations will be impacted by values of the time series far away from the time in question. This is almost never desirable.

Filtering in the Frequency Domain

a — This is what we want

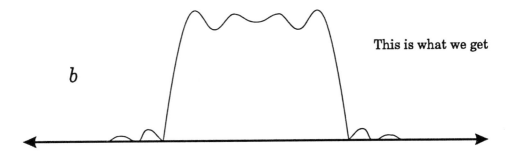

b — This is what we get

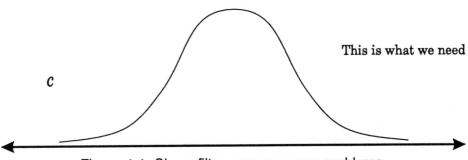

c — This is what we need

Figure 4.4 Sharp filter corners cause problems.

In order to reduce ringing, and more importantly, to shorten the length of the implied time-domain filter vector, we must round off the corners. The world's libraries contain thousands (really) of pages devoted to methods for optimally rounding the corners of filters. Every application has its own definition of *optimally rounded*. This text is most definitely not an appropriate forum for discussing even a few of the possibilities. One single method that has been proven effective in many time series applications will be presented. Interested readers should look elsewhere for alternatives.

Figure 4.4c illustrates a delightfully rounded filter function. The Gaussian function, shown here and defined in the next section, has its own excellent set of optimality criteria. Some applications, primarily in communications and electrical engineering, demand sharper corners than those shown in Figure 4.4c. However, most time series prediction applications, especially those in process control and econometrics, do very well with gentle curves. Moreover, it can be shown that, of all reasonable frequency-domain shapes, the Gaussian function has the shortest implied time-domain filter vector when several other important restrictions are imposed. (This is intimately related to the famous Heisenberg Uncertainty Principle. See Masters (1994) for a good discussion of the tradeoff between frequency-domain and time-domain shapes.) As will become clear when wavelets are discussed, short time domain filter lengths are immensely valuable in time series prediction.

Bandpass Filters

The most basic type of filter, the one from which practically all other filters are derived, is the *bandpass* filter. As the name implies, this filter passes a single band of periodic components, stopping all components having higher or lower frequencies. We initially focus on the simple and eminently useful style that has a single frequency of maximum response and that tapers smoothly to zero on both sides of this frequency. It resembles the filter shown in Figure 4.4c.

The bandpass filter predominantly used in this text is characterized by two parameters. The *center frequency*, called f_0, is the frequency maximally favored by the filter. It may be any value from 0.0 (DC) to 0.5 (the Nyquist frequency) cycles per sample. The *width*, called s, defines the width of the passband. Note that the literature contains many definitions of the width of a passband. This text defines the

Bandpass Filters

width in terms of the filter equation, which will be seen shortly. The width is specified in the same units as the frequency, and it typically ranges from 0.01 to 0.2 or so.

The frequency response function of the bandpass filter used here is a Gaussian function. This is the function that multiplies the DFT of the time series before transforming back to the time domain. It is shown in Equation (4.2).

$$H(f) = e^{-\left(\frac{f-f_0}{s}\right)^2} \qquad (4.2)$$

This function attains its maximum value of 1.0 at $f = f_0$, and it tapers toward zero on both sides of the center frequency. The rate at which it drops is determined by s, with larger values of s resulting in a slower fall. The effect of three different values of s, with f_0 fixed at 0.25, is graphed in Figure 4.5.

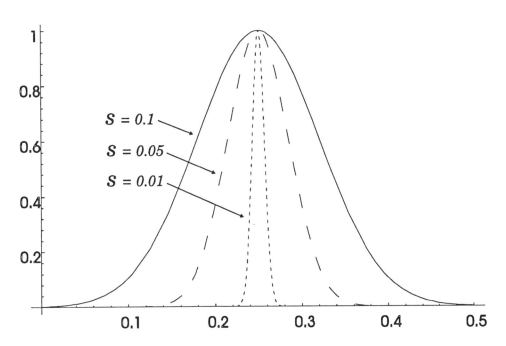

Figure 4.5 Three different filter widths.

Lowpass and Highpass Filters

A common reason for filtering a signal is to retain all periodic components on one side of a frequency limit and reject those on the other side. A *lowpass* filter removes all fast variation, keeping only slower components. This filter is often used to smooth a series, eliminating apparent noise so as to emphasize trends. A *highpass* filter removes slow trends and would be used when rapid changes are assumed to convey the important information.

It is trivially easy to modify a bandpass filter to turn it into a lowpass or highpass filter. To create a lowpass filter, discard the left tail of the bandpass filter, replacing it with the fixed value 1.0 at all points to the left of the center frequency. A highpass filter is created by replacing the right tail with the constant 1.0. Figure 4.6 demonstrates the technique with a lowpass filter. The dotted line is the discarded left tail of the Gaussian bandpass filter. The heavy solid line is the response of the resulting lowpass filter.

The literature is filled with alternative lowpass and highpass filters. Some of them may be superior to this choice for some applications. However, this style has many advantages. Ripple is low, and the corners, though not as sharp as many competitors, are sharp enough for the vast majority of time series prediction applications. Most importantly, the well rounded corners cause the implied time-domain filter to be relatively short. The value of this property in time series prediction cannot be emphasized enough. Figure 4.5 can be used to estimate the dropoff rate for different values of the width parameter.

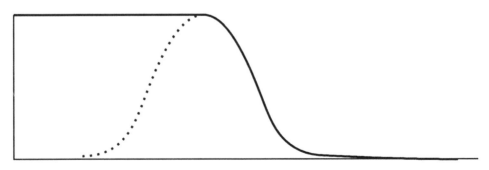

Figure 4.6 Modifying a bandpass filter for lowpass operation.

An Alternative Bandpass Filter

Sometimes the relatively narrow peak of the Gaussian bandpass filter shown in Figure 4.5 is just not right for the job. A filter with a wide, flat top may be needed. It is very easy to combine one lowpass filter and one highpass filter to form a flat-topped bandpass filter. To produce a bandpass filter of this style, apply a lowpass filter having a center frequency equal to the upper limit of the desired passband, and apply a highpass filter having a center frequency equal to the lower limit of the passband (either one first). The net effect is that of a bandpass filter whose top is flat between the two center frequencies. This is illustrated in Figure 4.7. The solid line is the response function of the lowpass filter, and the dotted line is that of the highpass filter. The net filter is their intersection.

The filters presented in this chapter are linear. This is the reason that the sequence of the lowpass and highpass filters does not matter. But there are even more profound implications of linearity. Filters can be combined in all sorts of imaginative ways. For example, suppose it is desired to remove a narrow band of frequencies from a time series, in essence using a filter that is constant at 1.0 except for a narrow notch at some frequency. One could apply a narrow bandpass filter to the series, then subtract the filtered output from the original series. The only limit is your imagination. Such versatility is difficult to achieve with nonlinear filters.

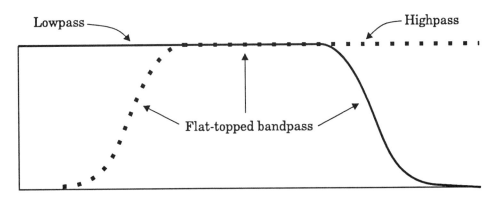

Figure 4.7 An alternative bandpass filter.

Implementation Details and Sample Code

We end this treatment of DFT lowpass, highpass, and bandpass filters with a discussion aimed primarily at programmers. The basic algorithm is surprisingly simple. However, by paying attention to a few subtle points, we can ensure correct and efficient operation.

The single most important implementation issue is padding, discussed on page 106. It is easy to forget about the need to pad the time series. Yet, in many applications, wraparound effects due to failure to pad adequately can introduce troublesome artifacts at the ends of the filtered series. If one applies a custom filter function to the DFT, computation of the required padding length involves specialized techniques. Luckily, if the filter function is based on the Gaussian function shown in Equation (4.2) on page 113, estimation of the padding length is trivial. Equation (4.3) is a conservative estimate of the minimum number of samples to be appended to the time-domain series before transforming. Further explanations of the padding issue, along with methods for dealing with arbitrary filter functions, are discussed in Masters (1994).

$$\text{minimum padding} = \frac{0.8}{s} \qquad (4.3)$$

Note that this is a minimum requirement. It is perfectly acceptable to use more padding. In fact, this brings us to a second subtle implementation detail. It is frequently in our interest to use more padding than the minimum shown in Equation (4.3).

Recall from Chapter 3 that the fast Fourier transform used to compute the DFT is most efficient when the length of the series is a power of two. The difference in computation time between a length that is a power of two versus a length that is a nearby prime number can be a factor of hundreds or even thousands. This fact inspires us to use whatever additional padding is needed to bring the length up to the next power of two. Intuition may balk at lengthening the series to shorten computation time, but that's the way it is. Do it.

The following code fragment illustrates the methods discussed so far in this chapter. It is edited for clarity. The complete listing is in the module FILTER.CPP on the accompanying code disk. Only the constructor of the Filter class and the single member function lowpass appear here. The other filter member functions are similar. A discussion of the code appears after each listing.

```
Filter::Filter (
   int npts ,                  // Number of points in series
   double *series ,            // Series to be filtered, not touched
   int pad ,                   // Pad with AT LEAST this to power of 2
   int use_qmf )               // Might qmf be called?  It needs an extra FFT.
{
   int i ;
   double mean ;

/*
   Push the user's n up to a power of two
*/

   length = npts ;             // Save original length to return correct length

   for (n=2 ; ; n*=2) {
      if (n >= npts+pad)
         break ;
      }

   halfn = n / 2 ;

/*
   Allocate the local storage and the FFT object(s)
*/

   xr = (double *) MALLOC ( 3 * n * sizeof(double) ) ;
   xi = xr + halfn ;
   workr = xi + halfn ;
   worki = workr + n ;

   fft = new FFT ( halfn , 1 , 1 ) ;
   if (use_qmf)
      qmf_fft = new FFT ( n , 1 , 1 ) ;

/*
   Compute and subtract the detrend line and pad with zeros
*/

   intercept = series[0] ;
   slope = (series[npts-1] - series[0]) / (npts-1) ;
```

```
  for (i=0 ; i<npts ; i++) {
    if (i % 2)
       xi[i/2] = series[i] - intercept - slope * i ;
    else
       xr[i/2] = series[i] - intercept - slope * i ;
    }

  for (i=npts ; i<n ; i++) {
    if (i % 2)
       xi[i/2] = 0.0 ;
    else
       xr[i/2] = 0.0 ;
    }

  fft->rv ( xr , xi ) ;            // Compute DFT
  }
```

  The user supplies to the constructor a pointer to the time series that will be filtered. The reason for this design is that many applications involve applying several different filters to the same series. There is no point in computing the forward DFT each time a new filter is to be applied. The most efficient approach is to compute the forward DFT just once and store it locally. Then, every time a new filter is applied, the program copies the saved DFT to a work area, multiplies by the filter function, and performs the inverse DFT. The potential time savings is substantial.

  The user also supplies the length of the series and the minimum padding to use. If npts is a power of two and pad is zero, no padding will be done.

  The constructor allocates a single array that will be used as storage for the DFT and as a work area for the inverse transform(s). The single array is broken into its components for programming clarity. The primary FFT object is allocated for only half the length of the series. This is because it can use the efficient *even real* method discussed on page 359 for all lowpass, highpass, and bandpass filters. If the user warns the constructor that quadrature-mirror filters (discussed later) may be used, an additional FFT object must be allocated to handle the full series length.

  The next step is confusing to those not familiar with the efficient FFT method used for real series of even length. The original series is alternated in the real and imaginary parts of the local storage vector.

Implementation Details and Sample Code 119

See page 359 for an explanation of this action. Simultaneously, the endpoints are used to detrend the series. This was discussed on page 108. The detrend line is computed so as to move both endpoints to zero. This value is used to pad the end of the locally stored series, still with alternation. The final step in the constructor is to compute the DFT in place. Code for the lowpass filter member function now appears.

```
void Filter::lowpass (
   double freq ,              // Highest frequency before dropoff starts (0-.5)
   double width ,             // Scale factor for exponential (in frequency units)
   double *out                // Filtered output here
   )
{
  int i ;
  double f, dist, wt ;

  for (i=0 ; i<halfn ; i++) {
     f = (double) i / (double) n ;    // This frequency
     if (f <= freq)                   // Flat to here
        wt = 1.0 ;
     else {
        dist = (f - freq) / width ;
        wt = exp ( -dist * dist ) ;
     }

     workr[i] = xr[i] * wt ;
     worki[i] = xi[i] * wt ;
  }

  dist = (0.5 - freq) / width ;               // Also do Nyquist in xi[0]
  worki[0] = xi[0] * exp ( -dist * dist ) ;   // Equation (4.2)

  fft->irv ( workr , worki ) ;

  for (i=0 ; i<length ; i++) {
     if (i % 2)
        out[i] = worki[i/2] + intercept + slope * i ;
     else
        out[i] = workr[i/2] + intercept + slope * i ;
  }
}
```

The user calls this filter function with three parameters: the frequency, the scale (Equation (4.2) on page 113), and a pointer to the array in which the filtered series is to be placed. This array only needs to be the length of the original series (npts in the constructor call). Storage for padding is taken care of internally.

The DFT in xr and xi is multiplied by the frequency function. Recall from Chapter 3 that the frequency (in cycles per sample) corresponding to each term in the DFT is i/n. Since this is a lowpass filter, the left half of the bandpass bell curve is ignored and is replaced by the constant value 1.0. This was illustrated in Figure 4.6 on page 114. As noted on page 359, the Nyquist term is stored in xi[0]. This term must also be multiplied by the filter function. The final step is to perform the inverse DFT and unpack the resulting time-domain series. We simultaneously reapply the endpoint detrending line that was removed in the constructor. Note that retrending is necessary only for lowpass filters. All other filters are so little affected by detrending that retrending adds distortion. The only exception to this rule is very-low-frequency bandpass filters in which a significant part of the left tail is in the lowest-frequency bin. *This situation is nearly impossible to deal with in an effective way.* If very low frequencies must be isolated, try to use a lowpass filter instead of a bandpass filter.

Summary of Basic Filters

The filters described in this section account for the vast majority of practical filter applications. They are easy to understand, easy to program, and easy to use. Moreover, they often provide valuable assistance in solving a problem. Here are some of the more important details that should be kept in mind when using these filters (as well as the more advanced filters that appear later).

- *Always* use at least the minimum padding shown in Equation (4.3). If the series remains roughly centered around its mean, pad with the mean. If it wanders about, use endpoint flattening.

- If speed is vital, use additional padding to bring the length to a power of two. Avoid large prime numbers!

- Never be afraid to use your imagination. Combine filters by adding and subtracting their outputs. This is always legal.

In-quadrature Filters

The filters discussed so far have a generally valuable property: They do not shift any periodic components in time. Each individual periodic component reaches its peaks, zero crossings, and troughs at exactly the same time before and after application of the filter. It should be apparent that this is a desirable property in most applications. The position of a waveform in time is called its *phase*. For this reason, the filters that have been presented so far are called *in-phase* filters.

Sometimes it can be useful to have a filter that shifts the phase of each periodic component by exactly one-quarter of its period. These filters are only occasionally useful alone, so we will not devote an inordinate amount of time to solo applications. However, they do have immense utility in combination with in-phase filters, so a brief examination of in-quadrature filters is in order.

A good introduction to nearly any filter is to examine its appearance in the time domain. Recall that the DFT method of filtering, done completely by multiplication in the frequency domain, always involves an implicit time-domain filter that is convolved with the time series. Convolving this implicit filter with the time series would produce exactly the same filtered output as the DFT method espoused in this text. Let's look at the time-domain representation of a pair of bandpass filters and some filtering examples.

Figure 4.8 is an in-phase bandpass filter. It corresponds to multiplication in the frequency domain by a Gaussian function. Think about the effect of computing the dot product of this filter with a series that oscillates at the same rate. Suppose we were to line up the filter with the series so that the high point at the center of the filter corresponds to a peak of the series. These high positive values would reinforce each other. The large negative values a half-cycle later would also reinforce. The dot product would be a large positive number. If we were to slide the filter a quarter cycle along in time, the peaks and troughs in one would correspond to the zero crossings of the other. Moreover, the places where a nonzero product occurred would exactly balance (positive and negative). The dot product would be zero. Finally, imagine the lineup being such that the filter and the series are offset one-half cycle. The highs in one would match the lows in the other. The dot product would be negative. In other words, the locations of the highs, lows, and crossings of the filtered output would match the highs, lows, and crossings of the input. Look at Figures 4.10 and 4.11.

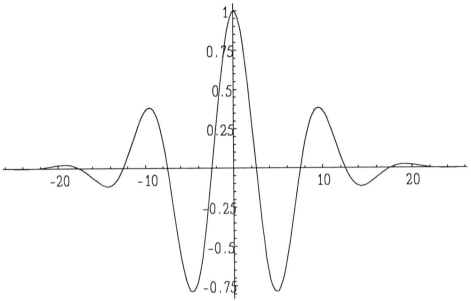

Figure 4.8 An in-phase bandpass filter.

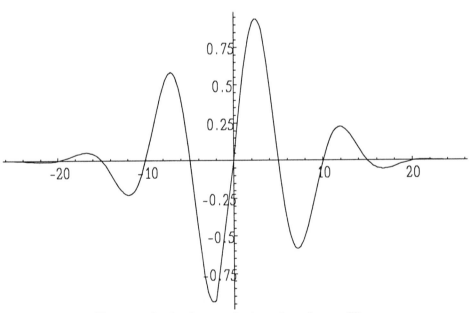

Figure 4.9 An in-quadrature bandpass filter.

In-quadrature Filters

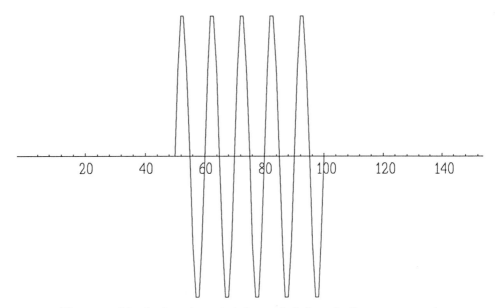

Figure 4.10 A signal containing a brief periodic component.

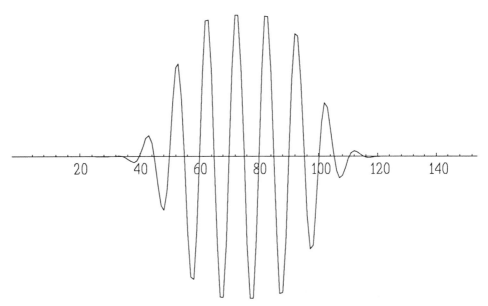

Figure 4.11 In-phase filtered signal.

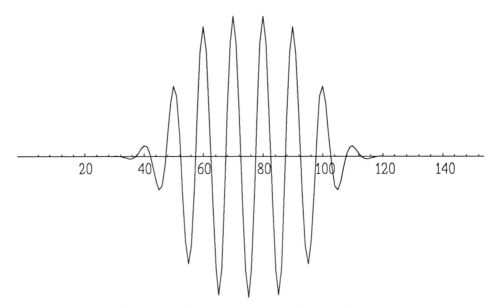

Figure 4.12 In-quadrature filtered signal.

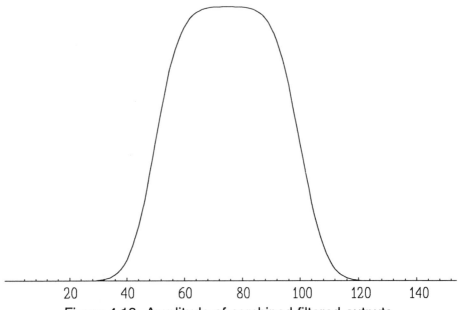

Figure 4.13 Amplitude of combined filtered outputs.

In-quadrature Filters

Figure 4.10 is a time series containing a short burst of a periodic waveform. Figure 4.11 illustrates the process just discussed. It is the convolution of the in-phase filter shown in Figure 4.8 with the time series shown in Figure 4.10. Observe that in both the original signal and the filtered signal, the rising zero crossings are at multiples of ten samples. No phase shift has occurred.

Now look at Figure 4.9. This is the time-domain representation of an in-quadrature filter having exactly the same passband as the in-phase filter just studied. Notice that while the in-phase filter is symmetric around its center, the in-quadrature filter is antisymmetric. One side goes up as the other side goes down in perfectly opposing symmetry. Imagine computing the dot product of this filter with the periodic series of Figure 4.10. Suppose the center of this filter lines up with a peak or trough in the time series. On one side of the center, the series and the filter move in the same direction, so they reinforce positively. On the other side of the center, they move in opposite directions, so they reinforce negatively. The net result is a dot product of zero. Now suppose this filter lines up with a zero crossing in the series. They both move in the same direction, so the net response is very large (positive or negative). In other words, the filtered output is large at zero crossings of the input, and it is zero at peaks and troughs of the input. The phase has been shifted one-quarter cycle. The filtered output is shown in Figure 4.12. Notice that the peaks occur where the zero crossings occur in the original signal.

Figure 4.13 is probably the most important one in this set. However, discussion of this figure is postponed for a short time. First, we need to mention one small but significant solo use for in-quadrature filters.

In-quadrature Filters for Change Detection

Think about the implication of zero-crossings in the input corresponding to peaks and troughs of an in-quadrature filtered output. This means that *the output of an in-quadrature filter is at its absolute maximum at the times when the input is changing most rapidly.* Conversely, at the times the input is holding steady (its peaks and troughs), the output is zero. An in-quadrature filter is sensitive to change. But more importantly, it is especially sensitive to changes that occur *at a particular rate*. Look at Figures 4.14 through 4.17.

126   Frequency-Domain Techniques II: Filters and Features

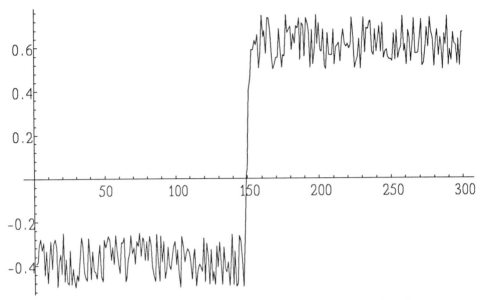

Figure 4.14   A series with an obvious change in level.

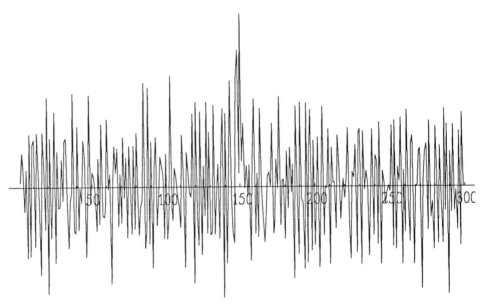

Figure 4.15   Adjacent differences.

Figure 4.16 In-phase bandpass filter.

Figure 4.17 In-quadrature bandpass filter.

Figure 4.14 is a time series that has an extremely obvious level change. Despite the fact that the human eye has no problem seeing such a blatant change, automated detection of this sort of feature is not always easy. That's not to say it can't be done; it certainly can. But there are subtle and often unexpected difficulties that must be avoided. For example, the naive approach is to compute adjacent differences. Intuition tells us that the differenced series will have a dramatic jump at the time of the level change. Figure 4.15 dashes that hope! If the level change had occurred all at once, there would be no problem. But since it is spread out across five contiguous observations, the individual differences are commensurate with the background noise. Counting on spikes in a differenced series to pinpoint level changes is dangerous.

Figures 4.16 and 4.17 are the results of applying in-phase and in-quadrature filters to the series. Both of these filters do a respectable job of detecting the level change. However, the in-phase filter is compromised by the fact that it does not clearly locate the change. In fact, at the time the change is in its exact center (slot 150), the in-phase filter is at zero. It is crossing zero in an upward direction at the same time the input series is crossing zero upward. This is a manifestation of its *in-phase* property. The in-quadrature filter is preferable because it peaks at the center of the level change.

This discussion is not meant to imply that in-quadrature filters are always the best way to detect level changes. For example, a well-chosen lowpass filter combined with differencing does a great job on the series in Figure 4.14. A carefully optimized template-matching filter does even better. But there is a notable set of conditions under which an in-quadrature filter is excellent. This is when the underlying process has inherent elasticity and momentum that cause shocks such as pulses or level changes to rebound and ring at a characteristic rate. This property is common in many physical processes, and it is not unknown in econometrics. In these situations, an in-quadrature filter can do an excellent job of pinpointing the impact.

We have devoted enough space to a relatively rare solo use of in-quadrature filters. Change detection is certainly an important subject, and in-quadrature filters are a powerful tool in any well-equipped arsenal. However, the principal use for in-quadrature filters is in combination with in-phase filters. This is the subject of the next section.

Quadrature-Mirror Filters

So far, we have used filters for one primary purpose: to separate different periodic components in a time series. These filters have been used to select certain components of interest while rejecting others. But sometimes we have a completely different goal related to periodic components. We may not care about getting our hands on a particular component. Instead, we may only want to know the answer to a simple question: Is it there? Suppose a characteristic (often brief) oscillatory pattern is known to appear in conjunction with an event that we wish to predict. In robotics, sudden contact between two parts may induce a brief ringing at a known resonant frequency. In econometrics, a rise or fall of an indicator at a roughly known rate, followed by a short period of damped ringing, may presage an interesting event. Whatever the case, detection of localized instances of periodic components can be valuable. Quadrature-mirror filters let us do that.

Why can't we just use an in-phase bandpass filter to detect an event? The problem (for this application) is that the periodic event appears in the filtered output. The output goes up and down, right along with the event in the input. Even though the event in question may be strongly present across a given time interval, the filtered output is zero twice every cycle. This is not much help. The solution appears when we remember that an in-quadrature filter does just the opposite: It is zero when the in-phase filter is at peaks, and it is at peaks when the in-phase filter is zero. But they are *both* zero when the event is not present. So we combine them. By simultaneously examining the two filtered outputs, we can detect a periodic event.

How do we combine the two filters? The standard method is to treat the pair of outputs as a single complex number and compute the absolute value of this number. The in-phase output is the real part, and the in-quadrature output is the imaginary part. As we saw in Chapter 3, the absolute value of a complex number is computed by squaring each of the two parts, adding them, and taking the square root of the sum. An additional advantage of this approach in some applications is that we can use the relationship shown in Figure 3.7 on page 70 to compute the phase of the event in the filtered output. This use will not be emphasized in this text, but readers should keep in mind that it is available if needed.

Turn back to Figures 4.10 through 4.13, starting on page 123. The series shown in Figure 4.13 is the pointwise square root of the sum of the squares of the series in Figures 4.11 and 4.12. This is the amplitude of the periodic event in Figure 4.10. Observe how the amplitude is zero when the event is absent, it rises smoothly when the event appears, it holds steady during the duration of the event, and it drops smoothly back to zero when the event terminates. *The amplitude of the complex filter pair is a feature detector optimized for the periodic event.*

When an in-phase and an in-quadrature filter pair having otherwise identical characteristics are used together this way, they are called a *quadrature-mirror (QM)* filter pair. This is an extremely valuable family of filters, not only in their original domain of electrical engineering, but even in econometrics and nonphysical sciences. Many economic and social systems behave surprisingly like a damped pendulum, responding to external forces by giving in to a degree, and then rebounding at a predetermined rate. The observed rate of giving in and rebounding can sometimes identify the system that is responding to the force. QM filters, with their ability to tune in to specific response rates, can make excellent feature detectors.

The algorithm for computing a QM filter is only trivially more complicated than the ordinary bandpass filter already described. On the other hand, the underlying mathematics is considerably beyond the scope of this text. Therefore, only a rough, intuitive justification of the technique will be presented. Readers who have a solid mathematical background should be able to follow it, and readers who do not have the necessary background may safely skip to the computer code. A more detailed explanation (though still without total mathematical rigor) can be found in Masters (1994).

The algorithm simultaneously computes both the in-phase and the in-quadrature outputs. It does this by means of a little trick: The in-quadrature output is not directly computed. Rather, the imaginary i times the in-quadrature output is computed. This allows use of a single FFT, with the in-quadrature output appearing in the imaginary parts of the inverse transform.

We already know how to compute an in-phase filter, because the lowpass, highpass, and bandpass filters studied so far are all in-phase. There are three steps: Compute the forward DFT of the time series, multiply by the filter function, then compute the inverse DFT. Recall from the discussion on page 68 that the DFT of the time series has conjugate symmetry around the Nyquist frequency. When we multiply the DFT by the filter function, we must multiply terms on *both* sides of the Nyquist point at the center of the transform. This fact is not directly reflected in the filter code listed on page 119. That code only multiplies the lower half of the transform. The reason is that the irv inverse transform routine takes advantage of the known symmetry to save time. Even though the symmetry is not directly addressed in the code, it is implicitly present. This point is important because when we compute a QM filter, we need to be aware of what's really going on under the surface.

Computing an in-quadrature filter involves the same three steps. The only difference is in the filter function. The filter function for an in-quadrature filter is pure imaginary (as opposed to pure real for an in-phase filter). So we could compute in-quadrature versions of the filters already studied by modifying the code to multiply the DFT by i times the Gaussian function in Equation (4.2) on page 113. This filter function is conjugate symmetric: The terms on opposite sides of the Nyquist frequency are equal in magnitude but opposite in sign. Unfortunately, this multiplication destroys the conjugate symmetry in the product (which was preserved with the in-phase filter). Thus, we can no longer use the efficient FFT method. We must resort to the full complex FFT. (There are efficient methods similar to the algorithm in irv, but we do not need them. Keep reading.)

Here is the trick: Do not directly compute the in-quadrature output. Instead, compute i times the in-quadrature output. This is done by multiplying the in-quadrature filter function by i. This function is pure imaginary, so when we multiply it by i, it becomes pure real. We can now double up by going one step further. Compute the in-phase output plus i times the in-quadrature output. Since the DFT is linear, we multiply the DFT of the time series by the *sum* of the in-phase and in-quadrature filter functions. They are both real. The in-phase function is positive and symmetrically equal in both halves of the DFT. The in-quadrature function is positive in the lower half and negative in the upper half (by conjugate symmetry), but equal in magnitude in both halves. When we add these two filter functions, they reinforce in the left half and cancel in the right half. This provides the key step in the QM filter: Multiply the DFT of the time series by the Gaussian filter function *in the left half only* (the terms below the Nyquist frequency). Set all terms in the right half equal to zero, since the two filter functions exactly cancel there. Then perform the inverse DFT to get back to the time domain. The in-phase outputs appear in the real parts of the transform, and the in-quadrature outputs appear in the imaginary parts.

This process is discussed more fully in Masters (1994). Several illustrations also appear there. Some readers may wish to consult that text for clarification. However, be aware that there is a slight difference between the approach there and the approach here. The 1994 text uses the strict mathematical definition of convolution, which reverses the filter in time relative to the input series. That is in deference to the assumption that readers of that text have a relatively strong mathematical or electrical engineering background. For this text, a more intuitive approach is taken by implicitly reversing the in-quadrature filter *before* convolution. The result is that the implied dot products move together in time. When the original series is rising, the in-quadrature output is positive (as opposed to negative in the 1994 text). *There is absolutely no difference in practical applications*. All that is involved is flipping the sign of the in-quadrature output. This point is stated only to avert potential confusion if a detailed comparison of the two algorithms is made.

Readers who are totally baffled by the preceding quasi-mathematical presentation should take heart. There is no need to understand that material. The algorithm is very simple. Here is an edited listing extracted from the module FILTER.CPP on the accompanying code disk. The constructor for this function was listed on page 117.

```
void Filter::qmf (
   double freq ,              // Center frequency (0-.5)
   double width ,             // Scale factor for exponential (in frequency units)
   double *real ,             // Filtered real output here
   double *imag               // Filtered imaginary output here
   )
{
   int i ;
   double f, dist, wt ;

   for (i=1 ; i<halfn ; i++) {            // Ignore DC, Nyquist for now
      f = (double) i / (double) n ;       // This frequency
      dist = (f - freq) / width ;
      wt = exp ( -dist * dist ) ;         // Gaussian weight function
      workr[i] = xr[i] * wt ;
      worki[i] = xi[i] * wt ;
      workr[n-i] = worki[n-i] = 0.0 ;     // Causes QMF outputs
      }
```

Quadrature-Mirror Filters

```
      workr[0] = 0.0

      dist = (0.5 - freq) / width ;          // Nyquist is in xi[0]
      workr[halfn] = 0.5 * xi[0] * exp ( -dist * dist ) ;
      worki[0] = worki[halfn] = 0.0 ;        // By definition of real transform

      qmf_fft->cpx ( workr , worki , -1 ) ;

      for (i=0 ; i<length ; i++) {
         real[i] = workr[i] / halfn ;
         imag[i] = worki[i] / halfn ;
         }
   }
```

    This routine is almost identical to the lowpass filter listed on page 119. The main loop passes through the DFT, computing the frequency and weight function for each term. The weighted terms are placed in the work arrays in preparation for the inverse transform. However, the previous filters used only the lower half of the work arrays, counting on irv to fill in the upper halves implicitly by symmetry. This routine explicitly zeroes the upper halves in order to implement the simultaneous in-phase and in-quadrature filtering. Also, the DC and Nyquist terms are handled individually. Recall from page 68 that the imaginary parts of the DC and Nyquist terms are always zero. Finally, the full complex inverse FFT is computed by calling cpx with the sign set to −1. Note that the FFT object in qmf_fft was allocated for the full series length. The other FFT object, used for the previous filters, was allocated for only half the length because the efficient symmetric method could be used for those filters. This allocation is seen in the constructor listing on page 117.

    Equation (4.1) on page 105 says that we must divide by n to compute the inverse DFT. This division is done in irv for the previously discussed in-phase filters. Since here we are using the full complex inverse transform, which does not include the division, we must do it now. But why do we divide by halfn instead of n? Recall from the mathematical presentation several pages ago that the in-phase and in-quadrature filter functions reinforce below the Nyquist frequency. This induces a factor of two that we put in place now.

    It is sometimes useful to examine directly the in-phase and in-quadrature filtered outputs. However, these individual components are not often interesting. The problem is that they reflect the underlying periodic component, rising and falling themselves. The amplitude of the combined outputs, as illustrated in Figure 4.13 on page 124, is usually the ultimate goal of QM filtering. The phase angle relating the outputs is occasionally of interest, as it may help to pinpoint the precise location of an event at a resolution finer than the original time-domain sampling rate. Even though computation of the amplitude and phase is trivial, in the interest of completeness, a code sample is provided. This is extracted from QMF_SIG.CPP on the accompanying code disk.

```
for (i=0 ; i<n ; i++)
   amp[i] = sqrt ( real[i] * real[i]  +  imag[i] * imag[i] ) ;

   if ((fabs(real[i]) > 1.e-40) || (fabs(imag[i]) > 1.e-40))
      phase[i] = atan2 ( imag[i] , real[i] ) ;
   else
      phase[i] = 0.0 ;
   }
```

The only item of note in this code is in the phase computation. It is always good programming practice to head off runtime complaints. By checking the magnitudes of the real and imaginary parts, we avoid the possibility of calling atan2 with an indeterminate pair of arguments. Some compilers respond to such an event by unceremoniously dumping the user back to the operating system.

5

Wavelet and QMF Features

- Feature presence versus feature state

- The width dilemma

- A comparative example

- Wavelet features

- The Morlet wavelet

Chapter 4 defined quadrature-mirror (QM) filters and discussed their mathematics and computational details. It was seen that they are a natural extension of the bandpass filters already thoroughly presented. For the sake of readers who wish to avoid the relatively complex mathematics of Chapter 4, and for those who desire more information regarding their practical applications, this chapter backs off and presents QM filters from an intuitive standpoint. A small amount of material must be repeated in order to ensure a smooth exposition. However, since QM filters are both complex and useful, readers will surely be forgiving.

After QM filters are under control, it will be shown how families of them, defined in special ways, can be immensely useful when taken as a group. These families, called *wavelets*, are similar to the DFT representation of a time series (Chapter 3) in that they express the information contained in the series using an alternative domain. However, the wavelet domain is different from the DFT domain in that features are located in terms of both frequency (like the DFT) *and* time (like the original series). This chapter also contains a very brief look at competing wavelet families and explains why some of the popular wavelet families not treated in this text are actually almost useless for time series prediction with nonlinear models.

Before proceeding, all readers should review the section on filtering in Chapter 1 (page 29), then thoroughly study the example at the end of Chapter 2 (page 48). Especially focus on the bandpass outputs shown in Figures 2.14 through 2.19. These figures are the intuitive foundation on which much of this chapter is built. Finally, look back at Figure 4.13 on page 124 and read the material leading up to that figure (page 121). Even though that material is relatively complicated, it provides a feeling for the direction in which this chapter is moving.

Feature Presence versus Feature State

When a bandpass filter is applied to a time series, there are two entirely different uses for the filtered output. We may supply both types of information to our prediction model, but often we will choose one or the other. In any case, it is crucial that we understand the distinction between the two types of information and the implications of the decision concerning whether to supply one or both of them to the model.

When we apply a lowpass filter to a time series, the filtered output has an obvious meaning: The value of this output at any time indicates the average value of the input series. This is clearly demonstrated in Figures 1.19 and 1.20 on page 36. The output of a highpass filter is almost as clear. Figure 1.22 shows that unusually high or low values of the highpass output correspond to similarly unusually high or low sudden jumps in the input series. In other words, these filtered outputs embody a special aspect of the *state* of the input series. High or low values in the input (subject to the constraints imposed by the filter) correspond to high or low values in the output. This effect is not as clear in bandpass outputs like that shown in Figure 1.21, but it is still there.

Now turn to the bandpass outputs in Figures 2.14 through 2.19 starting on page 56. With the possible exception of the fundamental output, the relationship between the original series and these bandpass outputs is not at all obvious. But it *is* there, and it may contain vital information. The point-by-point value of a periodic component that repeats a specified number of times per year might provide useful information. The knowledge that it is at a high or low value, and the knowledge of that precise value, might be important.

Examination of these figures, especially Figure 2.19, reveals another type of information. The rapidity of motion combined with the low resolution of the printed figure portrays an outline surrounding the total variation. Imagine sketching a smooth line along the top half of this outline. The height of this sketched line depicts (loosely) the amount of periodic variation at each time slot. It does not tell us whether the variation at a given moment happens to be high, low, or crossing the zero axis. It simply tells us the amplitude of the variation. In other words, rather than indicating the status of the variation, it indicates the *presence* of the periodic variation. When the sketched line is low, this particular variation is not strongly present, and when the line is high, the degree of variation is high.

It should be apparent that the smooth sketch along the outline of the filtered output contains an entirely different sort of information than the actual values of the output series. The general (but not total) thrust of this chapter will be toward capturing and utilizing the amplitude data as opposed to the individual points. The entire final section of Chapter 1 dealt with using the actual outputs, and this carried over into Chapter 4. Here and there throughout this chapter we will discuss how to make direct use of the outputs. However, we will focus primarily on the amplitude right now.

How do we compute the amplitude depicted by the sketched outline of the filter output? Some readers have already plowed through the mathematics and computer code in Chapter 4 (starting on page 129). For those who have not, be content to know that the algorithm is simple, the mathematics is difficult, and you may be better off not trying to pursue the subject any further than is necessary. QM filter amplitudes can be used in practical applications without knowing anything about their mathematics.

An Introductory Example

We begin our exploration of QM filters and the comparison of ordinary bandpass outputs versus QMF amplitude by studying a time series containing a diversity of features. Figure 5.1 plots long-term government bond rates from 1920 through 1994. In a great many applications, it is advisable to apply a lowpass filter so as to be able to supply the prediction model with general tendency information. The output of a lowpass filter having a cutoff frequency of 0.0 and a width of 0.03 is shown in Figure 5.2.

The easiest way to recover the high-frequency information that was removed by the lowpass filter is to subtract the lowpass output from the original series. This residual is graphed in Figure 5.3. In order to reduce numerical problems, as well as all sorts of artifacts due to large movements, all subsequent operations (bandpass filters and QMF amplitude computations) are performed on this residual series. This is not mandatory, but it is a good habit.

It is always advisable to consider the possibility of seasonal components or weak but significant narrow-band periodic components being present throughout the series. The power spectrum shown in Figure 5.4 confirms that there is nothing unusual in the residual. Do not be mislead by the large peak at the left side of the spectrum. This is nothing more than a manifestation of the fact that the original series is strongly skewed toward low frequencies. A spectrum of the original series would have a very strong peak at the far left. But we applied a lowpass filter, and this series is the residual after removal of the low frequency movement. The spectrum of the residual picks up where the lowpass filter left off.

Figure 5.5 is the output of a bandpass filter centered at a frequency of 0.15 cycles per sample. (Since there are twelve monthly samples per year, this is equivalent to 1.8 cycles per year.) The width of this filter is a relatively narrow 0.03. The point-by-point amplitude of the bandpass output is graphed in Figure 5.6. Observe how these two series, though closely related, convey extremely different types of information at each time slot. During time periods when the amplitude is consistently high, the bandpass output is rapidly changing from very positive, then to zero, then very negative, then to zero again, and so forth. Nonlinear prediction models treat these two sources differently, and the relative importance of each depends entirely on the application.

Figures 5.7 through 5.10 are bandpass outputs and QMF amplitudes at two other frequencies, one intermediate and one almost at the Nyquist limit of 0.5 cycles per sample. These filters have a width of 0.03, just like the first filter. At first glance they seem similar to the first pair of filters (Figures 5.5 and 5.6). This is primarily because the burst of energy around 1980 pervades all filters. However, on close examination, it is apparent that the three bands contain quite different information. It is possible that they comprise nicely complementary inputs to a prediction system.

Summary of Filter Outputs

We have studied three different types of filter outputs. The most common and most generally useful is an ordinary bandpass filter. In Chapter 4, this was referred to as an *in-phase* filter because the phases of the components passed by the filter are not affected. When our goal is simply to separate the information in a series into several different bands of interest, we use a bandpass filter.

The second type of filter output, not discussed in this chapter, but briefly treated in Chapter 4, is the *in-quadrature* output. This series is almost identical to the ordinary bandpass output. The only difference is that its phase is shifted by one-quarter cycle, causing it to indicate rates of change rather than actual values. Most time-series prediction applications have little direct use for this output alone, although it can be valuable in conjunction with the in-phase output.

The third type of filter output is the amplitude. We use this output when we care about the *presence* of a periodic feature, as opposed to the point-by-point state of the component. This is a very different type of information from that of the first two outputs.

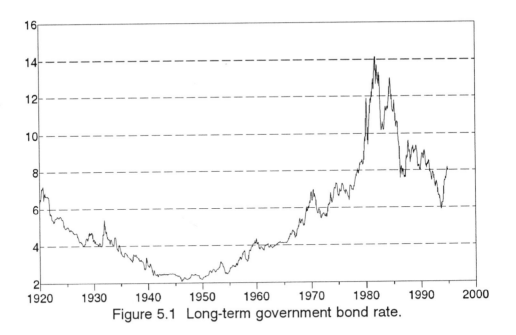

Figure 5.1 Long-term government bond rate.

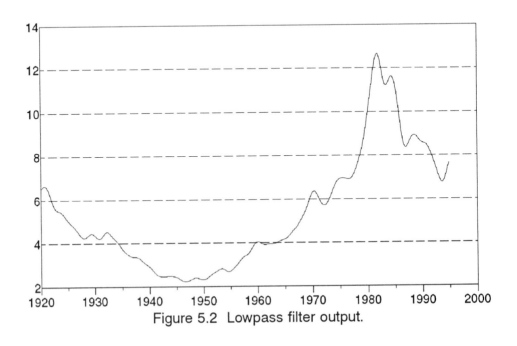

Figure 5.2 Lowpass filter output.

Feature Presence versus Feature State

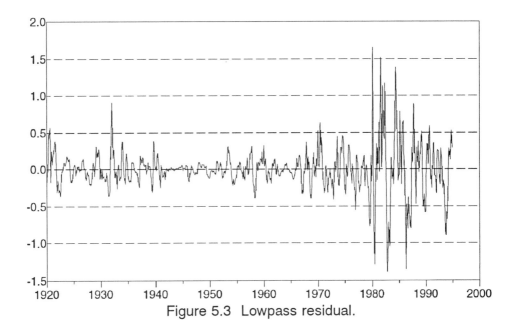

Figure 5.3 Lowpass residual.

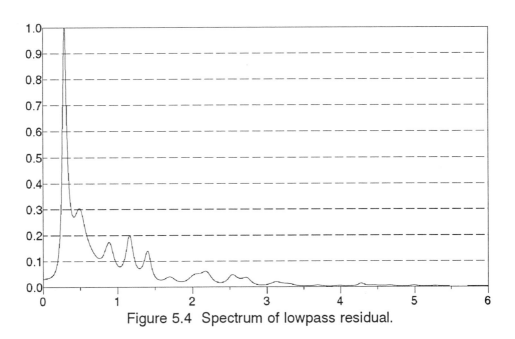

Figure 5.4 Spectrum of lowpass residual.

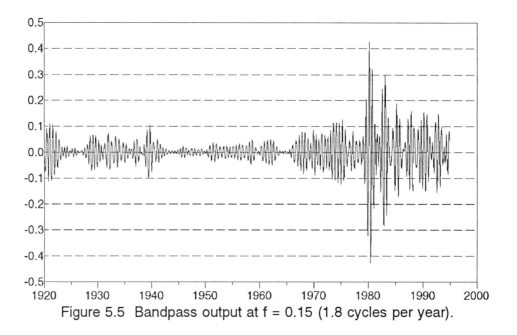
Figure 5.5 Bandpass output at f = 0.15 (1.8 cycles per year).

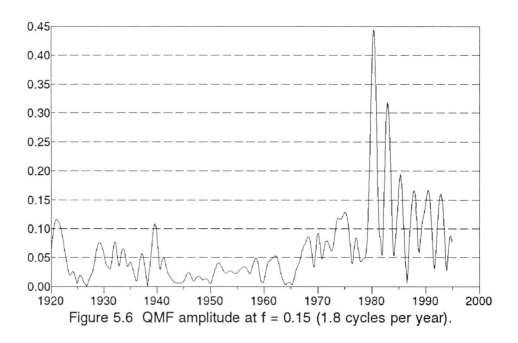
Figure 5.6 QMF amplitude at f = 0.15 (1.8 cycles per year).

Feature Presence versus Feature State 143

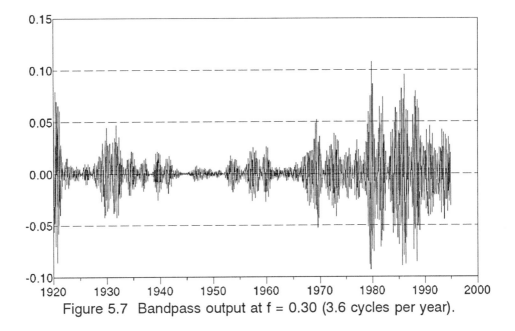

Figure 5.7 Bandpass output at f = 0.30 (3.6 cycles per year).

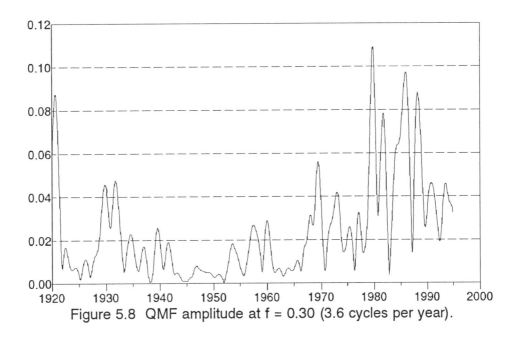

Figure 5.8 QMF amplitude at f = 0.30 (3.6 cycles per year).

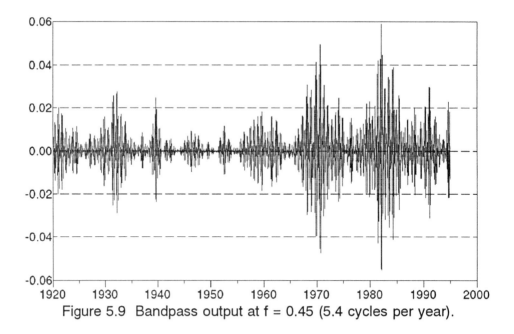
Figure 5.9 Bandpass output at f = 0.45 (5.4 cycles per year).

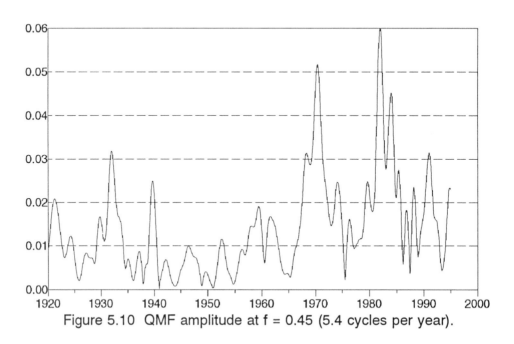
Figure 5.10 QMF amplitude at f = 0.45 (5.4 cycles per year).

The Width Dilemma

This section discusses the filter width parameter. This may seem to be a rather mundane topic, as it has been liberally treated in several previous chapters. However, the choice of a bandpass or QM filter width has deep and potentially serious ramifications. Moreover, the precise nature of these ramifications is the dominant motivation for the development of wavelets, both historically and in this text. In particular, *the entire intuitive motivation for wavelets that will be presented later derives from the material in this section.* Readers who are truly interested in wavelets should study this section carefully.

Let us begin with a quick review. The legal range of frequencies that we may observe and filter is from 0.0 to 0.5 cycles per sample. The width of a filter specifies a range of frequencies that will be passed on either side of a center frequency. This was illustrated in Figure 4.5 on page 113. The filter passage is significantly decreased at the center frequency plus and minus the width, and passage is almost nil at a distance of twice the width. In other words, suppose we have a filter with a center frequency of 0.2 (a period of 1 / 0.2 = 5 samples per cycle) and a width of 0.05. Frequencies from 0.15 to 0.25 will be passed (with some attenuation at the extremes). Frequencies below 0.1 and above 0.3 will be almost totally stopped.

When we want to zoom in on very narrow bands of interest, we need to use a small width parameter. But we have seen that there is a high price to pay for a narrow frequency-domain width. The heuristic estimate $0.8 / s$ has been presented as a way to gauge how far away (in the time domain) observed samples affect filter outputs. For example, with a frequency-domain width of $s = 0.05$, we compute the maximum time extent as $0.8 / 0.05 = 16$. This means that the filter output in each time slot is affected by observations as distant as 16 slots earlier and later. This is an annoyance in the interior of the series, and it is a disaster at the ends where the historical and future observations are unknown. Thus, we have a dilemma: Do we use a narrow frequency domain width in order to be able to focus on particular bands of interest, even though the price is large width in the time domain? Or do we opt for wonderfully narrow width in the time domain, with the price being a wide filter response in the frequency domain? The answer depends on the application, but some useful guidelines exist.

If several different filters are used, as is usually the case, the width tradeoff dilemma can be severe if we naively limit ourselves to choosing one common width for all of the filters. But the breakthrough comes when we realize that we do not need to abide by any such restriction. Moreover, it turns out that there is an excellent method for trading off resolution in the two domains in such a way that we win in one domain when we can most easily afford to lose in the other domain.

First, let's explore the need for resolution in the frequency domain. The period (in samples per cycle) of a periodic component is the reciprocal of its frequency (in cycles per sample, 0.0–.5). The components that reside in the upper half of the frequency range (0.25–0.5) have periods ranging from two to four samples per cycle, quite a narrow range. The components in the lower half (0.0–0.25) have periods from four all the way up to one long cycle spanning the entire measured time interval. The lower half is obviously where most of the action is. If we continue to subdivide the frequency range, this effect continues. The closer we come to the bottom of the frequency range, the greater the number of different periods that are packed into an interval of any fixed width. It is clear that we benefit most from narrow resolution when we are dealing with low frequencies.

The world contains many manifestations of this phenomenon. One of the most obvious examples is in music. As is commonly known, the basic musical interval is the *octave*. The relationship between pairs of notes has a consistent sound when transposed by octaves. If a guitar is playing a song in the key of G, a melody instrument will sound correct in any octave as long as it is also in the key of G. The interesting fact is that octaves are not linear. Instead, they involve factors of two. To raise a musical note by one octave, double its frequency. The frequency difference between A and B in the middle octave of the piano is exactly twice the difference between A and B one octave below the middle. As the frequency of a musical sound increases, differences that are comparable to the ear correspond to physical differences that are actually greater in absolute terms.

This logarithmic distribution of information in the frequency domain inspires the following rule: *The width of a filter should be proportional to its center frequency*. As an illustration of this fact, look at Figure 5.11. The three filters shown there all encompass the same amount of information. (Note that this argument applies in a general theoretical sense. A particular application may, by its nature, focus information differently. This possibility is somewhat glossed over here, but the principles still apply to some degree even when that is the case.) The filters in that figure have widths proportional to their center frequencies.

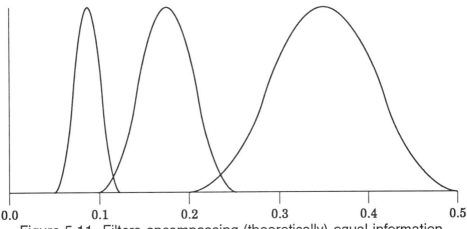
Figure 5.11 Filters encompassing (theoretically) equal information.

We are now motivated to base filter widths on their center frequencies. But what about the time domain? What does this rule do to us there? An example may help.

Suppose we are measuring the thickness of a wire as it emerges from a manufacturing device. Our goal is to predict whether a mechanical breakdown is immanent (or perhaps has already occurred). We know from the construction of the machine that variations in wire thickness are caused by two different underlying processes. One process produces slow changes in the thickness, while the other process causes rapid variation. In order to separate the information about these processes, we use two filters. One filter passes the low-frequency information that primarily describes the state of the first process, and the other filter passes the high-frequency information that is responsive to the second process. Ignoring the frequency-domain width for now, think about the widths we desire in the time domain. By definition, the low-frequency component changes its value slowly. When the low-frequency output is computed for some time slot, it does not matter very much if that computation is influenced by measured thicknesses a relatively large distance away. Since the component on which this filter focuses varies at a slow rate, distant observations will not introduce unseemly artifacts into the result.

The situation is reversed for the high-frequency filter. If we wish to have an accurate picture of the status of the high-speed variation at any given time slot, it is necessary that the filter output be influenced only by observations in the immediate neighborhood of the time slot under consideration. Since this component changes at a high rate of speed, distant observations have relatively little to do with the current state of the process and thus should be excluded from computation. *A high-frequency filter requires narrow width in the time domain.*

Let's compare this rule with the frequency-domain rule that was presented earlier. That rule said that we need narrow (in the frequency domain) filters at low frequencies, and we could get away with wide filters at high frequencies. This new rule says that we need narrow (in the time domain) filters at high frequencies, and we can tolerate wide filters at low frequencies. Now remember the heuristic rule that relates time-domain extent with frequency-domain width (s): *extent* = 0.8 / s. These quantities are inversely related. We are in luck! When we need narrow width in one domain, we can tolerate increased width in the other domain. So these two rules are in agreement. Observe them.

One More Implication of Time-Domain Extent

The tradeoff between time-domain extent and frequency-domain filter width is important, since this tradeoff is the motivation for wavelets. However, there is another aspect of time-domain extent that must be taken into account when filters are used to provide input data for prediction models. Remember that filter outputs become distorted when part of the filter extends past the end of the known bounds of the series. Since these distorted values can wreak havoc with a prediction model, we cannot make use of filter outputs near the end of the series.

Figure 5.12 illustrates two alternative prediction methods involving a filter having a time-domain extent of three samples. In practice, this would usually be unrealistically short, since a frequency domain filter width of 0.8 / 3.0 = 0.267 is implied. In both examples, the prediction distance is 12 samples (months, perhaps) ahead of the current time. The upper half of the figure depicts immediate use of the filtered output at each time slot. For the sake of this illustration, assume that the known data expires at month 12, somewhat to the left of the center of the time line.

Since this filter is influenced by samples up to three time slots on either side of the current output, the filter outputs will be distorted for months 10 through 12 by virtue of the filter (shown as a black triangle) extending into the unknown future. The last time slot for which fully valid filter outputs can be supplied to the prediction model is month 9. Since the prediction distance is (by design) 12 months, the most distant future prediction that can be made is the value for month

The Width Dilemma

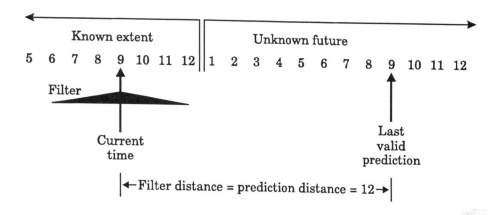

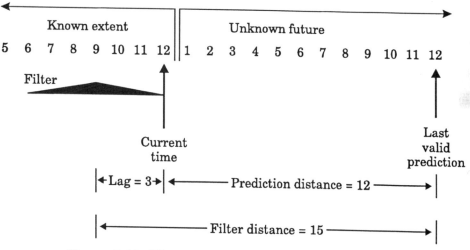

Figure 5.12 Filter lags increase prediction distance.

9 of the next year. In other words, even though the model was designed to predict 12 months in advance, and we possess data through month 12 of this year, we can only predict 9 months ahead. The filter has cost us three months of predictions.

The lower half of Figure 5.12 shows an alternative way of using the filter outputs. This method shifts the cost of the filter width, gaining prediction distance while probably losing a little accuracy.

Instead of using the current value of each filter output as a model input, we use a lagged value. To simplify this example, a lag equal to the filter width is used. In practice, any intermediate lag can be used to achieve a compromise between the two extremes shown here.

The last valid position for the filter is still month 9. This is unavoidable. But if the model explicitly uses the filter output at a lag of three months, the model can be applied right through the last known month. This allows valid predictions to be made through month 12 of the following year, thus attaining the full designed prediction distance.

There's no such thing as a free lunch. By using the filter output at a lag of three, the total distance across which predictions must be made is 15 months, as opposed to 12 months for the previous method. This can sometimes result in a significant loss of accuracy. Is it worthwhile? In a simple example like this one, the worthiness of lagging the filter is debatable. Certainly, there is a degree of dependence on the particular application. However, in only trivially more complex models, the answer is often an indisputable *yes*. The evidence comes down hard on the side of lagged outputs when one or more useful and valid inputs *for the current time slot* exist. We may, for example, believe that the current (unfiltered) value of the series contributes useful information. More likely, we may have one or more concomitant variables whose current values are deemed to be good predictors. In this situation, direct use of the filtered output would be extremely wasteful. As can be seen from the top half of Figure 5.12, we would be forced to ignore the most recent three months of perfectly valid supplementary measurements, a disgraceful waste. *Only by lagging the filtered outputs by a distance at least equal to the filter's time-domain width can we make use of current values of other variables.*

A More Realistic Example

The QM filter example that started on page 140 used the same width for all filters, an action that we now know to be a generally poor approach. Equal filter widths made the amplitude plots pretty, and the description of the filters was simplified. But we almost always want to vary the filter width according to the center frequency. Another practical problem with that example is that all operations were performed on the raw data. That's fine when the series stays roughly the same height throughout its extent. However, when the series wanders about as much as that one does, we are almost always better off working with differences rather than raw values. The subject of inducing stationarity by differencing will be covered in depth in Chapter 7. For now, take this as a given fact.

An interesting task is to use historical information in the long term government bond rate to predict future values of the rate. Examination of Figure 5.1 on page 140 indicates that this will be an extremely difficult task due to the tremendous variety of patterns visible in the data. Nevertheless, let us try. Actually, rather than predicting raw rates, an easier and probably more useful goal is to predict future *changes* in rates. In particular, the goal of this example is to predict the rate change exactly one year beyond the current date.

The dataset used for this example is LTGOVTBD.DAT, found in the FINANCE directory. This file contains monthly rates from January 1919 through December 1994. The 12 observations from 1919 will be used as a starting reference for computing the year-ahead differences. All displays will therefore start with the year 1920. The NPREDICT commands to read the data and display the raw values are now listed. The complete control file is QMF_DIFF.CON in the EXAMPLES directory.

```
NAME = , data
READ SIGNAL FILE = ltgovtbd.dat
NAME = actual
COPY = -900 data
DISPLAY ORIGIN = 1920
DISPLAY RATE = 12
DISPLAY = actual
```

The first value in each record is the date, so a comma appears first in the NAME command as a placeholder for the date. The total length of the file is 76 years times 12 months, or 912 observations. The last 75 years (900 cases) are copied to a signal named *actual* for display purposes. The origin is set to 1920, and a sample rate of 12 months per year is established. The display of the raw data is shown in Figure 5.13 on page 161.

The year-ahead rate difference is computed using the entire dataset. By starting at 1919 instead of 1920 (as was done for the raw data display), the time of each series point is kept consistent. When each new point is computed as the difference between the current observation and the observation one year earlier, the first year (which has no prior year) is lost. Explicitly keeping track of this offset can be irksome, so it is always preferable to use the technique shown here to ensure that the relative times of all series are comparable. In this example, by starting the seasonal differencing in January of 1919, we cause the first point in the new series to correspond to January of 1920. The one-month differences are computed with the same basic method. The only modification is that we start with December of 1919 because we need only one extra month, not 12. This is done by keeping the last 901 cases. These two differences are graphed in Figures 5.14 and 5.15, respectively. Note that because the one-month difference series will be used for neural network inputs, it is standardized.

```
NAME = year_ahead_diff
COPY = data
SEASONAL DIFFERENCE = 12 year_ahead_diff
DISPLAY = year_ahead_diff

NAME = diff
COPY = -901 data
DIFFERENCE = 1 diff
STANDARDIZE = diff
DISPLAY = diff
```

The first data-processing step taken here is a very common first step. A lowpass filter is applied to the data to reveal slow trends, and this smoothed series is subtracted from the unfiltered series to procure the remaining information concerning rapid local variation. Since the series to be filtered is centered around a fairly constant value, it is most appropriate to pad with its mean (as opposed to endpoint flattening). Finally, it is always advisable to check the power spectrum of the lowpass residual to verify that no deterministic periodic components are present. The results of these three operations are shown in Figures 5.16, 5.17, and 5.18, respectively. Note that because a display rate of 12 was specified earlier, the horizontal axis of the spectrum is labeled as 0–6 cycles per year, rather than the common 0–.5 cycles per sample.

```
PADDING = MEAN
NAME = low
LOWPASS = 0.0 0.05 diff

NAME = lowsub
SUBTRACT = diff AND low
```

```
STANDARDIZE = low
STANDARDIZE = lowsub
DISPLAY = low
DISPLAY = lowsub

NAME = lowsub_spec
MAXENT = 4001 50 lowsub
DISPLAY = lowsub_spec
```

An arbitrary decision is made to employ two QM filters. The lower-frequency filter has a center frequency of 0.15 cycles per sample (period of $1 / 0.15 = 6.67$ months) and a width of 0.1. Since our bandpass filters have a response as far out as twice the width, it would seem that this filter would be impacted by any constant offset (frequency = 0) of the input series. However, the QM filters in NPREDICT set the DC term in the transform equal to zero, forcing all QM filter outputs to be centered around zero. Later in this chapter, we will see a method that approaches zero at the DC point more gradually. However, this chopping at zero generally has no ill effect and is almost always beneficial, so there is no need to worry.

The higher-frequency filter has a center frequency of 0.35 and a width of 0.2. This simultaneous increase of width with frequency is commendable. We compute the in-phase, in-quadrature, and amplitude outputs for both filters. The phase angles are not computed (and, in fact, are almost never explicitly needed). These quantities are standardized and displayed in Figures 5.19 through 5.24.

```
NAME = p1 , q1 , a1
QMF = 0.15 0.1 lowsub
STANDARDIZE = p1
STANDARDIZE = q1
STANDARDIZE = a1
DISPLAY = p1
DISPLAY = q1
DISPLAY = a1

NAME = p2 , q2 , a2
QMF = 0.35 0.2 lowsub
STANDARDIZE = p2
STANDARDIZE = q2
STANDARDIZE = a2
```

```
DISPLAY = p2
DISPLAY = q2
DISPLAY = a2
```

The data is ready. We will now use four different subsets of this data, training a neural network on each subset and comparing the results. First, set up the global parameters for training and testing the networks. It is almost always easiest on the nerves if training progress is regularly reported. Also, a good habit is to strive for zero error and abort the training when patience expires. A small but reasonable number of restarts are requested. When so few restarts are done, it is most efficient to set the number of pretries equal to one more than the number of restarts, so that only one expensive refinement is performed. (More detailed explanations of these parameters can be found in Masters (1994).)

```
PROGRESS ON
ALLOWABLE ERROR = 0.0
ACCURACY = 5
REFINE = 3
MLFN RESTARTS = 4
MLFN PRETRIES = 5
```

We choose to use an MLFN model having one hidden layer of five neurons. Linear output activation is almost always best. The AN1_LM learning algorithm is usually the fastest in small-to-moderate networks.

```
NETWORK MODEL = MLFN
MLFN HID 1 = 5
MLFN OUTPUT ACTIVATION = LINEAR
MLFN LEARNING METHOD = AN1_LM
```

The decision is made to withhold the last three years (1992–1994) of known data to serve as a test set. It should be obvious that this will be a tough test, since even casual examination of the data indicates that several extremely different patterns appear. It turns out that a neural network can predict rate changes significantly better than the naive approach of assuming constant rates. However, the widely different results attained with the four subsets of variables that we are about to study indicate that more careful experimentation would be worthwhile. Also, it is never good to try to predict any series based on a single observed variable. Other econometric information would be valuable. The purpose of this test is simply to compare the utility of several families of filter outputs.

Several aspects of the command sequence are common to all four tests. If we were doing only one test, or if we were not doing any actual prediction into the future, we could use the *year_ahead_diff* series for the training and test sets. But since we will be extending this series into the future, and since we need to reuse the exact same data for each of the four tests, we cannot tamper with this series. Thus, the first step in each test is to copy the output series into a private scratch series that will be used throughout the test.

There are a total of 900 observations. The maximum lag of the model is 24 months (by our arbitrary choice), and the lead is 12 months. Therefore, the first 36 months are useless because predictions for those time slots involve inputs prior to the start of the known series. So we only have the final 864 months as valid data. We have chosen to withhold the last 36 months as a test set. Therefore, the training set will contain 828 cases.

With these ideas in mind, let us construct the training and test sets for the first test. This test avoids all use of filters. Only the actual data is used. There are 14 variables: the current value, all lags up to a year back in time, and the value exactly two years ago. The commands for building the training and test sets are as follows:

```
NAME = extend1
COPY = year_ahead_diff

CLEAR INPUT LIST
CLEAR OUTPUT LIST

INPUT = diff 0-12
INPUT = diff 24
OUTPUT = extend1 12

CLEAR TRAINING SET

CUMULATE INCLUDE = 828
CUMULATE EXCLUDE = 0
CUMULATE TRAINING SET
```

CLEAR TEST SET
CUMULATE INCLUDE = 999999
CUMULATE EXCLUDE = 828
CUMULATE TEST SET

The network is trained, saved to disk, written to an ASCII file for the user's edification, and tested on the final three years of known data that were withheld from the training set. Readers who wish to duplicate this process are warned that the training time is very long. This is a large model with a large training set.

TRAIN NETWORK = temp

SAVE NETWORK = temp TO qmf_dif1.wts
PRINT NETWORK = temp TO qmf_dif1.wpr
TEST NETWORK = temp

There are two interesting things that can be done with a neural network that has been trained for prediction. The simpler of the two is to compute a point-by-point prediction based on the known data. This is done by giving the predicted signal a name that is not in the current input or output list. The name *pred1* is chosen here. By asking for 900 predictions, we can see not only the network's performance within the duration of the training series, but its predictions for the final three years as well. Subtracting this predicted series from the true values gives an error series. For both of these series, it is best to start the display at month 36 so as to avoid the early predictions that are seriously erroneous by virtue of being based on inputs prior to the start of the known data.

NAME = pred1
NETWORK PREDICT = 900 temp

DISPLAY DOMAIN = 36 900
DISPLAY = pred1

NAME = error1
SUBTRACT = extend1 AND pred1
DISPLAY = error1
DISPLAY DOMAIN = ALL   ; Restore default to avoid confusion

A somewhat more complex prediction task is to extend the known series into the unknown future. This is done by giving the predicted signal a name that is in the input or output list. The easiest (and most common) choice is to use the same signal that was used for the training set. This is *extend1* in this example. It is also interesting to compute the historical confidence information so that confidence intervals can be displayed for predictions into the future. An important point should be made. Since the model we are employing does not reuse predicted outputs as recursive inputs, results will be badly skewed if an attempt is made to predict further into the future than the lead time. More distant predictions would be based on inputs that have no valid values, so errors can be serious. Since the lead for this model is 12 months, we will limit our future predictions and confidence computations to this extent. The predictions are most easily seen when the display starts near the end of the known series. By starting at month 888, we only share the screen with one year of known data. The last step is to erase this network so the name can be conveniently reused (not to mention saving memory). Note that these predicted series are not shown in this text, as they are not very exciting. Those produced by a later model, which does a good job, will appear.

```
NAME = extend1
NETWORK CONFIDENCE = 12 temp
NETWORK PREDICT = 912 temp
DISPLAY CONFIDENCE = extend1
DISPLAY DOMAIN = 888 911
DISPLAY CONFIDENCE = extend1
DISPLAY DOMAIN = ALL
CLEAR NETWORKS
```

The next model makes relatively primitive (but often very effective) use of filters. The outputs of the lowpass filter comprise one set of inputs, and the residual (which contains the complementary high-frequency information) is the other set of inputs. Also, just for the sake of having the most recent observation available, the current value of the unfiltered series is an input. Only the initial copy operation and the definition of the input list are shown here. All other operations are exactly the same as in the first test. There is one vital point to notice, though. Recall from the beginning of this example that the lowpass filter has a width of 0.05. Our heuristic rule tells us that the time domain extent of this filter is about 0.8 / 0.05 = 16 months. This might be a good time to reread the section starting on page 148. We chose a minimum lag of 6 months as a compromise between the two extremes illustrated in Figure 5.12. The implication is that predictions more than a few months into the future might be a little suspect. We can often get away with a lot in this regard, especially for lowpass filters. But extremely critical applications should employ the more conservative approach of starting with a greater lag.

```
NAME = extend2
COPY = year_ahead_diff
CLEAR INPUT LIST
CLEAR OUTPUT LIST
INPUT = diff 0
INPUT = low 6-12
INPUT = low 24
INPUT = lowsub 6-12
INPUT = lowsub 24
OUTPUT = extend2 12
```

The third test keeps the (usually vital) lowpass information, but replaces the residual with bandpass (in-phase QM filter) outputs. The low-frequency QM filter has a width of 0.1, so its time-domain extent is about 8 months. This heuristic is very conservative, so we can almost always get away with cheating a little in all but the most critical tasks. Starting this filter's lags at 6 months is perfectly reasonable. The high frequency filter has a width of 0.2, giving an extent of 4 months. Thus, a starting lag of 3 months is reasonable.

```
NAME = extend3
COPY = year_ahead_diff
CLEAR INPUT LIST
CLEAR OUTPUT LIST
INPUT = diff 0
INPUT = low 6-12
INPUT = low 24
INPUT = p1 6-12
INPUT = p1 24
INPUT = p2 3-12
INPUT = p2 24
OUTPUT = extend3 12
```

The last test uses the amplitude information instead of the in-phase outputs. This set of variables often goes to an extreme: Either it is very valuable, or it is totally worthless. The commands for building this input/output list are as follows:

```
NAME = extend4
COPY = year_ahead_diff
CLEAR INPUT LIST
CLEAR OUTPUT LIST
INPUT = diff 0
INPUT = low 6-12
INPUT = low 24
INPUT = a1 6-12
INPUT = a1 24
INPUT = a2 3-12
INPUT = a2 24
OUTPUT = extend4 12
```

It's time to look at the results of these tests. Please remember that the error surfaces of these models are covered with local minima of varying quality. Readers who duplicate these tests are likely to obtain significantly different results. However, the general relationships should remain fairly constant. The results are as follows:

	Training error	Test error
Test 1	0.4054	1.6557
Test 2	0.1813	0.8164
Test 3	0.2196	0.7454
Test 4	0.1473	1.1723

The obvious grand loser is the first test, in which only the unfiltered data is used. One common theme in these results is that the training and test errors are dramatically different. This is a familiar effect in econometric data. It is most often caused by the fact that patterns change across time. Rules that did an excellent job of prediction in 1930 probably do not mean a lot in 1990. One must always be careful to train the network with data that is representative of what will be presented to the network when it is put to work. In this example, the network expends the majority of its energy attempting to learn to predict the wild gyrations that occurred in the decade of 1980–1990, and in doing so, it largely ignores everything before and after this epoch! Nonetheless, some things never change, and these models are able to learn some tricks to improve on shots in the dark. Also, remember that inclusion of other econometric series usually improves performance. Using just one series to do prediction is foolish.

Since the third subset had the smallest errors on the three-year test set, its results are shown here. Figure 5.25 shows the pointwise predictions made by this model. Observe that (surprisingly!) it is able to predict the sudden jump late in 1994. It doesn't predict quite as much of a jump as actually occurred (see Figure 5.14), but it does a very respectable job. The errors of this prediction series are shown in Figure 5.26. Finally, the extension into the future, along with confidence lines, is shown in Figure 5.27. As is common with most prediction models, the predictions move toward the central tendency as the distance into the future increases.

Summary of this Example

This has been an extremely primitive attempt at prediction. The underlying series exhibits so many changing patterns that accurate prediction is probably impossible. Some sort of transformation to reduce the disparity between the decade of 1980–1990 and the rest of the series would be strongly advisable. Additional econometric inputs would surely improve performance. Nevertheless, useful general conclusions can be drawn. These include the following:

- When the variation of a series changes significantly across time, a prediction model will focus on improving performance in areas of large variation, neglecting areas of smaller variation.

- Splitting the raw data into different bands by means of filters can be very beneficial in terms of both training and test-set performance.

- There is not necessarily a positive correlation between training and test-set performance. A model having small training errors can have large errors in the test set.

- The three main families of supplementary filter outputs, (lowpass residual, bandpass, and QMF amplitude) can have very different capabilities.

The Width Dilemma 161

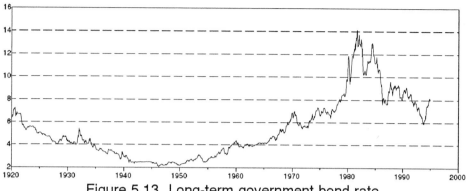

Figure 5.13 Long-term government bond rate.

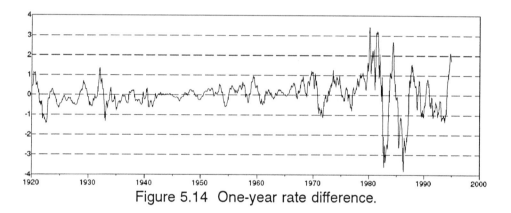

Figure 5.14 One-year rate difference.

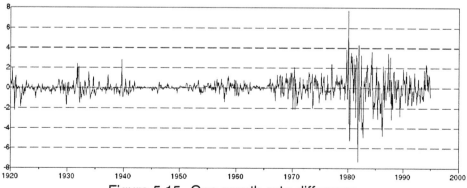

Figure 5.15 One-month rate difference.

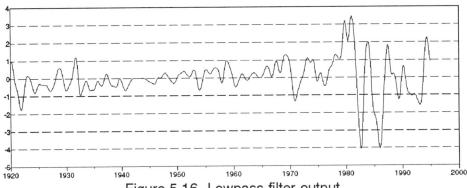
Figure 5.16 Lowpass filter output.

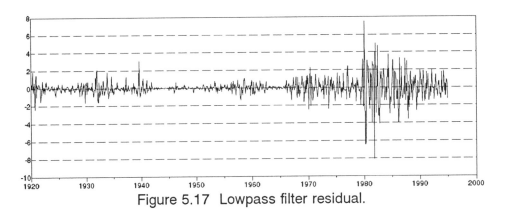
Figure 5.17 Lowpass filter residual.

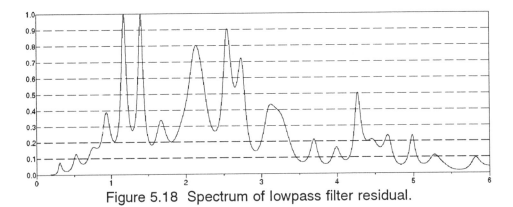
Figure 5.18 Spectrum of lowpass filter residual.

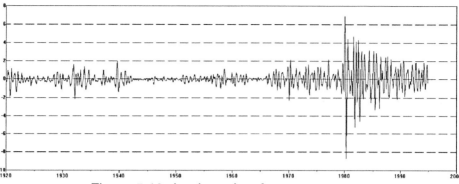
Figure 5.19 In-phase low-frequency output.

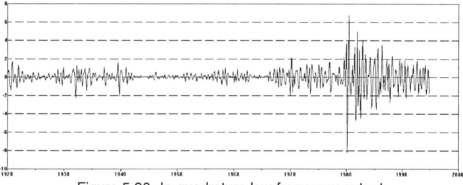
Figure 5.20 In-quadrature low-frequency output.

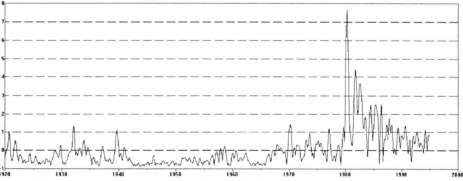
Figure 5.21 Standardized amplitude of low-frequency output.

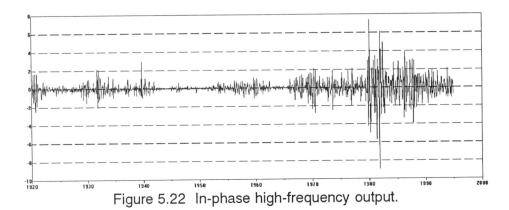
Figure 5.22 In-phase high-frequency output.

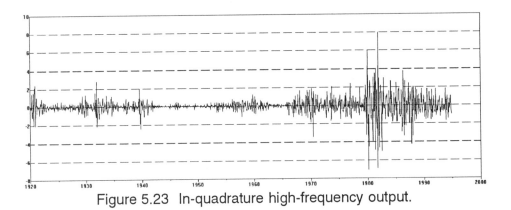
Figure 5.23 In-quadrature high-frequency output.

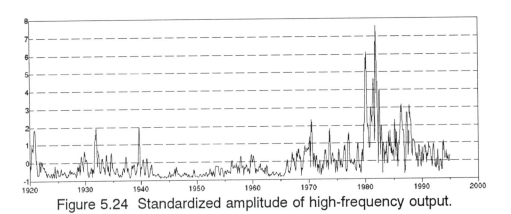
Figure 5.24 Standardized amplitude of high-frequency output.

The Width Dilemma

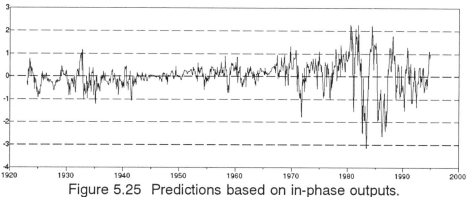

Figure 5.25 Predictions based on in-phase outputs.

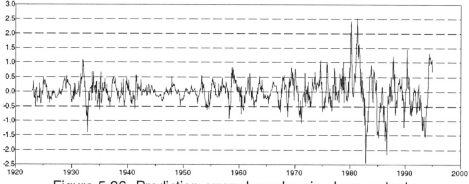

Figure 5.26 Prediction errors based on in-phase outputs.

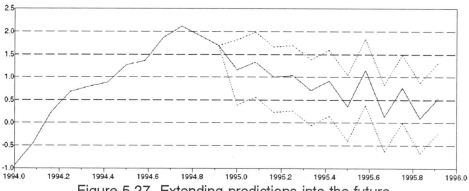

Figure 5.27 Extending predictions into the future.

Wavelet Features

We know that it can be useful to split a time series into several components, each corresponding to a different range of frequencies. The individual values of these new time series, taken together, can provide a more detailed picture of the state of multiple processes whose amalgamation generates the single measured series. But in many applications, knowledge of the current state of these processes may not be sufficient to solve the problem at hand. We may also require information about the recent history and near future of the process so that the current state can be placed in its proper context.

In the example presented in the previous section, the prediction model was supplied with contextual information by using filter outputs at many lags. This is acceptable, but a little more rigor would be nice. The filter frequencies, widths, and model time lags were chosen arbitrarily. With this naive approach, we may be feeding an excessive number of inputs to the model. Also, we may be accidentally discarding important information. There are sound mathematical principles that describe optimal methods for designing and spacing filters. These methods allow us to extract the maximum possible information from the original series while keeping the filters meaningful and asking the model to handle as few new inputs as possible.

As will be seen later, strict use of these principles is not only difficult from a logistical point of view, but it can also lead to an explosion in model inputs if care is not taken. The problem is that the principles of optimal filter design and placement involve three competing goals:

- We want to capture as much information as possible concerning the in-context state of the underlying processes as manifested in the observed time series.

- The filter outputs presented to the prediction model must have a simple, direct relationship with the underlying processes. An event must be evident in the filter outputs in a consistent and meaningful way.

- The prediction model cannot be overwhelmed by a vast number of filter outputs. The thorough and meaningful information capture must also be parsimonious.

Unfortunately, we are in a *pick any two* situation. It can be proven that all three of these goals cannot be simultaneously attained. Every application benefits from its own tradeoffs. This text suggests some compromises that are usually best for time series prediction. Be aware that everything is negotiable.

For the record, one inappropriate set of design goals will be briefly mentioned, then dismissed. Data compression tasks, as well as some types of optimal filtering and enhancement operations, are not concerned with filter outputs being meaningful or consistent. The second goal in the list can be ignored, since these applications do not care if events appear in the filter outputs in consistent ways. These tasks really care about only two things: Capture as much information as possible, and do it with as few filter outputs as possible. When this is the goal, the best wavelets are those in the famous *Daubechies* family. This family is discussed in a highly readable manner in Press *et al.* (1992), as well as in many other popular sources. It is vital for readers to understand that Daubechies wavelets violate the second goal in the list to an *extremely* serious degree. Identical events across the measured series can appear in Daubechies wavelets in so many different guises that most prediction models are unable to learn to recognize them well. No more will be said on this issue. The wavelets of choice in this text are those of the *Morlet* family. They have the weakness of generating more model inputs than Daubechies wavelets. However, their consistency in responding to underlying events makes them ideal for prediction applications. But we're getting ahead of ourselves. Let's back up and develop some more intuition before digging into Morlet wavelets.

Arranging Filters According to Width

Our goal is to be able to use a set of filter outputs to meaningfully describe the state of a signal-generating process. We also hope to increase the value of that state information by selecting the filter outputs so as to include as much contextual information as is reasonably possible. At the same time, in keeping with the admirable goal of parsimony, we want to use as few filter outputs as possible. How do we proceed?

We already know one useful fact: Outputs of correctly designed low-frequency filters embody information from a relatively long time duration. This is because these filters have (hopefully) been designed to have a narrow width in the frequency domain, which in turn implies large time-domain extent. This is in accord with the thought that low frequency components change slowly (by definition), so the output of a low-frequency filter at any given time slot is indicative of the state of that slow component in the recent past, in the present, and in the near future. This leads to the following important concept: *Just one single output of a narrow low-frequency filter is sufficient to describe the near past, the present, and the near future of the low-frequency component.*

Exactly the opposite is true of high-frequency filters. Remember that intelligent design principles dictate that high-frequency filters have a large width in the frequency domain, which implies a narrow time domain width. As a result, one single output of a high-frequency filter tells us almost nothing about the past or the future. Each output is primarily concerned with the present. Thus, we have the following guiding principle: *In order to obtain contextual information from high frequency filter outputs, we must supplement the present with values from the past and the future.*

We can make these rules even more precise. The inverse relationship between the frequency-domain width and the time-domain width is mathematically exact: When one width doubles, the other is cut in half. One implication of this fact is that whenever we double the frequency of a filter, we also (if we are smart!) double its frequency domain width, and hence implicitly halve its time-domain extent. In order for this higher-frequency filter to capture the same amount of contextual information in the time domain, we must examine its outputs twice as frequently, thus producing twice as many outputs as the lower-frequency filter. This is best illustrated with a picture. Look at Figure 5.28.

This figure illustrates the use of four filters having center frequencies and widths that are designed well. The tick-marks across the top of the figure represent the sampled time slots. The designer of this application requires context information within plus and minus 16 time slots. Four different frequency bands are chosen, and the width of each is defined to be one-half of the center frequency. The time domain extent and frequency-domain width of each filter are listed in the figure, with the time-domain extent of each filter also indicated by means of triangles. The fact that some triangles are solid and some are dotted has no significance whatsoever. This is simply a visual aid to more clearly show the overlapping filters. Notice that the time-domain extents and frequency-domain widths follow the familiar heuristic rule *extent* = 0.8 / *s*.

Wavelet Features

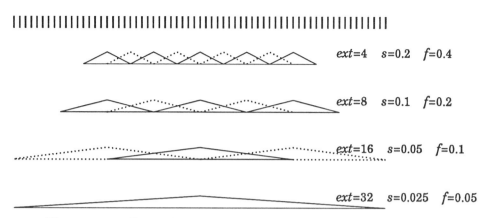

Figure 5.28 Capturing context information with multiple filters.

This set of filters deserves close examination. First, consider their spacing in the time domain. There are no response gaps between sampled filter outputs. We are in trouble if, for even one of the four filters, there is any time slot not well covered by the filter. When this happens, an event in the frequency band served by that filter may pass undetected if it occurs in the neglected time area. By closely spacing the sampled filter outputs (relative to the filter's time-domain extent), we ensure detection of all events. In this example, the top filter is sampled every fourth time slot, and its extent is four slots, with the responses significantly overlapping. An equivalent relationship holds for the remaining filters.

Similar overlap in the frequency domain is equally important if we want to be sure of capturing all events. Although the lack of corresponding triangles in this figure makes frequency-domain overlap a little more difficult to visualize than time-domain overlap, it can still be done. Remember that frequency response is significantly down at frequencies beyond the center frequency plus and minus the width, and response is almost totally gone by twice the width. The highest frequency filter, centered at 0.4 cycles per sample, has a width of 0.2. So, by a frequency of 0.2, where the next filter is peaking, its response is significantly down, and its response is nearly zero by a frequency of 0.0. The same relationships hold for the other filters. We appear to have good overlap in the frequency domain.

Finally, let's think about parsimony. We agreed to try to capture context information up to plus and minus 16 samples. This is a total of 33 measurements of the original raw data. Using the filter-bank approach, we sample the highest-frequency filter nine times, with each sample spaced four time slots apart. The other three filters are sampled five, three, and one times, respectively. This is a total of 18 measurements. We must use both the real (in-phase) and imaginary (in-quadrature) parts of each output, so we have a total of 36 new measurements representing a time extent covering 33 measurements. That's not too bad. However, there is one small problem that this calculation is ignoring. It's almost never serious in most prediction applications, but we should take a brief look anyway.

The trusting reader has been led to believe that by thoroughly overlapping measurements in both the time and frequency domains, all of the information in the original series is captured. Therein lies a long and sad tale. It's not quite so simple. It is certainly guaranteed that if any gaps appear in either the time domain (by spacing the samples too far apart relative to a filter's time-domain extent) or in the frequency domain (by spacing the filters far apart relative to their frequency-domain widths), information will be lost. However, the converse may not necessarily be true. What makes the situation even more troublesome is that the calculations necessary to ascertain the amount of information captured are brutally complicated. For a somewhat more rigorous presentation of this subject, consult Masters (1994). But even that source glosses over the most technical details. The ultimate reference is Daubechies (1990) or any of her subsequent works that appear in various compendia. Ingrid Daubechies is the reigning monarch in this subject, and anyone seriously interested in filter banks and wavelets should read everything she writes.

Luckily, all is not hopeless. The fact of the matter is that, as long as the filters and output samples substantially overlap in both domains, the amount of the information in the bands covered by the filters will be sufficient to satisfy the majority of practical applications. The point of diminishing returns comes quickly. In order to capture most of the relatively small amount of information missed by the example in Figure 5.28, the distance between output samples would have to be cut in half. It is almost never worthwhile doubling the number of variables to obtain a small amount of additional information.

What Is a Wavelet?

We have seen how a set of filters, arranged with geometrically increasing center frequencies, widths, and output sampling rates, can be used to capture the information in a time series in a meaningful and reasonably parsimonious way. The example in Figure 5.28 related the successive filters by a factor of two. However, there is no reason why this factor *must* be used. Narrower frequency-domain widths, combined with more closely spaced center frequencies, allow greater resolution in the frequency domain. Some applications even benefit from subgroups of filters called *voices* to increase frequency resolution while maintaining advantageous time-domain spacing patterns. We will not concern ourselves with these advanced techniques. The important point is that the relationship among filters is geometric: To progress from one filter to the next, we multiply the center frequency and the frequency-domain width by a constant, and we multiply the output sampling rate by the same constant. Naturally, limitations of discrete sampling mean that we cannot, in general, always increase the output sampling by the exact same quantity each time. But we do the best we can, rounding as necessary.

This geometric relationship implies an interesting fact. It turns out that the individual filters that are applied to the data are all identical in shape. They differ only in scale. Turn back to the filter shown in Figure 4.8 on page 122. The filters that we use in our filter bank all look something like that one. The only difference is that the high-frequency filters are obtained by pushing in on the sides of that filter, squeezing it in like an accordion. And the low-frequency filters are obtained by pulling out on the ends, stretching it. This operation of compressing and expanding along the time axis, without otherwise changing the shape, has precisely the effect we are looking for. It multiplies the time-domain width by the stretching factor, and it simultaneously divides the center frequency and the frequency-domain width by the same factor.

This leads us to a somewhat loose but perfectly respectable definition of a wavelet. *A wavelet is a function that defines a set of filters obtained by scaling this one function.* Not surprisingly, there are a host of mathematical properties that a filter function must satisfy in order to be called a wavelet. Readers of this text will not be burdened with any further mention of these properties. Masters (1994) provides a good overview of the important ones. All that we need to know here is that by choosing a good wavelet function (often called the *mother wavelet*), we can use this one single filter shape, compressed and expanded in the time domain, and sampled accordingly, to describe a time series in a meaningful and fairly parsimonious way.

Keep in mind that there is an infinite number of possible wavelet filter functions. They each trade off in different ways the three desirable traits listed on page 166. This text will present only one function that is known to be particularly effective for feature detection in time series. Keep an open mind in looking for others.

The Morlet Wavelet

We already know that the total time-domain extent of the filters used in this text can be approximated by $0.8/s$, where s is the filter width defined in Equation (4.2) on page 113. We also know that the filter response is significantly decreased at a distance of s from the center frequency, and response is almost zero by twice the width. It turns out that the time-domain response of this filter family is governed by exactly the same type of Gaussian function. Let t be the time-domain width analogue of s, and let j be the distance between an arbitrary time slot and the current output time slot. The sensitivity of each filter output to data in time slots other than the current slot is governed by Equation (5.1).

$$r(j) = e^{-\left(\frac{j}{t}\right)^2} \tag{5.1}$$

This equation says that the sensitivity of the filter is at its maximum (1.0) at the current time slot, and that it decreases as the distance from the current time slot increases. This is just what we expect. At a distance equal to t, the time-domain width, the filter's response is down to $e^{-1} \approx 0.37$. At twice the width, its response is down to $e^{-4} \approx 0.018$. This is exactly the same relationship that holds for the frequency domain.

It happens that s and t are related by a simple formula. In particular, $st = 1/\pi \approx 0.318$ for the filters used in this text. We are now in a position to justify our old heuristic rule stating that the time-domain extent of the filter is approximately $0.8/s$. Substituting $s = 1/\pi t$ into this heuristic, we can equivalently say that the filter's extent in the time domain is about $0.8\pi t \approx 2.5t$. By Equation (5.1), the filter's response this far away from the current time slot is $e^{-6.25} \approx 0.002$, small enough to disregard, and dropping fast.

Some readers may be wondering what influenced the decision to choose the particular family of filters used throughout this text. There is an infinite variety to choose from, many of which have interesting and useful optimality criteria. The reason for this choice has to do with the equation $st = 1/\pi$ that was just stated. It is obvious that the reciprocal relationship between the time-domain and frequency-domain widths is a nuisance. In a better world, we could make either (or both) of these widths as small as desired. It would be very nice if we could focus on a narrow frequency band while still having each output point being affected by only nearby neighbors. The fact that the width product is constant imposes a troublesome limitation. In view of this fact, a worthwhile goal is to choose a filter family for which the width product is as small as possible. This, in fact, is just what motivated the choice of the filter family used here. The *Heisenberg Uncertainty Principle* states that no filter can have a width product smaller than $1/\pi$, and the filters used in this text attain this theoretical limit. In other words, these filters are the best possible filters for locating events simultaneously in both frequency and time. For most time-series prediction applications, this is an excellent choice for an optimality criterion. (By the way, readers of Masters (1994) see that a different width product is used in that text. The reason is that widths in that text are defined in a different, more theoretical fashion.)

Chapter 4 explained that this family of filters is implemented in the frequency domain by multiplying the discrete Fourier transform of the original series by the Gaussian function shown in Equation (4.2) on page 113. We know from Chapter 4 that this action is equivalent to convolving the time series with a time-domain filter. An illustration of such a filter (in the time domain) appeared in Figure 4.8 on page 122.

Since this filter family has the nice property of optimally locating events in both time and frequency, it would seem that the filter function shown in Figure 4.8 would make an ideal mother wavelet. We could use that single shape, contracted and expanded, to define a family of filters for thoroughly and meaningfully presenting the information in a time series to a prediction model. In fact, this filter is *almost* an excellent mother wavelet. It has just one very small and easily solved problem: It does not quite sum to zero. It comes very, very close to summing to zero. It comes so close that when it is used in ordinary filter applications, the tiny offset is insignificant and can be safely ignored. But this is no ordinary filter application. When the mother wavelet is compressed along the time dimension so as to respond to the highest frequencies, we must ensure that it does so properly. Suppose that it does not quite sum to zero, and also suppose that the original time series wanders slowly away from zero. When these wanderings take the series far enough away from zero, the dot product of the filter function with a segment of the series will be biased away from zero by virtue of the offsets. The only way to keep series offsets from influencing filter outputs is to make sure that the filter sums to zero so that the offsets in the series cancel each other. If the filter does not sum to zero, low-frequency components (including constant offsets!) will find their way into high-frequency filter outputs. This is an extremely serious problem that cannot be tolerated.

Another way of viewing this problem is to examine the filter in the frequency domain. Look back at Figure 4.5 on page 113. The widest filter in this trio clearly (at least, with a magnifying glass) does not quite touch zero by the time a frequency of zero is reached at the far left side of the figure. In fact, Equation (4.2) on page 113 tells us that *no* filter, no matter how narrow, ever really reaches zero. We state without proof that the time-domain representation of a filter sums to zero if and only if the frequency-domain representation of the filter is zero at a frequency of zero. So what we need to do is to modify our filter in a way that makes its frequency-domain representation drop to a true zero. At the same time, we do not want to change the shape very much, since that would destroy the optimality property of the width product (*st*) attaining its theoretical minimum. Finally, we dare not simply use the exact filter function we have been using, but suddenly chop it to zero at the left side. The danger of sharp filter corners was explained in Chapter 4. What do we do?

The exact method for attaining this goal while minimally distorting the filter shape involves mathematics far beyond the scope of this text. The frequency-domain representation of the filter, analogous to Equation (4.2) on page 113, is shown in Equation (5.2).

$$H(f) = \frac{1}{\sqrt{\pi} s} \left[\frac{e^{\frac{2ff_0}{s^2}} + e^{\frac{-2ff_0}{s^2}} - 2}{2e^{\frac{f^2 + f_0^2}{s^2}}} \right] \qquad (5.2)$$

Even though this equation looks fierce, it is almost exactly like a (rescaled) Gaussian curve for almost all values of the center frequency (f_0) and width (s). The difference between this function and the functions graphed in Figure 4.5 on page 113 is visible only when a large width and a relatively low center frequency combine to cause significant response at a frequency of zero. When this happens, Equation (5.2) pushes the response peak slightly to the right and pulls the left side of

The Morlet Wavelet

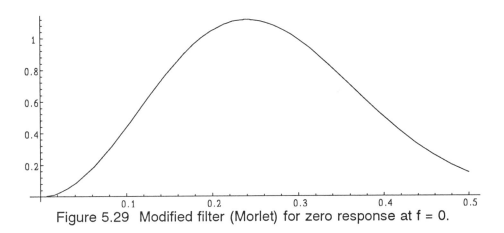

Figure 5.29 Modified filter (Morlet) for zero response at f = 0.

the bell curve downward, compensating by letting the right side rise. This rare and extreme case is graphed in Figure 5.29 using a center frequency of 0.2 and a width of 0.2.

The filter whose frequency-domain representation is given by Equation (5.2) is called a *Morlet wavelet* in honor of the person who first extensively used it to study geophysical phenomena. A much more detailed study of this immensely useful wavelet can be found in Masters (1994).

We will not bother graphing the time-domain representation of Morlet wavelets. Even in extreme cases, they are almost indistinguishable from the ordinary filter functions shown in Figure 4.8 on page 122. The centering modification, which only needs to be tiny, simply shifts the oscillations up or down by whatever minuscule amount is necessary to center the filter exactly. The distortion is minimal, and hence the time-frequency width optimality is, for all practical purposes, fully retained.

Code for the Morlet Wavelet

The code for implementing a Morlet wavelet filter is very similar to the code for an ordinary quadrature-mirror filter given on page 132. It would be good to review the discussion of that routine, as many of the comments made concerning that code apply to this routine as well. In order to highlight some subtle differences, the Morlet code will be presented and explained.

```
void Filter::morlet (
   double freq ,                  // Center frequency (0-.5)
   double width ,                 // Scale factor for exponential (in frequency units)
   double *real ,                 // Filtered real output here
   double *imag                   // Filtered imaginary output here
   )

{
  int i ;
  double f, wt, temp, widsq, numer, denom, wtmax ;

  widsq = width * width ;
  wtmax = 1.e-30 ;

  for (i=1 ; i<halfn ; i++) {
    f = (double) i / (double) n ;   // This frequency
    temp = 2.0 * freq * f / widsq ;
    numer = exp ( temp ) + exp ( -temp ) - 2.0 ;
    denom = 2.0 * exp ( (f * f + freq * freq) / widsq ) ;
    wt = numer / denom ;            // Equation (5.2)
    if (wt > wtmax)                 // Will rescale 0-1 so like qmf
      wtmax = wt ;
    workr[i] = xr[i] * wt ;
    worki[i] = xi[i] * wt ;
    workr[n-i] = worki[n-i] = 0.0 ; // Causes QMF outputs
    }

  workr[0] = 0.0 ;                  // Force DC to zero

  temp = 2.0 * freq * 0.5 / widsq ; // Do Nyquist in xi[0]
  numer = exp ( temp ) + exp ( -temp ) - 2.0 ;
  denom = 2.0 * exp ( (0.5 * 0.5 + freq * freq) / widsq ) ;
  wt = numer / denom ;
  if (wt > wtmax)
    wtmax = wt ;
  workr[halfn] = 0.5 * xi[0] * wt ;

  worki[0] = worki[halfn] = 0.0 ;   // By definition of real transform

  qmf_fft->cpx ( workr , worki , -1 ) ;
```

The Morlet Wavelet

```
      wtmax = 1.0 / (wtmax * halfn) ;
      for (i=0 ; i<length ; i++) {
         real[i] = workr[i] * wtmax ;
         imag[i] = worki[i] * wtmax ;
         }
   }
```

The Morlet wavelet is just a special QM filter that has been modified so that the weight function drops to zero at a frequency of zero. Otherwise, they are identical. One small implementation detail is that we keep track of the maximum value of the weight function in wtmax. The reason is that Equation (5.2) does not have a maximum value of 1.0 like the ordinary Gaussian function. Its maximum is a function of the width. It is nice (from a purely practical point of view) if the Morlet filter and the ordinary QM filter have comparable outputs. We effect this result by dividing the outputs by the maximum weight function. Observe that wtmax is initialized to a tiny positive number so that we do not inadvertently divide by zero at the end of the routine. This is cheap insurance against possible disaster.

Do not be confused by the line of code that explicitly sets workr[0] equal to zero. This would have happened anyway in the weight loop if this term were included. However, by starting the loop at i=1, we are able to set the upper half of the transform equal to zero to force QMF outputs. This was discussed in conjunction with the code on page 132. Also, we multiply the Nyquist term by 0.5 by virtue of the fact that the symmetric terms on either side cause them to have double value. If this vague statement is unclear, a few more details can be found in Masters (1994). Complete details are left to some of the more theoretical signal-processing texts.

Implementing Wavelets with NPREDICT

Turn back to Figure 5.28 on page 169. When we speak of implementing wavelets, we are not referring to the use of a single filter whose outputs are individually sampled. Rather, we are referring to an entire bank of filters spanning the range of frequencies in which we are interested. Moreover, we sample the filter outputs in proportion to their widths so that each filter's output set covers approximately the same total time duration. As shown in Figure 5.28, a complete wavelet implementation implies generation of a potentially large number of variables.

It would be impractical to use a single NPREDICT command to effect a complete wavelet implementation. The number of parameters to be specified in one command would be cumbersome, and the number of generated signals would be difficult to manage. Instead, a wavelet set is created by using a separate pair of commands for each filter. Then, when the filter outputs are presented to the prediction model, the lags are appropriately specified. This is a little extra work for the user, but the reward is greatly increased versatility. Also, by using this approach, the user is forced into a deeper understanding of exactly what is being done. *No pain, no gain* applies to wavelets.

The best way to clearly explain the method for building a wavelet set is by means of an example. Suppose we have a signal named *diff* for which we want to compute the wavelet decomposition shown in Figure 5.28 on page 169. We need eight signals: a real and imaginary pair for each filter. These are created by means of the following commands:

```
NAME = r1 , i1
MORLET = 0.4 0.2 diff
NAME = r2 . i2
MORLET = 0.2 0.1 diff
NAME = r3 , i3
MORLET = 0.1 0.05 diff
NAME = r4 , i4
MORLET = 0.05 0.025 diff
```

After the filter outputs have been computed, the context information in the time domain is captured by specifying appropriate lags in the input list. Since the lags are not contiguous, and hence must be specified with individual commands, a large number of commands are usually required. The layout shown in Figure 5.28 (the real parts only) is achieved with the following input list:

```
INPUT = r1 8
INPUT = r1 12
INPUT = r1 16
INPUT = r1 20
INPUT = r1 24        ; Center of first filter bank
INPUT = r1 28
INPUT = r1 32
INPUT = r1 36
```

The Morlet Wavelet

```
INPUT = r1 40
INPUT = r2 8
INPUT = r2 16
INPUT = r2 24          ; Center of second filter bank
INPUT = r2 32
INPUT = r2 40
INPUT = r3 8
INPUT = r3 24          ; Center of third filter bank
INPUT = r3 40
INPUT = r4 24          ; Center of fourth filter bank
```

The highest-frequency filter has a width of 0.2, which implies a time-domain extent of 0.8 / 0.2 = 4 points. We might be inclined to start the inputs with a lag of 4. However, we must consider *all* of the filters. The second filter has a width of 0.1, with a time extent of 8 points. The third filter has a width of 0.05, giving a time extent of 16 points. This leads us to consider using a minimum lag of 16. A tradeoff now comes into play. Only this one filter pushes us so hard. The first two filters have a much smaller time extent. The fourth filter has an extent of 32 points, but the center of this filter is itself 16 points further back in time than the closest approach of the other filters. By giving in to the extreme requirements of the third filter, we give up a lot of valuable recent information that can be obtained from the other filters. Is this a necessary price? Let's get down to the details. Look back at Equation (5.1) on page 172 and the explanation following that equation. For this third filter, $t = 1 / (0.05\pi) \approx 6.37$. The relative contribution of an observation 8 time slots beyond the current time is about $e^{-1.58} \approx 0.21$. Is this excessive? If the series is stationary enough for its mean to be innocuous when used to pad the ends, this is probably acceptable. If not, the minimum lag may need to be slightly increased by adding a constant to all of the lags in the input list. For most applications, cheating a little on the worst case is usually fine.

This example used only the real parts of the filters. Are the imaginary parts necessary? Unfortunately, when the time-domain spacing of the filters is pushed to the limit of overlap as is done here, the answer is usually *yes*. The real and imaginary parts contain complementary information. If this is not clear, reread the material starting on page 121. Their combined information is not so important if the filter outputs are sampled at tight spacing. But when the spacing is large, the extra information provided by the in-quadrature outputs is valuable. Therefore, the input list just given is only half complete. It should be duplicated for the imaginary filter outputs.

What about amplitude outputs? This series can be obtained from each filter by including a third signal name following the real and imaginary signals on the NAME command. Amplitude is a peculiar beast. Theoretically, it contains no information beyond that contained in the real and imaginary parts (taken together). However, it presents the information in a very different, nonlinear, fashion. For some applications it is valuable. For other applications it is worthless. The only way to find out for sure is to test it. You never know until you try.

6

Box-Jenkins ARMA Models

- Overview

- Predicting with predicted shocks

- Multivariate ARMA Models

- Designing and testing ARMA models

- Lagged correlations

One of the most venerable methods of time series prediction was pioneered by George Box and Gwilym Jenkins. Their classic 1976 text is the cornerstone of an entire branch of the science. Many improvements to the original method have been devised, but the underlying principles are as important and useful now as ever. This text will present a generalized multivariate version of the original model. The training algorithms have been completely revised to reflect the current state of numerical methods and computing hardware. But the underlying foundation remains the same.

Due to their inherent mathematical simplicity, the models discussed in this chapter lend themselves to much theoretical explication. Domains of stability, extrapolation properties, and a host of other topics are discussed in often gruesome detail in many other texts. Theoretical considerations will be largely glossed over in this text. The primary focus of this chapter is the practical utility of the models. Enough theory will be provided so that the reader understands the nature of the models. From there, each model's strengths and weaknesses will be shown, and examples will illustrate typical applications. Training algorithms will be discussed in detail. Readers who desire a full theoretical development are referred to the sources cited throughout this chapter.

Overview of the ARMA Paradigm

Box and Jenkins originally proposed that many time series could be explained by a relatively simple model. In its most basic form, this model is composed of one or both of two primitive models. One possible component model is called *autoregressive*. Most readers are familiar with the statistical technique of *regression*, in which one variable is predicted by using a linear combination of one or more other variables. Put an *auto* prefix on this word, and one gets a technique in which a variable is predicted by means of a linear combination of previous values of itself. An autoregressive (AR) model of a time series assumes that the current value of the series is equal to a weighted average of a finite number of previous values, plus a constant offset, plus a random noise term. These noise terms are often called *shocks*, and that word will be used throughout this text. By definition, the shocks are independent of each other *and* of the series itself. The shocks also are identically distributed and are assumed to have zero mean and finite variance. Some texts that rely heavily on traditional statistical tests also assume that the shocks follow a normal distribution, but that assumption is not made here.

The primitive AR model is expressed in Equation (6.1). Note that this formulation is trivially different from that usually found in the literature. Most traditional versions express the model as deviations from the mean rather than directly expressing the values as is done here. We use the constant offset ϕ_0 to allow direct modeling of the series itself. This simplifies the generalization to be presented later without affecting the power or spirit of the original model.

$$z_t = \phi_0 + \phi_1 z_{t-1} + \phi_2 z_{t-2} + \ldots + a_t \qquad (6.1)$$

When Equation (6.1) is used to model a time series, it usually involves relatively few parameters. Correlations between the parameters are quite high, leading to instability in their computation when many of them are to be found. Note that the constant offset, ϕ_0, is *not* the mean of the series. In fact, it will almost never even be close to the mean in actual applications. The current shock is a_t.

The second primitive model that plays a part in Box and Jenkins' approach to time series prediction is called a *moving-average* (MA) model. Again, the definition given here is trivially different from the usual definition so as to simplify later generalizations. The spirit and power of the original formulation are fully retained. The MA model says that the current value of a time series is equal to a weighted sum of a finite number of previous shocks, plus a constant offset and a current shock. This is expressed in Equation (6.2).

$$z_t = \phi_0 + \theta_1 a_{t-1} + \theta_2 a_{t-2} + \ldots + a_t \qquad (6.2)$$

The name *moving average* is not particularly appropriate for this model. To statisticians, this term implies that the weights are positive and sum to unity; but in fact, neither of these properties is required of MA models. Nevertheless, the term has stuck, so we will honor tradition.

Note that in the MA model, unlike AR models, the constant offset ϕ_0 is the mean of the series. This is because the shocks by definition have a mean of zero. Note also that the θ parameters of Equation (6.2) are the negative of the same parameters in Box and Jenkins' original formulation of the model. This is of no practical consequence and greatly simplifies later developments.

Duality Between AR and MA Models

On first glance, the AR and MA models may appear to be almost identical. One model bases the current value on previous values, while the other model bases the current value on previous shocks. Since the actual value at any given time and the shock at that time are closely related, it might be expected that the two models are nearly equivalent. Close, but no cigar. In fact, it turns out that they complement one another in a wonderful way.

It has been stated that pure AR and MA models typically employ very few parameters. Since the parameters have a large degree of interaction, reliably estimating them becomes difficult when more than a few are used. Unfortunately, many physical processes are best described by models whose terms, though simple in form, extend backwards in time for an infinite extent. Even though the weights eventually die out to insignificant quantities, they do not usually do so before an intractable number of terms have appeared.

This is where an interesting property of AR and MA models saves the day. It can be shown that a finite model of one type is mathematically equivalent to an infinite model of the other type. For example, Equation (6.3) illustrates the equivalence of a single-term AR model and an infinite MA model. Interested readers can easily derive this relationship by first using the AR model to write z_{t-1} in terms of z_{t-2}, then writing z_{t-2} in terms of z_{t-3}, and so forth, substituting each expression into the original model.

$$\begin{aligned} z_t &= \phi z_{t-1} + a_t \\ &= a_t + \phi a_{t-1} + \phi^2 a_{t-2} + \phi^3 a_{t-3} + \ldots \end{aligned} \quad (6.3)$$

There is no compelling reason to dwell on invertability. Box and Jenkins (1976) is an excellent source for more details. In practice, we rarely, if ever, need to convert one pure model to the other. The only possible exception is in the derivation of theoretical confidence intervals, but that approach is not taken in this text. The important point is to understand the relationship between AR and MA models and thereby be encouraged. If the physical process being modeled involves an infinitely long filter, whether it be in terms of values of the shocks or the series itself, there is hope that it can be explained by a model containing relatively few parameters.

ARMA Models

It is not necessary to limit the model choice to pure AR or pure MA expressions. Now that the reader understands the duality between the two pure processes, it should be apparent that tremendous versatility can be had by combining them. The ARMA model is shown in Equation (6.4).

$$z_t = \phi_0 + \phi_1 z_{t-1} + \phi_2 z_{t-2} + \ldots + \theta_1 a_{t-1} + \theta_2 a_{t-2} + \ldots + a_t \quad (6.4)$$

The primary reason why the model shown in Equation (6.4) has remained popular for decades is simple: It works well for a tremendous variety of real-life applications. A surprising number of physical processes can be comfortably modeled with very few ARMA parameters.

There is a standard notation that should be mentioned here. It is restrictive and does not generalize easily, so this text will not make much use of it. But it is so common in the literature that it should make at least a token appearance. The number of AR terms (not counting the constant offset) is traditionally called p, and the number of MA terms is traditionally called q. One can then speak of an ARMA(p,q) model. These terms are assumed to be contiguous, starting with a lag of one and extending to lags of p and q, respectively. It is possible to skip intervening lags, incorporating a single term at a distant lag. No simple notational provision is made for these *seasonal* models that will be discussed later. So, for example, an ARMA(2,1) model would make use of the prior two values of the series and one prior value of the shock. Pure processes are sometimes abbreviated AR(p) and MA(q).

Homogeneity, Periodicity, and Stationarity

Consider an AR(1) process defined by the model $z_t = 1.5 \, z_{t-1} + a_t$. It should be clear that a time series governed by this rule will not remain in one place for long. Once it breaks away from zero, it will head off into distant realms, never to return. Now consider the series defined by the weekly frequency of credit card use by the United States population. This series, though perhaps slowly trending upward, will always remain within finite bounds. But its appearance in the weeks before

Christmas will be very different from its appearance in July. How can these behaviors be characterized?

Let us start with a deliberately loose and vague definition. A time series is said (for the purposes of this text) to be *homogeneous* if its behavior across time follows the same pattern in some sense. What if a time series were created by sampling daily prices of hog futures for a randomly chosen time period, then sampling human birth rates for a while, and so forth. This is a rather extreme example of non-homogeneous behavior, but it is not difficult to imagine less extreme but equally troubling data. This chapter is concerned only with homogeneous data, the loose definition notwithstanding. ARMA models and their relatives are useful only for time series that follow a consistent rule of some sort.

Some time series are governed by rules that depend on time in a regular repeating pattern. These periodic components require special treatment if they are to be reconciled with ARMA models. Chapters 2 and 7 discuss general methods for detecting, removing, and restoring periodic components.

One concept that is vital to correct use of ARMA models is *stationarity*. A time series is said to be *strictly stationary* if its statistical distribution does not change across time. More specifically, suppose we have a set of m samples of the series made at times t_1 through t_m. These need not be contiguous times. Strict stationarity implies that the joint probability density function of those m samples is identical to the joint distribution of another m samples taken at times t_1+k through t_m+k. This must be true for all choices of m and k, as well as choices of the m relative sample times. There are other, less strict, definitions of stationarity. We will not be concerned with them here, and the term *stationary*, when used in this text, will always refer to the strict sense.

There are mathematical techniques for determining whether a given ARMA model is stationary. They are not difficult, involving the roots of polynomials defined by the model's parameters. Interested readers should see Box and Jenkins (1976) for details. For our purposes, the important concept is that we will be almost exclusively interested in time series that are stationary. The only exception of sorts is briefly discussed in the next paragraph and in detail in Chapter 7. We could use the basic ARMA model of Equation (6.4) to model a nonstationary series, but we usually will not. When that equation has parameters that define a nonstationary series, the model is unstable.

It has such strong tendencies to blow up out of control that it is nearly worthless for prediction.

Although stationarity obviously implies homogeneity in any sense of the word, the converse is not true. There are some commonly observed series that are homogeneous in that their local behavior is similar across time, but that are definitely not stationary. Stock prices are a classic example. An issue may bob around near a price of 40 for a while, then climb to 60 and hover there for another period of time. The relative behavior of the price may be similar at both levels, but the level change destroys stationarity. In such situations, it frequently happens that the *difference* between successive samples is stationary. Rarely, a second differencing is needed to achieve stationarity. This will be discussed in detail in Chapter 7.

When a homogeneous but nonstationary series can be made stationary by one or more differencing operations, it is quite possible that an ARMA model can be used to describe the differenced series. The inverse of differencing is *integrating*, so the process of differencing d times, fitting an ARMA model, then integrating d times to return to the original domain, is often given the label ARIMA(p, d, q). This term will not be used often in this text. Here we take a more general approach to the process of differencing, model fitting, and integrating. This is discussed in Chapter 7.

Differencing does not need to be restricted to adjacent samples. If a series exhibits periodic behavior, differencing across a distance equal to the period may yield a result that can be parsimoniously modeled by an ARMA process. These *seasonal* models play a vital role in many traditional econometric forecasting methods. Seasonal differencing is discussed in a more general context in Chapter 7.

Computing the Parameters

This text features a broad generalization of the ARMA paradigm. The relatively complex training algorithm for the general ARMA model is presented on page 211. However, it is instructive to discuss briefly the basic issues in traditional ARMA training, as many of them apply to the general model.

The most fundamental training consideration is the definition of the quantity that will be optimized. If we have a set of $p + q + 1$ weights for Equation (6.4), we can test their quality in a simple way. For the moment, ignore the fact that we do not have any observations prior to the first that was observed. This difficulty will be addressed soon. For now, use the first p observations (times 1 through p) to predict the value at time $p+1$. For lack of anything better, assume that all q shocks are zero, giving a pure AR model for this first prediction. The prediction error, which is the true value of the series at time $p+1$ minus the predicted value, is an estimate of a_{p+1}, the shock at time $p+1$. We then advance one point, predicting the series at time $p+2$. An MA term can come into play, as we now have a value to use for a shock. This procedure is illustrated in Figure 6.1.

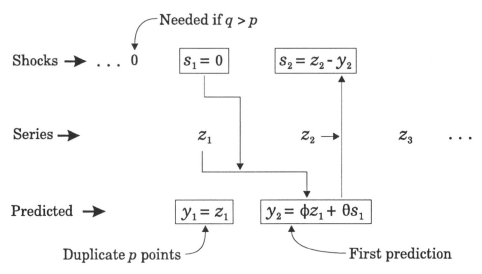

Figure 6.1 Computing the ARMA error.

This figure illustrates a simple model containing one AR term, one MA term, and no offset. Thus, $p=q=1$. The first p points are skipped, with prediction starting only when the full AR model can be used. As a consequence, the first p shocks are zero. If q exceeds p, the shocks prior to the start of the series are also assumed to be zero. After each prediction is made, it is subtracted from the true value to derive the shock for that time slot.

There are at least two reasons why the sum of squared prediction errors is a good measure to optimize when computing parameter estimates. First, it has obvious intuitive appeal. Minimizing squared error is a venerable old tradition in statistics. The second reason is more subtle and will be glossed over. If one assumes that the shocks follow a normal distribution, then minimizing the squared errors is equivalent to computing a maximum likelihood estimate of the parameters (subject to a condition that is best left to Box and Jenkins (1976)). Maximum likelihood estimators are paragons of virtue in the world of statistics.

The algorithm just given for recursively computing predictions and their errors demonstrates one technique for dealing with the fact that we do not have observations extending back to the dawn of time. The prediction does not start until time p so that all AR terms are based on known data. And the shocks for MA terms are assumed to be zero until they are explicitly calculated. This is easy, gives good results for all but very short series, and generalizes well. There are two alternatives. A very simple but usually inferior approach is to start right out predicting the first sample. Assume that all series values and shocks prior to the beginning are equal to the mean. In practice, the use of AR terms on such obviously faulty data leads to error transients that take a long time to die out. Except for very long series, this method should not be used.

There is a third method that provides results that have a slight statistical superiority to the method espoused in this text, at least for short series. However, this method is cumbersome for the basic ARMA model and nearly intractable for the generalized version presented later. For the reader's edification, here is a summary. Recast the ARMA model so that it predicts *backward*. Start with the last point in the series and work toward the beginning, predicting the series and the shocks in reverse order. Use this procedure to extend the series backward in time far enough that all series points and shocks that are required for the forward process are available. Since starting at the fixed end of the series introduced a transient that may not have died down before the beginning was reached, real fanatics will iterate again: Move forward through the series, extending the prediction beyond the original end. Then work back again. If the series is short relative to the larger of p and q, this may be worthwhile. But it is probably only rarely worth the effort, since series that are so short that it would make a difference are probably not long enough for reliable estimation anyway! In any case, there is no obvious way to incorporate backcasting in the generalized ARMA model featured in this text, so no more will be said on this issue.

A variety of methods have been proposed for computing parameters that minimize the prediction error. They nearly all have the same basic flow:

1. Use a fast deterministic algorithm to estimate the AR parameters.
2. (Optional) Use a fast iterative algorithm to estimate the MA parameters, conditional on the previously computed AR parameters.
3. Use a slow iterative algorithm to simultaneously optimize all parameters.

All early programs performed the first step, estimation of the AR parameters, by solving the *Yule-Walker* equations. This set of linear equations expresses the theoretical autocorrelation function of a series in terms of its AR parameters. If one computes the observed autocorrelation function, the resulting system of simultaneous linear equations is small and generally not too ill-conditioned. Solution of this system for the AR parameters is not difficult. When this algorithm was devised, computer memory was a scarce commodity. All of the required operations are economical in terms of storage. Now that memory is plentiful, a more straightforward approach may be better. Equation (6.1) defines a simple linear system whose least-squares solution can be obtained by the method of singular value decomposition. This method is described in detail starting on page 211.

At this point, some programs estimate the MA parameters, conditional on the existing AR parameters. There are at least two possible approaches to this task. Box and Jenkins (1976) define a system of nonlinear equations and use Newton's method to solve it. An alternative is to take advantage of the fact that an ARMA process can be thought of as a pure AR process whose noise follows a pure MA process. Treat the shocks from the AR step as a series in itself, and use an optimization algorithm to estimate MA parameters for that series. Most modern researchers believe that this MA step is a waste of time. Since the MA estimates are conditional on AR parameters that are almost certainly not very near their own optimal values, it is better to skip right to global optimization. Good minimization algorithms will have no trouble dealing with poor starting estimates for the parameters.

This leads to the choice of a global minimization algorithm. Most authorities agree that Powell's method is the best. The implementation used here is described on page 216, and its application to ARMA learning is presented on page 215, so nothing more will be said about it at this time. Be warned that some early writers, including Box and Jenkins (1976), espoused numerical estimation of derivatives and subsequent use of these estimates for quasi-Newton iteration. Numerical derivative estimation as part of a minimization scheme in which derivatives cannot be explicitly computed is abhorrent to responsible numerical analysts, and this method should be shunned. Finally, note that derivatives *can* be explicitly computed by ambitious programmers, but the effort is probably not worthwhile. Powell's method is nearly quadratically convergent, and its stability and reliability are impeccable.

Univariate ARMA Prediction

This introductory section ends with the topic that brought us all here: prediction. The majority of the groundwork has already been established. Look back at Figure 6.1 on page 188, which illustrates how the prediction error is computed within the known extent of the series. It is very easy to predict the series past its end: Don't stop. Just keep on going, using successively predicted values to recursively compute future values. There is only one catch. What should be used for shocks past the end? During the known extent of the series, the shock at each point is defined as the observed value there minus the predicted value. Unfortunately, this cannot be done when there are no known values! The only answer is that it must be assumed that the shocks are zero past the end of the series. The implication of this assumption is that the predicted series degenerates to a pure AR process after q predictions into the future. Recursive prediction is illustrated in Figure 6.2.

This figure illustrates computation of the first two predictions. It uses the same $p = q = 1$ model that was used in Figure 6.1. The first future prediction, shown with dotted lines at time $n+1$, is based on the last known term in the series and on the shock that was computed from that term. The next prediction, at time $n+2$, uses the previously predicted value for its AR component, and it must use a shock of zero for its MA component.

Even the availability of recursively predicted past values for use with AR terms is not magic. In virtually all cases of practical interest, the AR series will steadily decrease in variation, asymptotically converging to a constant.

The moral of the story is simple: ARMA models should *never* be used for predictions far into the future. Many references set forth elaborate formulas for computing confidence intervals for any future point by compounding noise effects. The sad truth is that in most

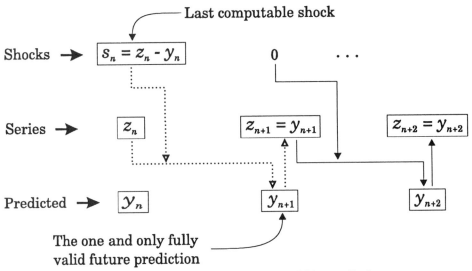

Figure 6.2 Ordinary recursive ARMA prediction.

practical applications, these formulas are worthless. They are based on assumptions of normality and independence that only in the rarest of occasions hold up in the real world. In practice, ordinary recursive prediction deteriorates so rapidly that it should be considered a last resort if significant prediction distances are required.

The prospect of distant prediction is not entirely hopeless. One method for predicting many samples into the future without using explicit recursion is to train the model that way. This may seem like a backhanded way of silently slipping implicit recursion into the mix, but the fact is that it often works well. For example, suppose one needs to predict the series four samples into the future. A model similar to that shown in Equation (6.5) might do the trick.

$$z_t = \phi_0 + \phi_1 z_{t-4} + \phi_2 z_{t-5} + \theta_1 a_{t-4} + a_t \qquad (6.5)$$

A second alternative, useful when explicit recursion is mandatory, is to use robust confidence intervals based on actual recursive predictions. The problem of deterioration as errors accumulate remains with us. But at least we can have an indication of how serious the deterioration has become for each future distance. This relatively complex subject is discussed in Chapter 8.

Predicting with Predicted Shocks

A traditional assumption of ARMA techniques is that the prediction errors, the implied shocks, are random. In many applications, this is only partially true. They may look random, and they may be demonstrably random as far as further ARMA modeling is concerned. But as anyone experienced with neural networks knows, not everything that looks random is random. The power of ARMA prediction may be significantly increased by the following hybrid technique:

1. Train an ARMA model on the series of interest.
2. Train a different model, such as a neural network, on the series defined by the ARMA shocks.
3. Use that alternative model to predict future values of the shock series.
4. Predict the original series with the ARMA model, but use the predicted future shocks instead of zeros. These should be used for both past values in MA terms and for the current shock of each predicted value. This is illustrated in Figure 6.3.

This algorithm is identical to ordinary recursive prediction through the last known point. After that, it changes in two ways. First, the predicted shocks are used for the MA terms instead of the usual zeroes. Second, the current predicted shock for each time slot is added to the prediction for that time. This action usually has greater impact on the predictions than use of predicted shocks for MA terms. Figure 6.3 illustrates two predictions. The dotted line shows the computation of the first future prediction. It is identical to the ordinary prediction of Figure 6.2 except that the first predicted shock is added in. The solid line is for the second prediction. It still uses recursion for the AR term, just like in Figure 6.2, but it makes use of the first predicted shock for the MA term. Also, it adds the second predicted shock to the second prediction.

Examination of a contrived example helps in understanding this technique. It must be made clear that this example is not representative of something that might normally be encountered in practice. The data has been constructed to emphasize the power of shock prediction. In particular, the entire experiment has been designed for visual effect. Practical applications will never present such obvious improvement, despite the fact that the actual improvement may be profound.

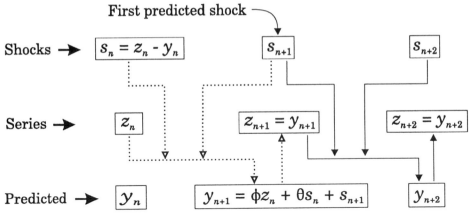
Figure 6.3 Predicting with predicted shocks.

The goal of this experiment is to work with a signal that is comprised of two components. One component, which will be called the *deterministic* component, is almost impossible to model with an ARMA model but is easily handled by a neural network. The other component, called *random* here, is a simple ARMA process. The deterministic part is the cosine wave graphed in Figure 6.4. The random part is an ARMA process with single AR and MA terms at a lag of ten. Their sum, graphed in Figure 6.5 on page 198, is the signal that will be predicted. The NPREDICT code for generating this data is shown below.

```
NAME = deterministic
GENERATE = 180 SINE 2.0 40.0 90
NAME = random
GENERATE = 180 ARMA  0 0 .2 0 0 .5
NAME = signal
ADD = deterministic AND random
```

Needless to say, no sensible person would consider fitting an ARMA model to a series that looks like the one shown in Figure 6.5. The obvious periodic component must be removed first, perhaps by filtering or by seasonal differencing. However, for the purposes of this demonstration, we need a component that has strong visual impact. This does the trick nicely. That said, the next step is to define an ARMA model having an AR term at a lag of ten and an MA term at a lag of ten. Note that a lag of ten was deliberately chosen, as using one-quarter of the period of the cosine wave causes minimal impact of the wave on the ARMA parameters. The model is defined and trained by means of the following NPREDICT commands:

```
INPUT = signal 10
OUTPUT = signal 10
TRAIN ARMA = temp
```

The training results in an AR weight of 0.005, an MA weight of 0.82, and a constant offset of −0.23. It is not surprising that these weights are not very close to the model that generated the ARMA component of the series. AR and MA weights at the same lag always exhibit high correlation, so one can take on the role of the other to a significant extent. Also, this series is short and is distorted by a deterministic component.

After the model is trained, the shocks must be computed. This is accomplished in two steps. First, do a point-by-point prediction within the series. This prediction is graphed in Figure 6.6. Then subtract that prediction from the true signal to get the shocks. NPREDICT code for these operations is shown below.

```
NAME = within_pred
ARMA PREDICT = 180 temp
NAME = shock
SUBTRACT = signal AND within_pred
```

The point-by-point prediction graphed in Figure 6.6 merits special discussion. Since the model has an AR term extending to a lag of ten, the first ten "predictions" are nothing more than a direct copy of the original data. The next ten predictions are almost equal to the constant offset. This is partly because the MA component has not yet taken effect. The shocks for the first ten duplicated points are assumed zero. The AR term controls these ten predictions, but recall that its weight is only 0.005, so it has little to contribute. If the training had computed an AR weight significantly different from zero, variation would start after the ten initial points had been skipped. This will become more clear later in this chapter when the actual computations are discussed in detail. Also, Chapter 11 rigorously states the rules by which predicted signals are created.

The next step is to train a neural network to predict the shocks. The network inputs are carefully chosen for this demonstration. Their lags are exactly a half and a full cycle. Many cases are excluded from the start of the shock series, since these are invalid for the reasons just discussed. NPREDICT commands for training the neural network are now shown.

```
CLEAR INPUT LIST
CLEAR OUTPUT LIST
INPUT = shock 19-21
INPUT = shock 39-41
OUTPUT = shock
CUMULATE EXCLUDE = 25
CUMULATE TRAINING SET
NETWORK MODEL = PNN
TRAIN NETWORK = temp
```

After the network is trained, the shock prediction is done. By using the same name for the predicted signal, the prediction is simply appended to the end of the series (as opposed to being a point-by-point prediction within the series). This is appropriate for subsequent ARMA prediction. The shock series, along with the network's prediction of the last 20 points, is shown in Figure 6.7. NPREDICT commands for the prediction are as follows:

```
NAME = shock
NETWORK PREDICT = 200 temp
```

The final step is to do the ARMA predictions. This example will demonstrate two approaches. The first, whose commands are shown below, is the traditional method in which all shocks beyond the end of the series are set to zero. This prediction is graphed in Figure 6.8. The first ten of the twenty predictions have valid shocks for the MA terms. The AR terms are negligible due to the small weight. The final ten predictions are nothing but the constant offset. Clearly, the quality of this prediction diminishes rapidly as the end of the series is left behind.

```
CLEAR INPUT LIST
CLEAR OUTPUT LIST
INPUT = signal 10
OUTPUT = signal 10
NAME = predict_a
ARMA PREDICT = 200 temp
```

Now switch to the method of using predicted shocks. Commands to accomplish this are shown below, and the prediction is graphed in Figure 6.9. The benefit of incorporating predicted shocks is obvious.

```
ARMA SHOCK = shock FOR signal
NAME = predict_b
ARMA PREDICT = 200 temp
```

The NPREDICT commands shown in this example have been slightly edited for clarity. Users who wish to duplicate this example should not type them in directly. The file ARMA_NN.CON in the EXAMPLES directory of the accompanying disk contains the complete and fully correct text of the control file that produced these figures. The differences between the commands listed here and those in the file are very minor, but they are important.

Once again, it should be emphasized that this example is a greatly exaggerated illustration of hybrid ARMA/neural network prediction. In a practical application, deterministic components that are so obvious should not be handled this way. They should be explicitly dealt with by seasonal differencing, filtering, or more careful modeling. The most appropriate way to employ the power of neural network prediction is to use it on components that follow no apparent patterns. In particular, we will learn later (and we have already partially seen in Chapter 3) that the shock series for a good ARMA model should be white noise. Several tests, including lagged correlations and power spectra, are useful for evaluating the shocks. Even though the shock series may pass all the white noise tests, it still may be good to attempt to train a neural network to predict the shocks. It is not unusual for a good neural network to find predictable patterns that are not only invisible, but that are undetectable by many common tests. When this happens, hybrid ARMA/neural network prediction can significantly improve on the abilities of one of these models working alone.

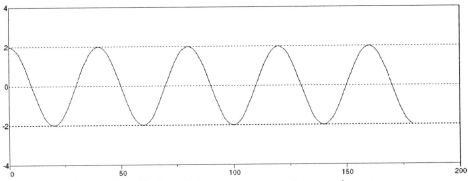
Figure 6.4 Deterministic component, a cosine wave.

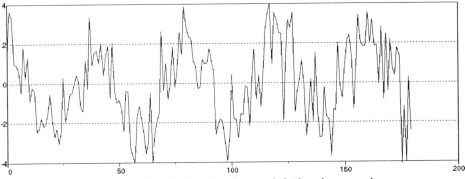
Figure 6.5 Final signal: deterministic plus random.

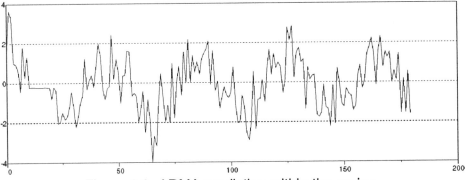
Figure 6.6 ARMA prediction within the series.

Overview of the ARMA Paradigm 199

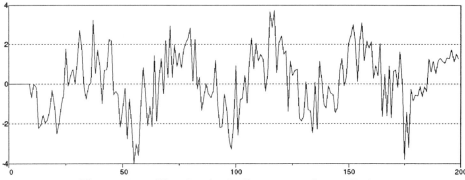

Figure 6.7 Shocks, including network prediction.

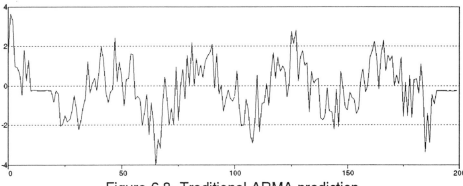

Figure 6.8 Traditional ARMA prediction.

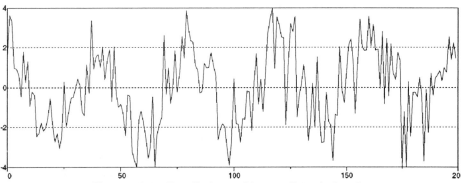

Figure 6.9 Prediction with predicted shocks.

Multivariate ARMA Models

So far, this chapter has considered only univariate ARMA models. Their original derivation was in the context of a single series, although Box and Jenkins (1976) did present a sort of bivariate version that they called a *transfer function* model. Since that time, several multivariate extensions of the basic univariate model have been introduced. The version presented in this text encompasses nearly all of the practical capabilities of the alternatives, and it has the additional bonus of being highly intuitive in its construction and use.

In real-life applications, it is reasonable to assume that the current value of a time series may be driven by five generic influences (other than the current shock, which is assumed random and unmeasurable at that time). The first three influences shown in the list below are accounted for in the univariate model already discussed. The remaining two are encompassed by the general multivariate model.

1. A central tendency, typically the mean
2. Historical values of this series
3. Historical values of the shocks that impacted this series
4. Historical values of other measurable series
5. Historical values of the shocks that impacted other series

The last item in this list is not often found in other multivariate ARMA models, at least not explicitly. Yet it can be extremely important. There are two ways of looking at this component. Recall from page 184 that a finite AR series is equivalent to an infinite MA series, with the converse also being true (under reasonable invertability conditions). So item five can be thought of as an economical means of incorporating the effect of a large number of historical values of the series. Alternatively, it is not difficult to imagine situations when the shock itself may be meaningful. Suppose that personal computer sales are being predicted. Sales of CPU chips are also monitored. In response to some event in the world, CPU sales are impacted by an unusual shock. The possibility certainly exists that the CPU shock affects future PC sales in a more predictable way than the CPU sales figure itself. Regardless of which view is taken, the fact remains that shocks from some series are often important predictors of future values of other series. Skeptics should sneak a look ahead at Figure 6.20 on page 233.

Multivariate ARMA Models

Including multiple predictor series complicates the model notation. Without loss of generality, we will write the model in such a way that it expresses prediction of only one output series. The fully general form of the model allows multiple simultaneous predictions. In fact, if there is even one MA term based on a different series than the one of interest, two simultaneous predictions must be made. (Predictions must be made if shocks are to be computed!) Moreover, the multiple predictions can be intertwined in fiendish ways. We may have two series, x and y. The predictions of x may be based in part on shocks from y, and predictions of y may be based on shocks from x. The potential for such interdependence implies that training and testing must be simultaneously done for all outputs. Despite these facts, it is still best to break up multiple prediction models into individual models for each output series. No versatility is lost, and we can get away with double subscripts instead of triple.

We will continue to use z_t to represent the value of the output series at time t, and we will use a_t to represent the shock at that time. Other predictor series need an additional subscript to identify themselves. Input series i (which may itself be an output series in a different equation that is part of the grand model) is called z_{it}, and its shock is a_{it}. The weights are similarly treated. Those relating the output series to itself have only one subscript, identical to the univariate model. Weights for the other series require one subscript to identify the predictor series and a second as the ordinal number of the weight. With this in mind, the multivariate ARMA model for predicting this single output may be expressed as shown in Equation (6.6).

$$\begin{aligned}z_t = {} & \phi_0 + \phi_1 z_{t-1} + \phi_2 z_{t-2} + \ldots + \theta_1 a_{t-1} + \theta_2 a_{t-2} + \ldots + a_t \\ & + \phi_{1,1} z_{1,t-1} + \phi_{1,2} z_{1,t-2} + \ldots + \theta_{1,1} a_{1,t-1} + \theta_{1,2} a_{1,t-2} + \ldots \\ & + \phi_{2,1} z_{2,t-1} + \phi_{2,2} z_{1,t-2} + \ldots + \theta_{2,1} a_{2,t-1} + \theta_{2,2} a_{2,t-2} + \ldots \end{aligned} \quad (6.6)$$

There is one small way in which the model expressed in Equation (6.6) does not precisely represent the model set forth in this text. The weight subscripts in this equation correspond to the time lag of the term that is multiplied by the weight. The original ARMA model was designed this way, and writing the model is easy when this convention is used. However, it is a significant and needless limitation.

There is often good reason to assume that all weights up to some physically meaningful lag are equal to zero. For example, a model may include only one MA term, that term being at a lag of four. This might occur with quarterly economic data. It should be apparent that generalizing the equation that way would severely complicate things while adding no effective versatility. Offended readers may imagine the equation written out in its full glory, with some coefficients fixed at zero and the other coefficients free to be optimized. In other words, this discrepancy is a programming issue only, and an easily handled one at that.

It may be easier to comprehend the multivariate ARMA model by means of a diagram. Consider a simple model that has one AR term and one MA term for the series of immediate interest, one of each for another series, and no constant offset. Prediction at time t is illustrated in Figure 6.10.

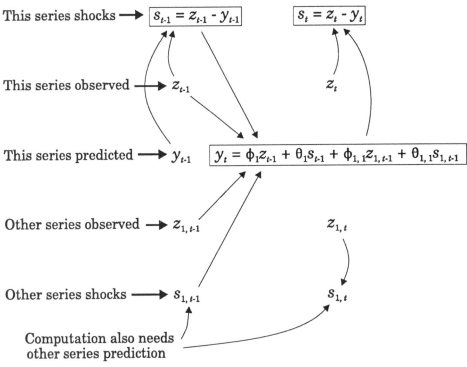

Figure 6.10 Multivariate ARMA prediction.

The top three lines in that figure are identical to the univariate model. The second line from the bottom is simply the observed values of another series. It is the bottom line that is potentially confusing. In order to compute those shocks, predictions for that other series are needed. The equation for that series is not shown. It has its own set of AR and MA weights, any of which may involve the series that is shown here! It should be clear to the reader why the multivariate ARMA model has been written for one output at a time. The complete representation would require another subscript to differentiate the weights according to their output. Despite this complexity, careful study will reveal that there is no ambiguity in the model. Computation of the predictions for each successive time slot may be made in any order because the mutual references are always to *past* values that have already been computed. The fact that the outputs can be mutually dependent adds tremendously to the power and versatility of the model without compromising its rigor.

Computing the Error and Predicting

This section presents an algorithm for computing the error of a set of multivariate ARMA weights. Complete C++ code for this algorithm can be found in the function predict in the module ARMA.CPP on the accompanying disk. The version in this section is a combination of C++ and pseudocode, having been substantially edited for clarity. Editing is indicated by italics.

The implementation given here makes the opposite tradeoff from most traditional ARMA programs. It is wildly extravagant in its memory usage in order to save a small amount of execution time. This reflects modern hardware characteristics. Users who are restricted to hardware with minimal memory relative to the series length will need to modify the code so that the shock storage rotates on an as-needed basis.

To simplify the life of the caller of this algorithm, the output series and their shocks are strung out into one long vector. All cases of the first output series come first, followed by the other outputs. Then, after all outputs have appeared, the shocks for the first output come, followed by the other output shocks. This layout is illustrated in Figure 6.11.

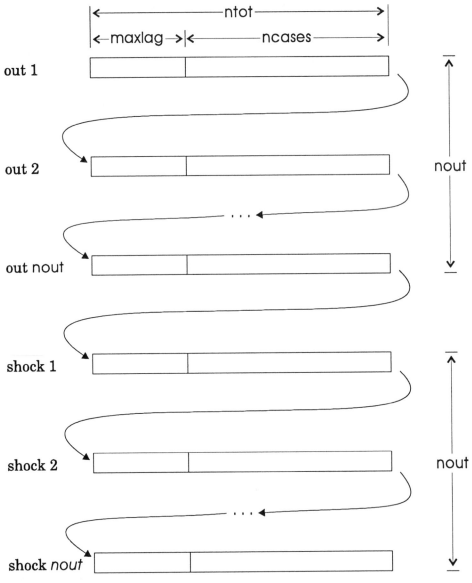

Figure 6.11 Layout of outvars vector.

Multivariate ARMA Models

There are many variables in this algorithm, and some of them are used in complex ways. Here is a list of the main variables along with a brief description of each.

ncases — Outputs for this many cases will be computed. This includes the point-by-point predictions within the known series, used for shock computation, as well as any predictions beyond the end of the series. It does *not* include the first maxlag cases, which are simply copied from the original series.

maxlag — This is the maximum AR lag in the entire model. This many cases at the beginning of each output will be copied rather than computed.

ntot — The sum of ncases and maxlag, this many outputs (and shocks) will be stored.

nin — There are this many inputs (AR terms) in the entire model.

nma — There are this many MA terms in the entire model.

nout — This many output series exist. An output series is one for which the user wants predictions. All series for which MA terms exist in the model are automatically considered outputs, regardless of user interest. These series must be predicted in order to compute their shocks for the MA terms.

outvars — This is a huge vector that holds all ntot*nout outputs and their shocks. If predictions are being made beyond the end of the known output series, those values should be initialized to a huge number as a flag (see the code below). Similarly, shocks that are not known should be initialized to a huge number.

nio — There are this many input and output series named in the model. Each one of these named signals falls into one of three categories: Either it is a pure output, or it is an output for which MA terms take part in the model, or it is an input (generalized AR). For the latter two categories, a range of lags is specified. Since a series may fall into more than one category, repetition is possible. The order of the weights in wts corresponds to the order of the nio names.

shocks_known — This flag tells the algorithm whether the shocks are already known, having been computed and probably predicted by another method. If this flag is false, all shocks not based on actual predictions within this algorithm are assumed to be zero.

fixed — This flag indicates whether the constant offsets are fixed at zero. If not, they will be estimated along with the weights.

var[nout] — This vector holds the error (shock) variance for each output series. The sum of its components will be minimized during learning.

optr — As each case's prediction begins, optr points to this case's output in the first output vector. Adding ntot to this pointer advances it to the next output.

shocks — As each case's prediction begins, shocks points to this case's shock for the first output. Adding ntot to this pointer advances it to the next output.

wts — All of the model's weights are here. Those for the first output series come first, followed by those for the second output, and so forth. Within each output, the order of the weights corresponds to the order of the nio named series, with the optional constant offset last.

pred — For each case in each output series, that individual prediction will be cumulated here.

```
ntot = ncases + maxlag ;          // First maxlag outputs remain at original values

for (i=0 ; i<nout ; i++)          // Will sum error variances here
   var[i] = 0.0 ;

for (icase=0 ; icase<ncases ; icase++) {
   optr = outvars + icase + maxlag ;   // Keep first maxlag at original vals
   shocks = optr + nout * ntot ;       // Shocks go here (if MA terms)
   for (iout=0 ; iout<nout ; iout++) { // For each predicted output
      wptr = wts + iout * nvars ;      // Weights for this output here
      pred = 0.0 ;                     // Will cumulate predicted here
```

```
for (i=0 ; i<nio ; i++) {                        // Check all ins and outs
   if (this is an input) {                       // If this is an input
      dptr = this series + icase + maxlag ;      // Point to this case
      for (j=min lag ; j<=max lag ; j++) {       // For its lag range
         if (icase+maxlag-j < this series' length) // Common: within signal
            pred += *(dptr-j) * *wptr++ ;        // Sum prediction
         else if ((k = index of output series) >= 0) { // Recursive?
            if (j <= icase)                      // Normally true, but insurance
               pred += optr[k*ncases-j] * *wptr++ ; // Sum prediction
            }
         else  // Rare situation: beyond end of unpredicterd signal
            pred += *(this signal's last point) * *wptr++ ; // Dup
         }
      }
   else {     // This is an output.  Handle MA term.
      k = index of output series ;               // Output vector for this MA term
      for (j=min lag ; j<=max lag ; j++) {
         if (! j)                                // Only lags are MA terms
            continue ;                           // This must be pure output
         if (j <= icase)                         // If this shock is available
            pred += shocks[k*ntot-j] * *wptr++ ; // Sum prediction
         else                                    // Shocks before the first prediction
            ++wptr ;                             // Are assumed zero
         }
      } // This is an output
   } // For all nio named series
if (! fixed)
   pred += *wptr ;                               // Last weight is the constant offset
if (pred > 1.e30)                                // Prevent runaway
   pred = 1.e30 ;
if (pred < -1.e30)
   pred = -1.e30 ;
if (optr[iout*ntot] < 1.e35) {                   // If this is still known signal
   err = optr[iout*ntot] - pred ;                // Then this makes sense
   var[iout] += err * err ;                      // Cumulate error
   }
else {                                           // We are now beyond the end
   err = 0.0 ;                                   // Assume no errors after known
   if (var[iout] >= 0.0)                         // True the first time end passed
      var[iout] = -var[iout] / icase ;           // Find the mean for this out
   }                                             // And flag to not divide again
```

```
      if (nma || shocks_known) {               // If there is a shock vector
        if (shocks[iout*ntot] < 1.e35)         // If this shock is known
          pred += shocks[iout*ntot] ;          // Use it to improve prediction
        else if (nma)                          // Unknown.  If there are MA terms
          shocks[iout*ntot] = err ;            // Must compute the shocks
        }
      optr[iout*ntot] = pred ;                 // Record the prediction
      } // For all predicted output signals
    } // For all cases

  error = 0.0 ;                                // This will be grand mean error
  for (iout=0 ; iout<nout ; iout++) {          // For each predicted output
    if (var[iout] < 0.0)                       // If there were predictions past end
      var[iout] = -var[iout] ;                 // Mean was already computed
    else
      var[iout] /= ncases ;                    // Compute its mean error
    error += var[iout] ;                       // And grand error
    }
  error /= nout ;
```

Several sections of this code need special comments, so the code will be explained from top to bottom. There are ntot output cases altogether. The first maxlag of them remain at their original values, while ncases of them will be computed. This may or may not include predictions beyond the end of the series.

Before this algorithm is begun, the output series must have been initialized to their true values. If any of the outputs go beyond their known extents, as would be the case for predictions into the future, these output series points must be initialized to a huge number, typically 1.e40. If shocks are in use, as would be the case if there are any MA terms, or if the shocks are being input as known values, then any elements of the shocks that are unknown must also be initialized to a similar huge value. These huge numbers serve as flags, and their use will become clear later.

The error variance of each output series is cumulated separately. However, we will see later that the total error is minimized when the model is trained. These variances are initialized to zero.

The outermost loop passes through all cases, computing them one at a time. For each case, set optr to point to this case in the first output series, and set shocks to point to the corresponding position in the shock vector for the first output series.

For this case, compute all nout output series predictions. The dot product of the data points with the weight vector will be cumulated in pred, so initialize it to zero. Make wptr point to this output's weight vector in the grand weight array.

The next loop is potentially the most confusing part of the whole algorithm. There is a total of nio series that take part in this procedure. Some of them may be inputs (AR terms), some may be outputs for which MA (shock) terms are needed, and some of them may be pure outputs (no MA involved). It is possible for some series to play multiple roles. It is also possible for some series to have disjoint sets of lags. For example, we may be using a series at lags of 1, 2, and 7. Thus, we must pass through all nio items and handle each separately. Note that the order of the weights in wptr corresponds to the order of AR and MA terms as they are encountered in this loop. The constant offset for this output, if any, is the last weight.

If this item is an input term, point to its current value. Remember that everything is offset by maxlag because that is where the actual predictions start. Then process each lag in the user-specified contiguous range. For each lag, one of three things will be true. By far the most common event is that this element is within the known extent of this series. If so, just multiply it by the corresponding weight and cumulate the sum. If it is outside the known extent, the user is probably doing recursive prediction into the future and we have passed into that realm. The variable k serves dual duty as a flag and as an index. If this input is not also an output, k will equal −1. In the more likely event that this is recursive prediction, k indicates which output series corresponds to this input. In this case, multiply the previously computed value by the appropriate weight and cumulate. Note that for the sake of clean programming, we include insurance against the practically impossible situation of the output series being ridiculously short. Careful users of this algorithm may safely remove that check.

The last of the three possibilities for this input term is rare, but conceivable. This may be an input series that is not recursively predicted and that expires before the end of the predictions. Probably the best response is to duplicate the last point in the series, although some would argue that a better response would be to use the series mean. In any case, this generally leads to such poor results that responsible users will almost never let it happen in practice.

If this term is not an input, it must be an output, pure or MA. Set k to the index of this output series, then pass through the contiguous lag range of this term. A lag of zero indicates that this is a pure output, so skip it. Right now we are interested only in MA terms. Shocks before the first actual prediction are assumed to be zero, so only include this MA term if its shock has been computed.

This prediction is almost complete. If the constant offset is not fixed at zero, add it in. Finally, as part of a continuing campaign for safe computing, limit this prediction to an arbitrarily defined valid range. It is frighteningly easy to come up with weight sets that are so nonstationary that predictions rapidly blow up out of control. At best, this will generate nan outputs, and, at worst, some systems may generate error conditions that cause less than graceful exits from the program. This simple expedient prevents these unhappy occurrences while still giving the user plenty of warning that something has gone amiss.

The next step handles error (shock variance) computation. Recall that the user has initialized outputs beyond the known extent of this output (if any) to a huge number. If we are still within the known extent, compute the shock as the true output minus the prediction and cumulate its variance. But if we have passed into the future, we are done computing the error. Set the error for this prediction to zero in case we need it later. The first time this happens (the first future prediction), divide the sum in var by the number of predictions that went into this sum to get the variance, then flip its sign as a flag to avoid doing this division again.

Now we need to deal with the shock vector. This vector will not even exist if there are no MA terms and if the shocks are not being provided. Thus, we only worry about the shocks if at least one of these conditions is true. Recall that the user has previously initialized all unknown shocks to a huge number. If this shock is not huge, the user has supplied its value, so we add it to the prediction. Otherwise, this shock is not being provided for us. But we may not need to keep it anyway. We need to preserve the shocks only if there are MA terms using them. If so, save this shock.

At this point, the prediction for this case in this output series is complete. Copy it into the correct output slot.

After the predictions have been done, all that remains is the trivial computation of the shock variance for each output series and the grand mean error. If an output had predictions beyond its known extent, its variance is negative as a flag that the division has already been done. Otherwise, we must now divide the sum of squared shocks by the number of cases to get the variance. The grand error is the mean shock variance across all outputs.

Training the Multivariate ARMA Model

There are two distinct phases in multivariate ARMA training. First, the constant offset and the input (AR) weights are estimated, with all MA weights set to zero. If there are no AR terms, the constant offset of each output is set equal to the mean of that output. If there are no AR terms and the offset is to be fixed at zero, there is nothing to do in the first phase.

The second phase is needed only if there are one or more MA terms. In this phase an iterative algorithm is used to simultaneously optimize all of the weights. Iteration is necessary because the effect of the MA weights on the predictions is nonlinear.

Training will be presented in three sections. The first topic is initial estimation of the AR parameters. The second topic is iterative refinement of all parameters. The final section puts everything together into the complete algorithm.

Computing Initial AR Parameter Estimates

The traditional method of estimating the AR weights is to compute the autocorrelations and solve the Yule-Walker equations. A more recent excellent alternative is Burg's algorithm (page 364). The approach taken here is to solve the linear system directly by using singular value decomposition. This method, though far more expensive in terms of memory usage, is slightly more accurate and much more straightforward. Moreover, it certainly generalizes cleanly.

Let us start with a brief review of ordinary linear regression, for this is the technique that will be used to compute initial AR parameter estimates. The fundamental assumption of linear regression is that a dataset can be explained by a linear model. In particular, suppose that for each case, several *independent* variables are measured. Call these x_1, x_2, and so forth. A *dependent* variable, called b, is also measured. The hope is that b depends on the x variables in a linear manner. For each case, this relationship can be written as shown in Equation (6.7).

$$b \approx \phi_0 + \phi_1 x_1 + \phi_2 x_2 + \ldots \qquad (6.7)$$

The goal is to estimate values of the ϕ coefficients that optimize this equation's performance over a known dataset. The regression equation can be written to encompass a collection of cases by using matrix notation. Let **b** be a column vector containing the dependent variable for each case. Let **A** be a matrix, each of whose rows is a single case. Its columns are the measured values of the independent variables. An additional column, corresponding to the constant offset ϕ_0, is set to 1.0 for all cases (rows). The weights are in the column vector Φ. The linear model is written for all cases as shown in Equation (6.8).

$$\mathbf{b} \approx \mathbf{A}\Phi \qquad (6.8)$$

There are many mays of computing the weight vector. No space will be wasted here repeating topics that are discussed in nearly every good text on numerical methods. Suffice it to say that the method generally recognized as the best is called *singular value decomposition*. The module SVDCMP.CPP provided on the accompanying disk implements this algorithm. All we need to be concerned with is construction of the **A** matrix and the **b** vector.

Each element of the **b** vector is an observed value of a series of interest. The corresponding row in **A** is the lagged values of the input series. If more than one output series is to be predicted, there will be a separate **b** vector for each output series. The same **A** matrix applies to all predicted outputs.

Nothing beats a picture. Suppose we have two series. They will be used for both input and output. To clarify the picture, the values of one of the series will be 0, 1, 2, and so forth. The values of the other series are 10, 11, 12, and so forth. The model uses as inputs the values of series one at lags of 1, 2, and 4. It uses series two at the single lag of 6. The **A** matrix and the two **b** vectors are shown in Figure 6.12.

After the **A** matrix and the **b** vectors have been constructed, the matrix equation is solved for each output. First, function svdcmp() is called once to compute the singular value decomposition of **A**. Then, function backsub() is called for each **b** vector. Each time it is called, it returns the set of ϕ weights for predicting that output series.

The following edited code fragment demonstrates how **A** and the **b** vectors may be constructed from the series. This code is extracted from the module ARMA.CPP on the accompanying disk. See the discussion starting on page 203 for an explanation of the variables used here. There is one important difference to note between that discussion and this code. Here we have no use for the maxlag cases at the start of the series, so they are omitted. Only ncases cases are stored in the output vectors.

Training the Multivariate ARMA Model

Series 1 lags: 1, 2, 4   Series 2 lag: 6

Series 1:   0   1   2   3   4   5   6   7   8   9
Series 2: 10 11 12 13 14 15 16 17 18 19

A

5	4	2	10	1
6	5	3	11	1
7	6	4	12	1
8	7	5	13	1

b1

6
7
8
9

b2

16
17
18
19

Figure 6.12 Preparing for singular value decomposition.

```
aptr = A matrix
optr = b vectors here (outvars in Figure 6.11)

for (icase=maxlag ; icase<ncases+maxlag ; icase++) {   // Only ncases done
    optr = outvars + icase - maxlag ;                   // Do not offset by maxlag
    for (i=0 ; i<nio ; i++) {                           // Check all ins and outs
        if (this is an input) {                         // If this is an input
            for (j=min lag ; j<=max lag ; j++)          // All lags for this input
                *aptr++ = this series[icase-j] ;        // Get this lagged input value
            }
        else {                                          // This is an output
            if (this is the first appearance of this output) {   // Just once each
                *optr = this series[icase] ;            // Get this value into the b vector
                optr += ncases ;                        // Cases strung out as vectors here
                }
            } // This is an output
        } // For all inputs_outputs
    if (! fixed)                                        // If offset not fixed at zero
        *aptr++ = 1.0 ;                                 // Last column is for the constant offset
    } // For all cases
```

After the data has been set up as just shown, two steps remain. The linear system(s) must be solved, and the AR weights must be unpacked into the grand weight array, filling in zeros for all MA weights. The equation solution starts by computing the singular value decomposition of **A**. Then, one at a time, the **b** vector for each output series is copied into the appropriate location and the linear system is solved. This set of weights is expanded into the grand weight array. These operations can be accomplished with the following code:

```
svdptr->svdcmp () ;                                   // Decompose A (the slow part)
for (iout=0 ; iout<nout ; iout++) {                   // For each output series
   memcpy ( svdptr->b , outvars+iout*ncases , ncases * sizeof(double) ) ;
   svdptr->backsub ( 1.e-8 , work ) ;                 // Solve for this output's AR weights
   wptr = work ;                                      // The computed AR weights are here
   optr = wts + iout * nvars ;                        // Move them to grand weight array
   for (i=0 ; i<nio ; i++) {                          // Check all ins and outs
      if (this is an input) {                         // If this is an input
         for (j=min lag ; j<=max lag ; j++)           // For all lags of this input
            *optr++ = *wptr++ ;                       // Copy the AR weights for these terms
         }
      else {                                          // This must be an output
         for (j=min lag ; j<=max lag ; j++) {         // For all lags of this output (MA terms)
            if (! j)                                  // Only lags are MA terms
               continue ;                             // Skip pure outputs
            *optr++ = 0.0 ;                           // MA weights are zero for now
            }
         } // MA term
      } // For all inouts
   if (! fixed)                                       // If the constant offset not fixed at zero
      *optr = *wptr ;                                 // Copy the computed constant term too
   } // For all outputs
```

If there are no AR terms but the constant offset is not fixed at zero, weight initialization is much easier than what we just dealt with. Set all MA weights to zero and set each offset equal to the mean of the series. Finally, if the offset is fixed, the only initialization is to set all MA weights to zero. Code for these situations is tedious busywork, so the interested reader is directed to the routine learn in module ARMA.CPP on the accompanying disk.

Iterative Refinement of All Parameters

The AR parameters affect the predictions in a linear fashion. As a result, optimal estimates of these parameters can be computed by explicit solution of a linear system. The same is not true of the MA parameters. These weights have a nonlinear influence on the predictions, so they can be estimated only by iterative methods. Furthermore, the MA parameters interact with the AR parameters, so both sets of parameters must be optimized together. Fortunately, unless there is a very large number of parameters, these interactions usually are not severe. Standard optimization algorithms are adequate for finding good weight sets. This is in marked contrast to many neural networks. As an example of the degree and nature of interaction of AR and MA parameters, look at Figure 6.13. This contour plot shows the error for a simple model having one AR parameter (horizontal axis) and one MA parameter (vertical axis), both at a lag of one. Note that the contours, though skewed, are not excessively narrow. This is typical.

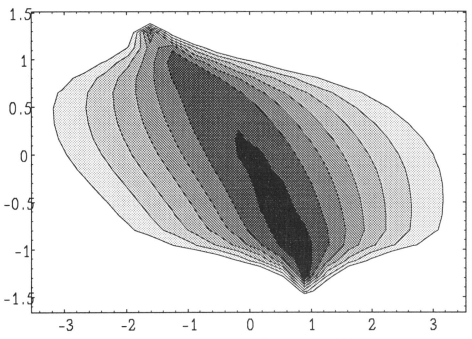

Figure 6.13 Error involving one AR and one MA parameter.

Since this is not a particularly difficult optimization problem, many solutions are possible. In fact, many different solutions have been used throughout the years. Box and Jenkins (1976) propose a modification of Marquardt's algorithm in which the gradient is numerically estimated, and a variation of Newton's method is employed. Modern numerical experts frown on this approach, and its use is not recommended. For simple models, the gradient can be explicitly computed, and that is certainly a viable option. However, a simple and straightforward method that is known to work well is Powell's method. This algorithm is documented in most good books covering numerical optimization, so details are omitted here. Figure 6.14 is a rough flow chart of Powell's algorithm.

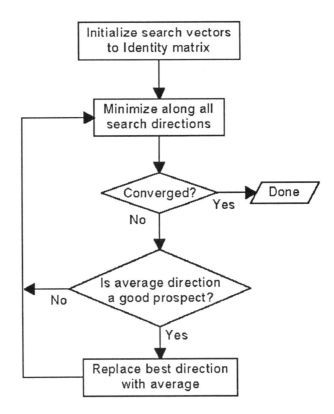

Figure 6.14 Powell's algorithm for minimization.

The simplicity of the algorithm belies its effectiveness. An array of search vectors is initialized to the identity matrix, and the line minimum is found in each of those directions. In other words, when the algorithm starts, one parameter at a time is optimized, with the others held fixed. After a pass through all search directions (parameters, initially), the average direction of movement is computed. This direction may be a good direction to pursue because there is a significant chance that it points toward the global minimum. Its quality is tested. If this direction is nothing special, another pass is made with the same set of search directions. But if the average motion vector shows good promise for the future, it replaces the search direction that provided the most improvement. The exact nature of the quality test, and the reason for replacing the former best vector, are subtle and beyond the scope of this text. See any good reference, such as Press *et al.* (1992), for more information.

One practical warning is in order. If the model is overfitted (more parameters than are needed to model the data), there will be large interactions among the parameters. This may even be true to a lesser degree if the model is not overfitted. The result is that the error will decrease rapidly on the first few iterations, then continue dropping at an exceedingly slow rate. *There is no reason to pursue accurate parameter estimates.* Remember that only on the rarest of occasions do we have a hard reason for being interested in the actual values of the parameters. The vast majority of the time all we really need is *some* set of parameters that provides a reasonably good fit. It is a waste of time to pursue improvements out in some distant significant digit.

The Complete Training Algorithm

The preceding two sections have discussed initial estimation of the AR parameters and iterative refinement of all parameters. It is time to put everything together into a complete multivariate ARMA training algorithm. Its many memory allocations and safety checks make the entire subroutine too long and complex to list here. It can be found in function learn in module ARMA.CPP on the accompanying disk. On the other hand, the core algorithm is really too simple to justify a flow chart. As a compromise, its salient features will be given as a combination of C++ code and comments. This allows the reader to comprehend the technique without getting lost in the details. In order to understand this material, it is necessary that the reader be familiar with the definitions found on page 205.

One potentially confusing aspect of this code is the variable *trained*. Sometimes it is useful to train the model partially, then resume training at a later time. This variable flags whether the model has already been trained.

```
/*
   Compute the number of cases: Length of the shortest signal minus
   the longest input lag.
*/

   ncases = MAXPOSNUM ;
   for (i=0 ; i<nio ; i++) {   // Check all ins and outs
     if (length of this signal < ncases)
       ncases = length of this signal ;
   }

   ncases -= maxlag ;                      // This many cases used in training
   if (ncases < (nvars * nout))            // If the linear system is underdefined
     return -2 ;                           // We should not try to do anything

/*
   Allocate storage for each output variable.

   If there are any MA terms, allocate outvars twice what is needed.  Get_shocks will
   use the far half of the vector to save the shocks for computing predictions based on
   the MA terms.  Technically, we only need to allocate for the MA terms that are
   present.  But memory is cheap and time is expensive.  The logic for decoding
   addresses is too slow to justify a little memory savings.
*/

   if (nma)
     outvars = (double *) MALLOC ( 2 * nout * ncases * sizeof(double) ) ;
   else
     outvars = (double *) MALLOC ( nout * ncases * sizeof(double) ) ;

/*
   There are three possibilities.  If there are inputs and the ARMA is not already
   (partially) trained, use regression to estimate the input weights and set any MA
   weights to zero.  Also do this if there are no MA terms, as in that case learning is
```

Training the Multivariate ARMA Model

```
    not iterative.  Else if it is pure MA and not trained, set all MA weights to zero and
    set the constant offset to the mean.  Finally, regardless of status, we need the
    outputs in outvars.
*/

   if ((! nma) || (nin && ! trained)) {  // Estimate just input weights

/*
   Allocate the SingularValueDecomposition object for equation solution
   for the input terms.  Also allocate short work vector for solutions.
*/

      svdptr = new SingularValueDecomp ( ncases , fixed ? nin : (nin+1) , 0 ) ;
      work = (double *) MALLOC ( (fixed ? nin : (nin+1)) * sizeof(double) ) ;

/*
   Fill in the A matrix in svdptr and the output vectors.  Start at the case offset by the
   longest lag.  The columns of A correspond to the weights, with the last column
   being the constant (hence set to 1.0).  For each input, the value is the lagged
   signal.  We skip MA terms for now.
*/

      aptr = svdptr->a ;
```

> This code is on page 213.

```
/*
   Decompose the matrix and solve for all weights.  For each output, copy the output
   signal to the RHS (b), then put the solution in work.  Finally, move the computed
   weights from work to wts, simultaneously filling in any MA weights with zero.
*/
```

> This code is on page 214.

```
      delete svdptr ;
      FREE ( work ) ;
      } // If nin, solve for input weights
```

```
/*
   If there are no inputs and it is not trained, set all the MA weights to zero and set the
   constant offset to the mean.
*/

   else if (! trained) {

      for (iout=0 ; iout<nout ; iout++)            // Borrow variance vector
        var[iout] = 0.0 ;                          // For summing means

      for (icase=maxlag ; icase<ncases+maxlag ; icase++) {
        optr = outvars + icase-maxlag ;            // Offset outputs
        k = 0 ;                                    // Count unique outputs

        for (i=0 ; i<nio ; i++) {                  // All of these are outputs (pure MA!)
          if (This is the first) {                 // The first appearance of this output?
            *optr = This series[icase] ;           // Get this value
            var[k++] += *optr ;                    // Sum means
            optr += ncases ;                       // Cases are strung out as vectors here
            }
          } // For all inputs_outputs

        } // For all cases

      wptr = wts ;                                 // Set MA wts to 0, constants to mean
      for (iout=0 ; iout<nout ; iout++) {          // For each predicted outut
        for (i=0 ; i<nma ; i++)                    // Each MA weight
          *wptr++ = 0.0 ;                          // is zero
        if (! fixed)                               // If constant offset is not fixed
          *wptr++ = var[iout] / ncases ;           // It is the mean
        }

      } // Else pure MA and not trained
```

/*
   Else already trained. Just put outputs in outvars.
*/

   else
      This code is identical to A, b initialization, except A is not initialized!

```
/*
   Initialization is practically done.  The weights are in wts.  All input weights have
   been found by regression.  All MA weights are zero.  The constant offset has been
   estimated.  The true outputs are in outvars.  Note that outvars[0] is for time
   'maxlag'.  This offset is because earlier predictions cannot be made.  Complete
   initialization by computing the errors (shocks).  Put these in the far half of outvars.
*/
```

This code is nearly identical to the prediction code found on page 206.

```
/*
   If there are no MA terms, we are done.
*/

   if (! nma) {
      trained = 1 ;
      goto FINISH ;
      }

/*
   Iteratively minimize the grand mean error.
*/

   work = (double *) MALLOC ( nvars * nout * sizeof(double) ) ;
   err = error ;

   memcpy ( work , wts , nvars * nout * sizeof(double) ) ;
   error = powell ( ... , work , ... ) ;
   memcpy ( wts , work , nvars * nout * sizeof(double) ) ;

   if ((error >= 0.0)  ||  (-error < err))   // This is pretty subjective.
      trained = 1 ;         // Consider it trained if not interrupted or improved
```

Compute the final errors and pointwise predictions here

```
   FREE ( work ) ;
```

The preceding code contains many omissions, but these are always for the sake of increasing clarity. Let us now work through the code. It starts by computing the number of cases that will contribute to the training process. It would be counterproductive to introduce errors by duplicating short series beyond their end, so the shortest series imposes a limitation. Also, it is dangerous to omit AR terms, so maxlag cases must be skipped at the beginning. If the resulting number of cases is less than the number of variables to be estimated, the system is underdetermined and solution should not be attempted.

A temporary storage area is needed for the point-by-point predictions that are made as the error is evaluated. Also, if there are any MA terms, the shocks must be stored. Memory for these vectors is allocated. Note that, technically, there is no need to allocate memory for the entire series. A much shorter allocation could be done, with the elements being rotated as needed. But the logic for such rotation would slow operations, and most users will have more than enough memory available for this worthy extravagance.

The initial estimation phase encompasses three possibilities. If there are no MA terms, iterative refinement will not be needed, so the AR computation must be done regardless of the previous training status. Otherwise, we will estimate the AR weights if there are any to estimate, and if the model has not yet been trained. Code for that process was presented earlier in this chapter, so it is not repeated.

The second initialization possibility, when the conditions just described do not apply, is that it is a pure MA model that is not yet trained. In this case we still must get the outputs, as they will be needed in the optimization that comes later. Also, cumulate the means, just in case they will be needed to initialize the constant offsets. Then set all of the MA weights to zero and set the offsets if needed.

If neither of the above conditions applies, then the model must be trained already and include one or more MA terms. In this case we do not tamper with the weights. Just fetch the outputs. This code is so similar to that shown on page 213 that it is not repeated here. The only difference is that there is no A matrix to construct.

Initialization is now complete. If there are no MA terms, we are done. Otherwise, prepare for iterative refinement of all parameters. The preliminary ARMA weights are copied to a work vector that will be optimized by the powell subroutine. The error before refinement is also preserved. After refinement, the optimized weights are copied back to the ARMA weight area. The powell routine returns the negative of the error if the user pressed ESCape. An easy way to determine whether the model can legitimately be considered trained is based on two tests. If the returned error is positive, the optimization was not interrupted, so convergence was attained. If the user did interrupt, check to see if any improvement at all was had. If so, say the model is trained. This is admittedly an arbitrary method, but it should be good for most applications.

Finally, note that powell does not necessarily leave everything neat and tidy. Now that the optimal weights are back in the ARMA model's member array, the errors associated with that weight set should be computed. This code is nearly identical to that shown on page 206, so it is omitted here.

Designing and Testing ARMA Models

Thus far, this chapter has focused on computing the parameters of multivariate ARMA models and predicting with these models. Several important aspects of the total problem have been omitted. How does one decide on the number of AR and MA parameters for the model? What lags should be used? How can the adequacy of the trained model be checked? These closely related questions are crucial to good performance. Unfortunately, they are difficult to answer. A lot of intelligent guesswork, combined with a fair amount of trial and error, is needed. There is no substitute for experience, as no simple magic formulas exist. However, this section will attempt to provide some basic guidelines.

Two powerful tools for answering these questions expose the relationship between current values of a series and lagged values. The *autocorrelation* of a time series is itself a time series, each of whose elements is the correlation between the current value and a lagged value of the original series. The first point in an autocorrelation series is the correlation between the current value and the previous value. The second point is the correlation between the current value and the lag-two value, and so forth. Unless there is a strong seasonal component, the autocorrelations rapidly drop to small values as the lag increases. Therefore, unless seasonality is being investigated, there is rarely any need to examine many autocorrelations.

A closely related tool is the *partial autocorrelation*. This is similar to the ordinary autocorrelation in that it describes the relationship between current and lagged values of a time series. But there is an important and subtle difference. The partial autocorrelation at lag i is the correlation between the current value and the lag-i value *after the influence of all lesser lags has been accounted for*. This has major implications for model designers. As a general example, suppose we are considering using two variables, X and Y, to predict a third variable, Z. In the context of time series, Z may be the current value, and X and Y may be lagged values. Perhaps we observe that both X and Y are strongly correlated with Z, so we are tempted to include both of them in the model. But what if X and Y are highly correlated with each other? In this case, there is little point in including them both in the model, as they are largely redundant. We would conclude that this is the situation if we discover that the partial correlation of Y with Z, after the effect of X is accounted for, is nearly zero. That piece of information tells us that Y may be safely omitted from the model as long as X is included. The converse may also be true.

Let's look at this in the context of a real time series. We will consider four different series. The first is an autoregressive series in which the current value is equal to a positive number times the previous value, plus a random shock. The second is identical to the first except that the weight is negative, so the series will tend to alternate in sign. The third is an MA series with a single positive weight. In other words, the current value is equal to a positive weight times the previous value's shock, plus a shock of its own. The last series is identical to the third except that the MA weight is negative. Again, this series will tend to alternate in sign. This set of four series encompasses the four major simple serial relationships.

The upper-left graph in Figure 6.15 shows the theoretical autocorrelations of the first series. As expected, the first element of the autocorrelation series is large and positive, reflecting the fact that the current value is heavily based on the previous value. The previous value is similarly influenced by the value before it, so the current value is correlated with the lag-two value as well, though not so much, because of the dilutive effect of the shocks. This extends to infinity, with the correlations decaying exponentially. This is probably the most commonly seen pattern in real-life data.

The upper-right graph in that figure shows the autocorrelations when the parameter is negative. The first (lag-one) correlation is negative, reflecting the negative influence of the previous value. But that value is negatively influenced by its earlier neighbor, so the end result is that the current value and the lag-two value are positively correlated. The same exponential decay occurs as in the case of the first series.

Designing and Testing ARMA Models 225

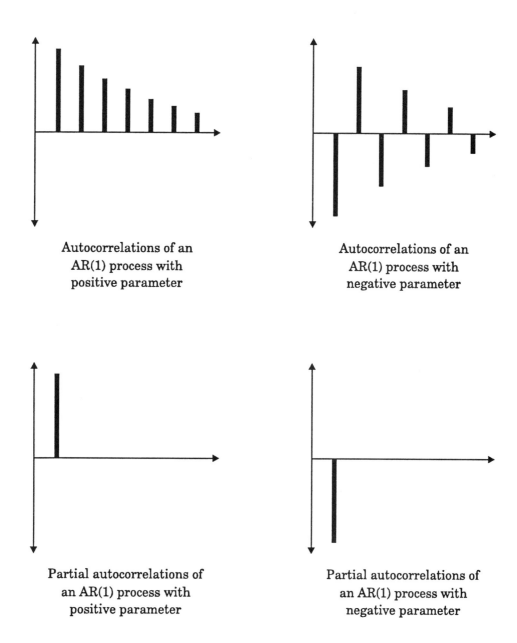

Figure 6.15 Correlations in an AR process.

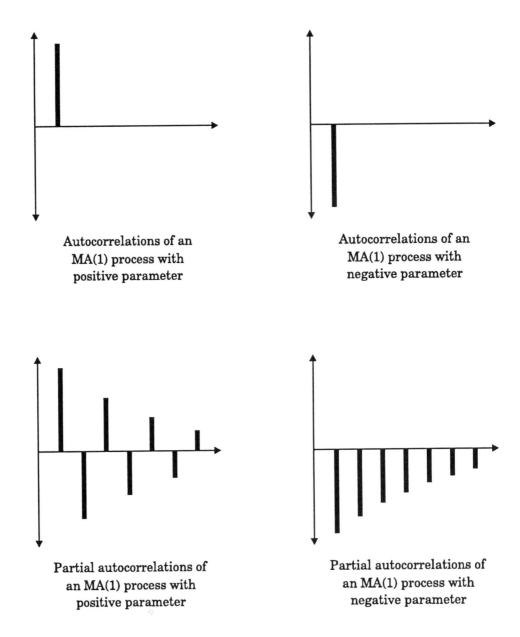

Figure 6.16 Correlations in an MA process.

Designing and Testing ARMA Models

These same two series clearly illustrate the nature of partial autocorrelation. Look at the lower-left graph in Figure 6.15. The first partial autocorrelation is exactly equal to the first autocorrelation because there are no earlier lags whose effects are removed. This is the only nonzero correlation. Recall that this AR series is defined in terms of the prior value only. When the effect of the prior value is removed, all other correlations vanish. This is expressed in the following general rule:

- *When a time series is purely autoregressive and involves only lags up to some maximum lag, all partial autocorrelations beyond that maximum lag are zero, and the ordinary autocorrelations decay exponentially.*

Now let's look at two pure MA series as illustrated in Figure 6.16. Suppose the current value of a series is equal to a positive multiple of the previous value's shock, plus a shock of its own. Each shock affects the series value in that time slot and the next, but this effect does not propagate any further. This is reflected in the upper-left graph in Figure 6.16. The autocorrelation at a lag of one is positive due to the double effect of each shock. But all other autocorrelations are zero. A similar pattern appears in the upper-right graph in which the MA weight is negative.

The partial autocorrelations are a little more difficult to fathom. By definition, the first partial autocorrelation is equal to the ordinary autocorrelation. The second requires a little thought. To compute it, the effect of lag-one terms on current terms must be removed. But there is no such effect in a direct sense. The effect is indirect, through the single common shock. Therefore, when we remove the effect of the value itself, we remove too much. The result is a negative partial autocorrelation at a lag of two. A similar thought process explains the lower-right graph in that figure. The following rule applies:

- *When a time series is purely moving-average and involves only lags up to some maximum lag, all autocorrelations beyond that maximum lag are zero, and the partial autocorrelations decay exponentially.*

Unfortunately, these two rules concerning correlations are not nearly as valuable as might be thought. It should be obvious that even a relatively simple series consisting of one AR component and one MA component can have ordinary and partial autocorrelations that extend for considerable distances. The exponentially decaying correlations due to one component make it difficult to appraise the sharply truncated correlations due to the other component.

Another complicating factor is that these rules apply only to the theoretical correlations of series having infinite length. In practice, finite length series give rise to nonzero correlations everywhere. So we cannot simply look for the lag at which correlations become tiny. We must be wary of the fact that random sampling generates correlations that may be large enough to fool us into thinking something is there when really there is nothing. The traditional approach to this problem is to compute the statistical significance of correlations. This is a dangerous approach, though. There is a big difference between *statistical* significance and *practical* significance. Especially if the series is long, trivial influences in the process that generates the data can easily produce correlations that have statistical significance. If we attempt to make the model account for every statistically significant correlation in the series, we will likely create a monster. A frequently better approach is to generate a relatively long series of regular and partial autocorrelations and let our eyes be the judge.

For the sake of readers who absolutely demand some statistics, and with considerable fear and trembling, here is a little information about judging the significance of autocorrelations. For a time series consisting of n independent random samples following a normal distribution, its regular and partial autocorrelations are approximately normally distributed (asymptotically as n increases). Their mean is zero and their variance is about $1/n$. For this reason, it may sometimes be reasonable to draw a pair of lines on a plot of the correlations. One line could be two standard deviations ($2/\sqrt{n}$) above zero, and the other, two standard deviations below zero. These lines could provide rough guidance about the significance of correlations.

There are numerous pitfalls in using these statistical guidelines. Please take them seriously, as they can easily invalidate conclusions. Among the dangers are the following:

- *This variance estimate might be too large.* If the samples are not independent, the variance of low lags can be much smaller than $1/n$. Also, if the series is the residuals from an ARMA model, the variance of early lags can be smaller than $1/n$ as a result of the effect of parameter fitting.

- *This variance estimate might be too small.* If the samples are not independent, the variance of large lags can exceed $1/n$. (See Figure 7.3 on page 257 for a good example of this.)

- *Even if the variance and normality assumptions are correct, this method can still be misleading.* Basic probability tells us that even if all of the correlations are truly zero, random sampling leads us to expect about 5 percent of the computed autocorrelations to exceed two standard deviations in magnitude.

It should not surprise readers to learn that the best method for finding an effective yet parsimonious model is iterative. Start by examining the regular and partial autocorrelations. In the unusual event that a large correlation stands out, an AR or MA term at that lag is definitely indicated. Otherwise, look at the largest correlation. If it is an autocorrelation, try an MA term at that lag. If it is a partial autocorrelation, try an AR term at that lag. Compute the model and look at the regular and partial autocorrelations of the residual (shock) series. Repeat this process until all correlations appear reasonably small. This is not a perfect method, but it seems to work well.

The only real caveat to iterative model building is this: *Add just one term at a time.* It is possible to find parameter sets for which the model is ill-defined. For example, an ARMA(1,1) model whose weights are equal in magnitude but have opposite signs is just white noise! (This fact is not obvious, but ambitious readers can derive it with some work.) Geometrically, the error function has a flat ridge along the line of equal weights. A good parameter estimation program will not become trapped, but why tempt fate? By adding just one parameter at a time, guided by correlations of residuals, the likelihood of including unnecessary parameters that may lead to degeneracy is decreased.

Examining the Power Spectrum

The power spectrum of a series (and of the residuals as an ARMA model is iteratively constructed) can sometimes reveal important information. Readers should be familiar with the material in Chapter 3, especially that discussed starting on page 70, before continuing in this section. Also, the material on the cumulative power spectrum (page 83) is vital.

The goal in creating an ARMA model is straightforward: End up with a residual series that is essentially white noise. The implicit premise is that if the residuals are not random, more detailed modeling is possible. Naturally, this is usually an unattainable goal. Not much real-life data can be *fully* explained by an ARMA model operating on white noise. Nevertheless, it is a worthy goal.

How can the power spectrum help us work toward this goal? The answer is that white noise has a flat spectrum. So we simply examine the spectrum and look for anomalies. If any suspicious humps appear, something is amiss. Unfortunately, it is not quite that easy. Random sampling and the laws of probability guarantee that even white noise has suspicious humps all over the place. It is vital that the smoothing methods and the statistical significance tests discussed in Chapter 3 be employed. For the purposes of this chapter, we will take the expedient of focusing on the raw spectrum in hopes that the reader will see both the general principles and the dangers of that approach.

Figure 6.17 is the power spectrum of a sample of white noise. It's not exactly flat! In fact, one or two of the peaks could easily be mistaken for indicators of actual periodic components. The moral of the story is that the techniques of Chapter 3 should be considered before jumping to conclusions based on raw power spectra.

In most situations, the primary use for power spectra in assessing randomness is the general trend. Figure 6.18 is the spectrum of an ARMA signal having a moderate positive weight at a short lag. Note the clear concentration of power in the low frequencies. This is very typical. A spectrum that looks like this invariably indicates that an AR or MA term at a short lag should be tried.

When the weight is negative, exactly the opposite effect occurs. Figure 6.19 is the spectrum of an ARMA process having a moderate negative weight at a short lag. The concentration of power at high frequencies is obvious. An unbalanced spectrum like this indicates the need for an AR or MA term at a short lag.

In summary, we should not be overly concerned with local peaks in the power spectrum. These often occur due to nothing more than random sampling effects. In Chapter 7 we will discuss the only significant exception to this rule. The general principle is that the overall distribution of the power is far more important than local artifacts. If the general tendency is not toward a horizontal layout, alarms should be triggered and short-lag ARMA terms should be considered.

Designing and Testing ARMA Models

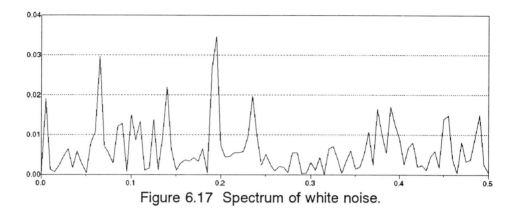

Figure 6.17 Spectrum of white noise.

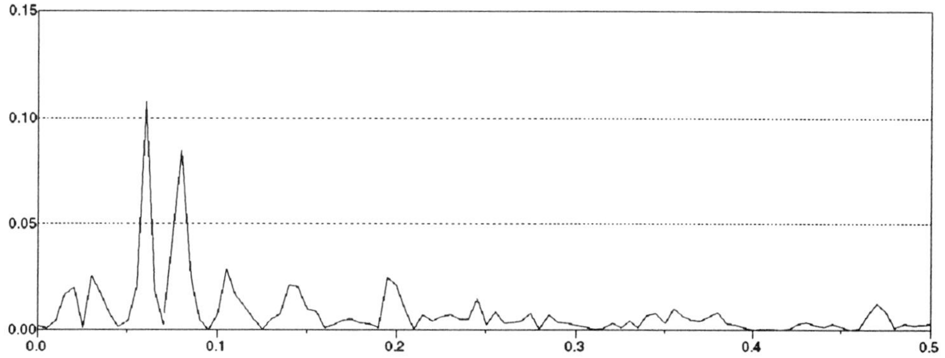

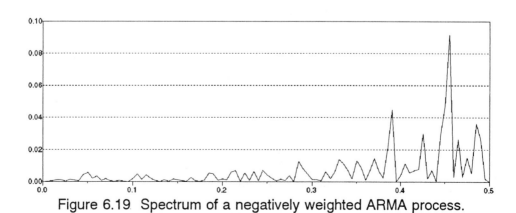

Figure 6.18 Spectrum of a positively weighted ARMA process.

Figure 6.19 Spectrum of a negatively weighted ARMA process.

Multivariate Models

The situation is more complicated when a multivariate model is needed. A potent pair of tools are the *crosscorrelation* and the *partial cross correlation*. These are similar to the lagged correlation tools described earlier. The difference is that these tools relate two different series to each other, as opposed to relating a single series to itself. The cross correlation is a series each of whose values is the correlation of the current value of one series with the lagged value of another series. Note that some applications treat crosscorrelation in a symmetric sense in that both lags and leads are considered. We will not need that approach in this text, so only the asymmetric case of positive lags will be considered.

Partial crosscorrelation is like ordinary crosscorrelation except that the effect of lesser lags is removed from each successive correlation. In other words, the partial crosscorrelation of one series with respect to another series at lag i is defined as the correlation of the current value of the first series with the lag-i value of the second series, *after the effect of all lesser positive lags of the second series has been removed*.

The nice relationships between ARMA terms and lagged correlations shown in Figures 6.15 and 6.16 do not apply to cross correlations. The situation is complicated by autocorrelations within the individual series. But we can still make effective use of these tools. Peaks in the crosscorrelation suggest promising lags. If the partial crosscorrelation drops to nearly zero and stays there past a certain lag, there is reasonable assurance that terms that far back are not needed. Far more guesswork and testing is needed than when there is just one series to worry about. The basic principle is the same, though. Examine the crosscorrelations, looking for prominent peaks. Construct a model with one term (probably AR is the best first try) at the largest peak. Compute the crosscorrelations of the residuals and look for another peak. Try replacing an AR term with an MA term to see if improvement is had.

Multivariate MA terms are a nuisance because each one adds another weight to iteratively optimize, a slow operation. But they can sometimes be very useful. Think about the marginally stable series defined by Equation (6.9) and graphed in Figure 6.20.

$$z_t = .999\, z_{t-1} - .4\, a_{t-1} + a_t \tag{6.9}$$

Designing and Testing ARMA Models

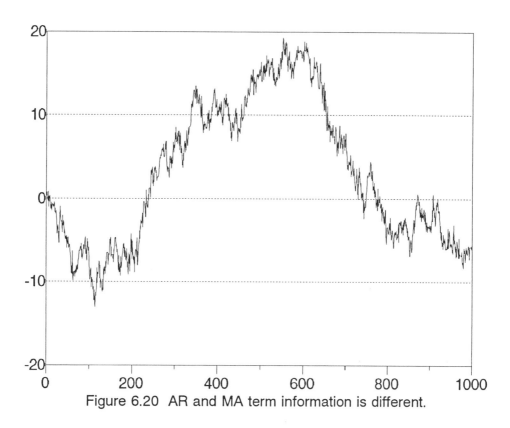

Figure 6.20 AR and MA term information is different.

This series tends to wander far and wide due to the large AR weight. The negative MA weight causes rapid local oscillation. It should be clear that as far as predicting another series goes, z_t and a_t convey very different information. An AR term (a lagged value of the actual series, z_t) provides the slowly varying general neighborhood of the series, while an MA term (a lagged value of the computed shock, a_t) provides information about the outside forces acting on the series at each time slot.

When a supposedly final multivariate ARMA model has been formulated, the last step should be to check the crosscorrelations of the residual with respect to every input series. They should be reasonably small. Otherwise, another term may be needed. The confidence intervals plotted on an NPREDICT display may provide a rough indication of the significance of nonzero correlations. However, a visual examination of a long set of lagged correlations is usually the best indicator, especially to a practiced eye.

Computing Lagged Correlations

This section presents a set of methods for computing regular and partial auto and crosscorrelations. Be advised that the methods given here are not the only possibilities. There is a surprising variety of contenders for even the seemingly straightforward operation of computing autocorrelations. These are the methods used in the NPREDICT program that accompanies this text, and experience indicates that they are effective and efficient in a wide variety of applications. Complete source code for all of these algorithms can be found in the module AUTOCORR.CPP on the accompanying code disk.

Autocorrelation

The simplest lagged correlation is autocorrelation. The most popular method of computing it is shown in Equation (6.10). The top line of that equation defines the autocovariance, and the bottom line shows that the autocorrelation is equal to the autocovariance divided by the variance (c_0).

$$c_k = \frac{1}{n} \sum_{i=k+1}^{n} (z_i - \bar{z})(z_{i-k} - \bar{z})$$

$$r_k = c_k / c_0 \tag{6.10}$$

Several points about this definition should be mentioned. It shows the mean being subtracted from each point. This will nearly always be appropriate. However, some processes are such that the mean is known to be zero. In this situation, the raw values should be used to avoid the possibility of error due to random sampling causing a nonzero mean in the sample.

Alert readers may be surprised that a factor of n is always used to divide the sum of crossproducts. It might seem better to divide by $n - k$, the number of terms in the sum. The definition used here exhibits less bias than the alternative, so it is generally considered preferable.

For the sake of completeness, it should be mentioned that there is another algorithm for efficiently computing autocorrelations when a large number of them are desired. Although this algorithm is relatively simple, its details are beyond the scope of this text. Basically, it goes like this: Compute the fast Fourier transform of the series. Perform some simple operations on the result, then transform back. If the length of the original series is a power of two, this can be very fast. However, since the applications covered in this text rarely require many autocorrelations, this fast method will not be pursued.

Crosscorrelation

Crosscorrelation is a trivial extension of autocorrelation. There are only two differences. First, the crossproducts involve two separate series rather than one series. Second, instead of dividing by the variance of the single series, one divides by the standard deviations of both series. With this in mind, look at Equation (6.11). In this equation, s_x is the standard deviation of x, and s_z is the standard deviation of z.

$$r_k = \frac{1}{n s_z s_x} \sum_{i=k+1}^{n} (z_i - \bar{z})(x_{i-k} - \bar{x}) \qquad (6.11)$$

Note that this definition encompasses only positive lags of x with respect to z. This is all that is required for the applications here. Some applications involve the full symmetric version, in which the correlation is defined at a lag of zero as well as negative lags (lagging z relative to x).

Partial Autocorrelation

There are many algorithms for computing the partial autocorrelation of a time series. One method is to compute the ordinary autocorrelations, arrange them in a correlation matrix, and use any of the traditional algorithms for computing partial correlations from ordinary correlations. Readers who have ready access to standard statistical libraries may want to use this straightforward approach. However, this method is not recommended because it fails to take advantage of the band structure of the correlation matrix derived from autocorrelations. Also, it is slow, and numerical problems may hinder its operation.

Another method is to make direct use of the following fact: *The partial autocorrelation at any lag is equal to the weight at that lag in a pure AR model containing all lags up to that lag (but no further).* In other words, to compute the partial autocorrelation at a lag of five, compute the weights of an AR model involving lags of one through five. The computed weight at a lag of five is the partial autocorrelation at a lag of five. It doesn't take much thought to realize that naive application of this method is an abominably slow way to compute partial autocorrelations. Having to compute a full set of AR weights for each and every partial autocorrelation is an extravagance that few can afford.

The method of choice is somewhat of a hybrid of the two methods just described. This method, called *Burg's Algorithm*, is a little too complicated to present in detail in this text. Elliot (1987) provides an excellent discussion, along with a lot of interesting related information. (But beware of his flow chart; it is not excellent.) Press *et al.* (1992) provides an implementation (called memcof) that has an absolutely ingenious way of conserving scratch memory by shifting successive lags. See page 364 of this text for details on using BURG.CPP which appears on the disk accompanying this text. This implementation is heavily based on the program in Press *et at.* (1992).

Here is a rough overview of Burg's algorithm. It starts by computing the ordinary autocorrelation at a lag of one. By definition, this is also the partial autocorrelation at a lag of one. Then, two steps are taken. First, the partial autocorrelation at a lag of two is computed by means of the traditional formulas that can be found in any good multivariate statistics text. However, the clever part is that the algorithm takes advantage of the special band structuring of the correlation matrix, which is derived from autocorrelations. This dramatically reduces the amount of computation needed. The second step is that the lag-one correlation computed earlier is modified to be the lag-one AR weight. At this point we have the two weights of an AR(2) model. For our immediate purposes, all we care about is the lag-two weight, as that is the partial autocorrelation at that lag.

Burg's algorithm then iterates the procedure just described. The lag-three partial autocorrelation is computed, then the first two weights are adjusted to become the AR(3) weights. This pair of operations is repeated for as many partial autocorrelations as are desired. This recursive building of successive models is tremendously faster than computing each model from scratch. It also discourages buildup of rounding errors that can occur when one massive model is constructed.

Partial Crosscorrelation

Sad to say, this discussion of computation of lagged correlations must end on a note of inefficiency. There does not appear to be a readily available bivariate analogue of Burg's algorithm. Such an algorithm could surely be devised by an ambitious student. Perhaps this has already been done. Unfortunately, this algorithm, if it exists, has not made its way into the NPREDICT program. Instead, the slow route of direct calculation is taken.

This method, though slow, is certainly straightforward. To compute the partial crosscorrelation at a given lag, use the method described on page 211. Put the independent (lagged) series in the **A** matrix, using lags from one through the required lag. Put the current values of the dependent series in the **b** vector. Solve the linear system. The partial crosscorrelation at this lag is the computed weight for this lag. This terribly slow process must be repeated for each partial crosscorrelation. This method is obviously not useful for computing a long partial crosscorrelation series.

A Multivariate Example

We conclude this chapter with a relatively long and complicated example of multivariate ARMA prediction. The task chosen for this demonstration is extremely difficult: prediction of the monthly total precipitation a month in advance. As all of us who listen to weather reports know, this appears to be an almost impossible job. Nevertheless, it makes a good demonstration. The complete text of this example can be found in the file PREDPREC.CON in the \EXAMPLES directory of the accompanying disk.

The first step is to read the data. We will use the temperature in the ten regions of New York (TEMP.NY) to predict the precipitation (PREC.NY) in the first region. The first 83 years serve as a training set, and the remaining 12 years are a validation set. The signal splitting for the first temperature series is shown. That pattern is repeated for the remaining nine regions and for the precipitation. Also, we use the log transform to tame the wild precipitation tails.

```
NAME= , tmp1,tmp2,tmp3,tmp4,tmp5,tmp6,tmp7, tmp8, tmp9, tmp10
READ SIGNAL FILE = temp.ny   ; Skip the first data value:  the date
```

```
NAME = , prec        ; Again, the first field in each case is the date
READ SIGNAL FILE = prec.ny
LOG = prec           ; Log transform tames heavy tail

NAME = t1            ; This is for the training set
COPY = 996 tmp1      ; Keep the first 83 years
NAME = v1            ; This is for the validation set
COPY = -144 tmp1     ; Keep the last 12 years
... Repeat for nine more temperature series, and the precipitation
```

Just for fun, let's start out with a very naive prediction model. Use the precipitation exactly one and two years ago to predict the current value. This should pretty well handle simple annual cycles. The audit log file tells us that the error in the training set is 0.434, and the error in the validation set is 0.478.

```
INPUT = t_precip 12        ; One input is precipitation a year ago
INPUT = t_precip 24        ; And also two years ago
OUTPUT = t_precip          ; Predict the current precipitation
TRAIN ARMA = naive         ; Train this simple model
CLEAR INPUT LIST           ; Must start new input and output lists
CLEAR OUTPUT LIST
INPUT = v_precip 12        ; Now use the validation data
INPUT = v_precip 24        ; Which is the final 12 years
OUTPUT = v_precip          ; The t_precip was the first 83 years
NAME = v_naive             ; Any name will do
ARMA PREDICT = 144 naive ; This will tell us the validation set error
```

Ten temperatures is definitely too much data. Compute the principal components. The audit log informs us that the first three factors account for 99.91 percent of the total temperature variation. We decide to use just these three. Compute the model, then apply it to both the training and the validation series.

```
CLEAR INPUT LIST
CLEAR OUTPUT LIST
INPUT = t1
...     Repeat for t2 through t9
INPUT = t10
CLASS = dummy              ; Any name will do
CUMULATE TRAINING SET      ; This builds the training set
```

A Multivariate Example

```
ORTHOGONALIZATION TYPE = PRINCIPAL COMPONENTS
ORTHOGONALIZATION STANDARDIZE = YES
DEFINE ORTHOGONALIZATION = princo   ; Compute the model

NAME = t_fac1 , t_fac2 , t_fac3  ; Factors for the training set
APPLY ORTHOGONALIZATION = princo

CLEAR INPUT LIST            ; We now need a new input list
INPUT = v1                  ; to compute the validation factors
   ...   Repeat for v2 through v9
INPUT = v10
NAME = v_fac1 , v_fac2 , v_fac3 ; Factors for the validation set
APPLY ORTHOGONALIZATION = princo
```

The first two principal components of the training set, and the (log) precipitation in the training set, are graphed on page 244 in Figures 6.21 through 6.23, respectively.

We now define a multivariate ARMA model. This model is much larger (more terms) than we ordinarily want. We will use AR lags of 1–3, 12, and 13 months for all four series. The reason for using a lag of 13 is to allow the model to make an exact annual comparison of values one and 13 months ago. In addition, we will use a single lag-one MA term for each of the four series. This model is specified and trained as follows:

```
CLEAR INPUT LIST
CLEAR OUTPUT LIST
INPUT = t_fac1 1-3          ; AR inputs at lags of 1 through 3
INPUT = t_fac1 12-13        ; AR inputs at lags of 12 and 13
   ...   Repeat for t_fac2 and t_fac3
INPUT = t_precip 1-3        ; AR inputs at lags of 1 through 3
INPUT = t_precip 12-13      ; AR inputs at lags of 12 and 13
OUTPUT = t_fac1 1           ; MA terms at lag of 1
OUTPUT = t_fac2 1
OUTPUT = t_fac3 1
OUTPUT = t_precip 1
TRAIN ARMA = arma1          ; This is a slow training process!
```

Compute a pointwise prediction within the training set. Subtract this from the true values to produce the shocks. These quantities are shown in Figures 6.24 through 6.29.

```
NAME = t_pred_fac1 , t_pred_fac2 , t_pred_fac3 , t_pred_precip
ARMA PREDICT = 996 arma1        ; Pointwise prediction within tset
NAME = t_shock_fac1             ; Shocks for this factor
SUBTRACT = t_fac1 AND t_pred_fac1
...     Repeat for factors 2 and 3
NAME = t_shock_precip           ; Shocks for (log) precipitation
SUBTRACT = t_precip AND t_pred_precip
```

Do exactly the same pointwise prediction and shock calculation, except this time use the validation data. Visual examination of the graphs (not shown here) helps spot trouble areas, and the errors reported in the audit log file are valuable numeric results. For the record, the RMS training error for the precipitation series was 0.411, and the validation error was 0.437. This compares to 0.434 and 0.478 attained with the naive model employed at the beginning of this example. A little sophistication obviously helps.

```
CLEAR INPUT LIST
CLEAR OUTPUT LIST
INPUT = v_fac1 1-3
INPUT = v_fac1 12-13
...     Repeat all inputs and outputs as above, except use v_

NAME = v_pred_fac1 , v_pred_fac2 , v_pred_fac3 , v_pred_precip
ARMA PREDICT = 144 arma1        ; Pointwise within validation set
NAME = v_shock_fac1
SUBTRACT = v_fac1 AND v_pred_fac1
...     Repeat for other two factors and precipitation
```

Our whole goal is prediction, so let's do it. Use this independent validation set to compute honest confidence intervals. Go back to the raw precipitation domain (not logs) for the user's convenience. Then recursively extend the validation set into the future. These predictions are graphed in Figures 6.30 through 6.32.

```
CONFIDENCE LOG = v_precip       ; Inform confidence computer that
ARMA CONFIDENCE = 24 arma1      ; we will be undoing the log
NAME = v_fac1 , v_fac2 , v_fac3 , v_precip ; Predict all four series
ARMA PREDICT = 168 arma1        ; 144 + 24 (two extra years)
EXP = v_precip                  ; Undo log to return to real data
```

It is often shameful to ignore the potentially useful validation data when training the model. Now that we have an idea of performance, let's pool all of the data together and compute one grand model.

```
CLEAR INPUT LIST
CLEAR OUTPUT LIST
INPUT = tmp1                  ; This is the full 83+12 years
  ...     Repeat through tmp10
NAME = fac1 , fac2 , fac3 ; New factors (95 years)
APPLY ORTHOGONALIZATION = princo  ; Keep the same model

CLEAR INPUT LIST
CLEAR OUTPUT LIST
INPUT = fac1 1-3
INPUT = fac1 12-13
  ...     Repeat for other two factors and precipitation
OUTPUT = fac1 1
OUTPUT = fac2 1
OUTPUT = fac3 1
OUTPUT = prec 1
TRAIN ARMA = arma2
```

The usual next step is to compute and display the pointwise predictions and shocks within the series. These are not shown.

```
NAME = pred_fac1 , pred_fac2 , pred_fac3 , pred_prec
ARMA PREDICT = 1 140 arma2    ; Pointwise predictions within series
NAME = shock_fac1
SUBTRACT = fac1 AND pred_fac1
  ...     Repeat for other two factors and precipitation
```

Ideally, confidence intervals should be based on data that is independent of the training set. We did that earlier, but now we cheat a little. Compute the confidences and predict two years beyond the end of the series. The predictions are in Figures 6.33 through 6.35. Later in this demonstration, we will need the data again. The prediction will extend it, so we must make copies of all four series.

```
NAME = fac1_save
COPY = fac1
  ...     Repeat for other two factors and precipitation
```

```
CLEAR CONFIDENCE COMPENSATION    ; Get rid of the previous one
CONFIDENCE LOG = prec            ; It had a different name
ARMA CONFIDENCE = 24 arma2       ; Two years of confidence
NAME = fac1 , fac2 , fac3 , prec ; These are the predictions
ARMA PREDICT = 1164 arma2        ; 1140 + 24
EXP = prec                       ; Undo the log transform
```

Let's see if a hybrid approach helps. Train a neural network to predict the shocks. With such strongly seasonal data as this, it is probably worthwhile to include raw data in the model in addition to the shocks. However, we *cannot* ever let any unpredicted inputs reach their end during shock prediction. Repetition of the terminal point introduces tremendous error. To avoid the complexity of full recursive prediction, use the raw data at lags equal to the anticipated maximum future prediction, 24 months. This way, unknown data is never used.

```
CLEAR INPUT LIST
CLEAR OUTPUT LIST
CLEAR TRAINING SET
CLEAR CLASSES
INPUT = shock_fac1 1-2    ; The shocks, recursively predicted
INPUT = shock_fac1 6      ; Just for fun, try a different lag
...     Repeat for other two factors and precipitation
INPUT = fac1_save 24      ; Raw inputs to grab season info
INPUT = prec_save 24      ; They must be lagged to avoid expiring
OUTPUT = shock_fac1       ; Recursive prediction of shocks
... Repeat for other two factors and precipitation
CUMULATE EXCLUDE = 10   ; 24 (raw data) - 12 (shocks) + 10 = 22
CUMULATE TRAINING SET

NETWORK MODEL = MLFN
...     Specify other network parameters here
TRAIN NETWORK = temp
```

The trained neural network is now used to predict the shocks 24 months into the future. For kicks, let's also compute and examine the confidence intervals. These are shown in Figures 6.36 through 6.38.

```
NAME = shock_fac1 , shock_fac2 , shock_fac3 , shock_prec
NETWORK CONFIDENCE = 24 temp
NETWORK PREDICT = 1164 temp
```

A Multivariate Example

The final step is to use the hybrid model to predict the future. Note that the current version of NPREDICT does not allow computation of confidence intervals for hybrid models. This may be addressed in a future version. Also note that there is no such thing as pointwise prediction for a hybrid model. All prediction, even that within the known series, is based on the known shocks. Therefore, prediction within the series exactly duplicates the known data. (What else could it be, when we know the true shocks!) The implication is that even when we specify a different output name on the NAME command, the result is exactly as if the same name were specified to force preservation of known data. These predictions are shown in Figures 6.39 through 6.41.

```
CLEAR INPUT LIST
CLEAR OUTPUT LIST
INPUT = fac1_save 1-3
INPUT = fac1_save 12-13
    ...    Repeat for other two factors and precipitation
OUTPUT = fac1_save 1
    ...    Repeat for other two factors and precipitation

ARMA SHOCK = shock_fac1 FOR fac1_save    ; It is legal to use
ARMA SHOCK = shock_fac2 FOR fac2_save    ; any or all of the
ARMA SHOCK = shock_fac2 FOR fac3_save    ; known shocks.
ARMA SHOCK = shock_prec FOR prec_save    ; Here, we use all.

NAME = hybrid_fac1 , hybrid_fac2 , hybrid_fac3 , hybrid_prec
ARMA PREDICT = 1164 arma2
EXP = hybrid_prec
```

There is an extremely important lesson to be learned from this demonstration. Observe the deterioration starting in Figure 6.33. When the neural network comes into play (Figures 6.39 through 6.41), recursive predictions beyond a few months have terribly rapid error buildup. Why? The reason is *overfitting*. Both the ARMA models and the neural network employ far too many terms. They learn inappropriate ideosyncracies of the training set that fail them when attempts are made to use recursively computed outputs as new inputs. The sensible approach is stepwise: Pick one good term, examine the residuals, add another, and repeat as needed. And *always* be wary of extended recursive predictions. Twenty-four months is far too many!

244 *Box-Jenkins ARMA Models*

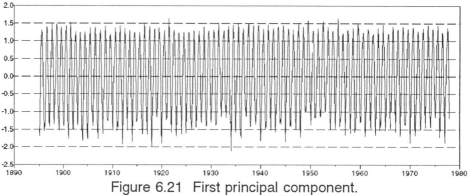

Figure 6.21 First principal component.

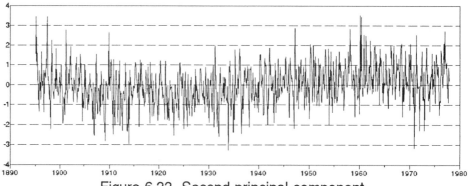

Figure 6.22 Second principal component.

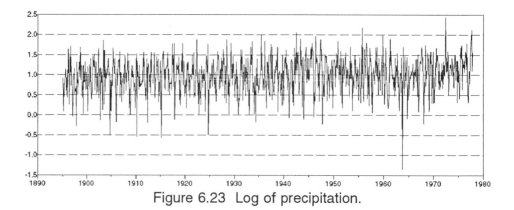

Figure 6.23 Log of precipitation.

A Multivariate Example

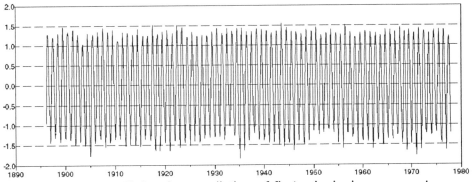

Figure 6.24 Pointwise prediction of first principal component.

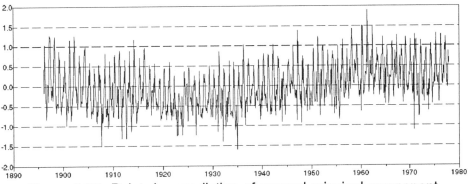

Figure 6.25 Pointwise prediction of second principal component.

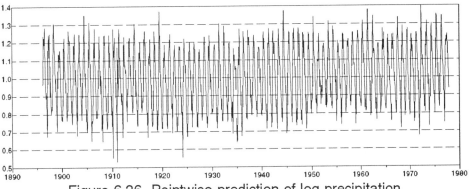

Figure 6.26 Pointwise prediction of log precipitation.

246 Box-Jenkins ARMA Models

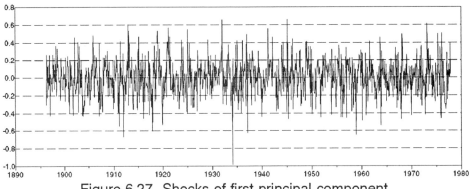
Figure 6.27 Shocks of first principal component.

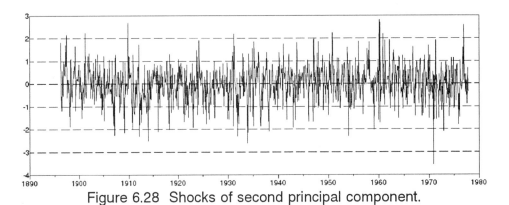
Figure 6.28 Shocks of second principal component.

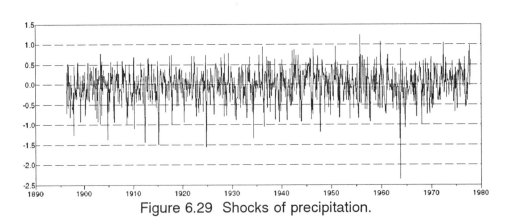
Figure 6.29 Shocks of precipitation.

A Multivariate Example

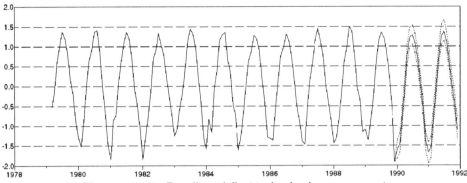
Figure 6.30 Predicted first principal component.

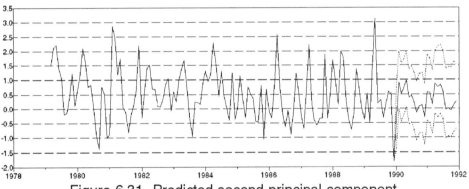
Figure 6.31 Predicted second principal component.

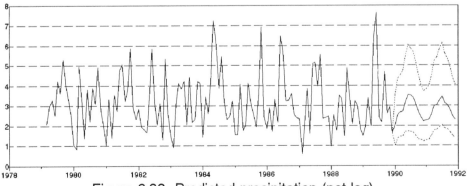
Figure 6.32 Predicted precipitation (not log).

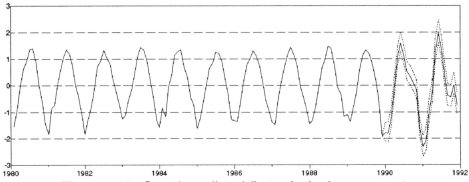
Figure 6.33 Grand predicted first principal component.

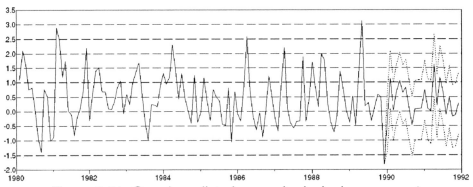
Figure 6.34 Grand predicted second principal component.

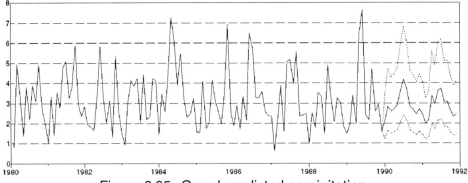
Figure 6.35 Grand predicted precipitation.

A Multivariate Example

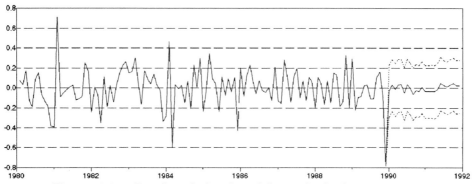

Figure 6.36 Predicted shocks of first principal component.

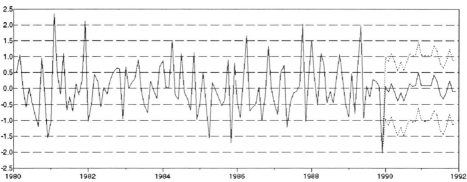

Figure 6.37 Predicted shocks of second principal component.

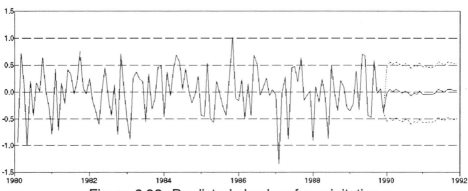

Figure 6.38 Predicted shocks of precipitation.

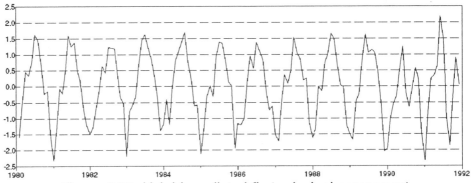
Figure 6.39 Hybrid predicted first principal component.

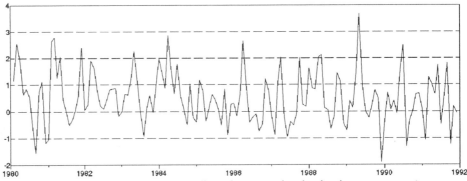
Figure 6.40 Hybrid predicted second principal component.

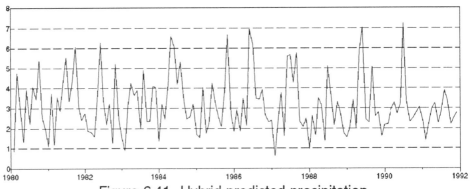
Figure 6.41 Hybrid predicted precipitation.

7

Differencing

- Identifying and removing nonstationarity

- ARIMA models

- Seasonal differencing

- A detailed example of multiple differencing

Sometimes we want to work with data essentially as it arrives from the measuring device, and sometimes we don't. There is a tremendous variety of data preprocessing techniques: linear rescaling, log transformations, wavelet decomposition, and an infinitude of other possibilities. Many examples appeared in Chapter 1. This chapter focuses on one of the simplest, yet most useful of them all. Differencing is such a basic operation that it is often taken for granted. It deserves more.

Two distinctly different types of differencing are considered here. The first is *adjacent-point differencing*. When the term *differencing* is used alone, this is what is meant unless the context clearly indicates otherwise. In this type of differencing, a new series is defined as the change (relative to the previous sample) of the original series at each time slot.

The second type of differencing is *seasonal differencing*. In this type, a new series is computed as the change relative to the sample at some fixed distance in the past. For example, it may be useful to compute the net profit of a company each quarter relative to the profit in that same quarter a year ago. In this case, the season length is four samples.

Why might we want to difference a signal by subtracting adjacent points? A few reasons are listed here. There are certainly more, but this is a representative collection. Note that many of these reasons are different ways of looking at the same situation.

- Sometimes the fundamental nature of the process asserts that it is the differences that are important, as opposed to the raw data. When we know this to be true, we should use differences.

- We may deliberately want to separate two types of information in a series. For example, the series defined by recent changes in the inventory level of some commodity probably conveys different information than the inventory level itself. Look at Figure 6.20 on page 233 for an example.

- Most practical methods for modeling time series assume that the series is stationary. These models are hindered by data that wanders far and wide as time passes. Differencing is an effective way of removing most types of nonstationarity.

- We may have reason to believe that the high-frequency information in the series is more important than the low-frequency information. Differencing is a simple means of applying a potent and easily reversible high-pass filter.

What about seasonal differencing? This is a separate topic that is individually dealt with later in this chapter. The key to this technique is in its name: *seasonal*. We consider seasonal differencing when there is something seasonal about the data. Monthly economic data is surely seasonal, and some applications might benefit from differencing at a lag of 12 samples. We will return to this topic later. For now, let us focus on adjacent-point differencing.

Stationarity

A mathematical definition of stationarity appeared on page 186. Intuitively, we say that a time series is stationary if the behavior of the series does not change with time. This implies that we consider not only the behavior of individual points, but the collective behavior of sets of points as well. For example, suppose the distribution of individual samples early in a series is identical to that late in the series. But suppose neighboring points early in the series have one sort of relationship, while neighbors late in the series have a different relationship. This series is not stationary.

The reason for featuring stationarity in a chapter on differencing is that differencing is a common method for inducing stationarity on a nonstationary series. Do not read too much into that statement. There is an infinite number of ways for a series to be nonstationary, and differencing works on only a fraction of them. However, it turns out that a great many series encountered in daily life exhibit a kind of nonstationarity that can be loosely called *homogeneous*. Apart from occasional changes in level or perhaps slope and level, these series exhibit generally uniform behavior across time. The classic example is stock prices. An issue may bob around a price of 40 for many months, then drop to 30 where it remains for a few more months. Its day-to-day behavior in the latter time period is nearly identical to its behavior in the former time period. Only its central tendency is different. This sort of homogeneous nonstationarity is common. Differencing can be effective in producing an essentially stationary series from such data. And there is a bonus to using this method, too. Differencing is easily reversed by summing. More on this will appear later.

Identifying Nonstationary Behavior

There is no magic formula for deciding whether a series is nonstationary, and, if so, whether it is at least homogeneous. In fact, there are no clear lines separating these possibilities. If one happens to be examining the exact mathematical formula responsible for generating a series, these questions are often easily answered. For ARMA models whose weights are exactly known, the roots of certain polynomials tell the whole story. But we never have the formulas. All that we workers in the field have is a pile of raw data. Given a time series and nothing more, we must be content with intelligent guesswork. Fortunately, in most applications, there is a lot of room for error. Unnecessarily differencing a series that falls into a gray decision area is rarely a crime. And when a series definitely needs or definitely does not need differencing, the methods of this section will almost always clearly indicate this fact.

The first step in any time series analysis procedure is looking at the data. It may be difficult for the eye to detect subtle changes in serial correlations. But changes in level or slope should be obvious. Clear changes in level, like that shown in the top half of Figure 7.1, indicate that differencing is required. In the relatively unusual event that the slope obviously changes as well, as shown in the lower half of that figure, two stages of differencing may be required. However, *do not blindly difference twice based on observed changes in slope*. Try differencing just once. If the result resembles the top half of Figure 7.1, go ahead and difference again. But be warned that it often happens that the transition periods during level changes are misinterpreted as changes in slope. Examination of the result of a single differencing helps one avoid being fooled.

The previous statement is so important that it bears repeating. Later in this chapter, and especially in Chapter 8, it will be seen that the operation of undoing a differencing has a strong tendency to be numerically unstable. Undoing a double differencing compounds the problem to a tremendous degree. Therefore, this guiding principle is vitally important: Difference once, if the need is obvious. Then examine the result. Only difference a second time if the result of the first differencing is obviously nonstationary. A third differencing is virtually never required. If the result of two differencings is still nonstationary, something peculiar is going on, and further investigation is definitely necessary.

Stationarity

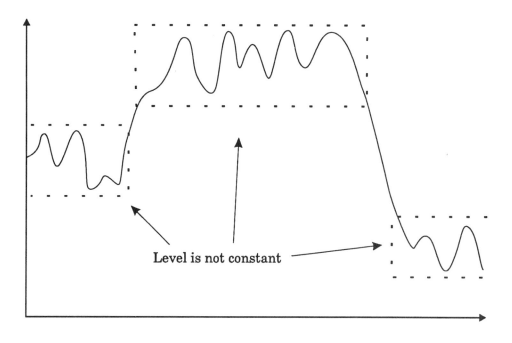

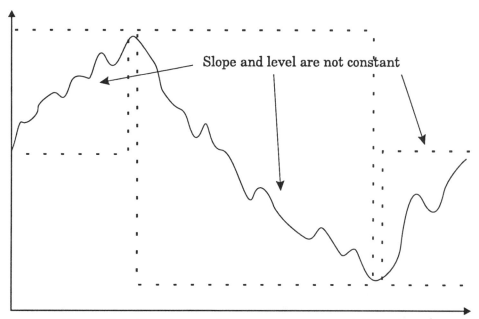

Figure 7.1 Two types of homogeneous nonstationarity.

Another useful tool for detecting nonstationary behavior is the autocorrelation of the series. (Review the material starting on page 223 if necessary.) The autocorrelation of a stationary series usually drops to zero quite rapidly, while that of a nonstationary series remains high for relatively large lags. Look back at Figure 6.15 on page 225. The two graphs in the upper portion of that figure show that a stationary AR series can have autocorrelations that taper slowly to zero. However, unless the AR weights are large (leading to nonstationary tendencies), the autocorrelations drop rapidly. If the series is nonstationary, it will spend extended periods of time away from its mean, as illustrated in Figure 7.1. This results in noticeable autocorrelations at lags commensurate with the length of its sojourns at different levels.

The effect of nonstationarity on autocorrelations can be profound. Figure 7.2 is a series that is very close to being nonstationary. It is AR(1) with a weight equal to 0.9. (An AR(1) series is nonstationary when the magnitude of its weight equals or exceeds 1.) Close examination of this series reveals that it regularly wanders away from zero, staying out there for significant amounts of time. Nevertheless, its autocorrelation (shown in Figure 7.3) drops to zero fairly rapidly. (The oscillation in the tail is a random sampling effect that clearly validates the warnings about judging significance given on page 228.)

Now look at the truly nonstationary series graphed in Figure 7.4. This series is AR(1) with a weight of 1. Its autocorrelations, shown in Figure 7.5, certainly suggest nonstationarity. It is especially interesting to see (in both the raw series and in the autocorrelations) the tremendous difference between a weight of 0.9 and a weight of 1.0. This makes the difference between wanderings that must always return to zero versus a true random walk.

Another tool that can help identify nonstationary behavior is the power spectrum of the series. A nonstationary series will virtually always have most of its spectral energy concentrated in the lower frequencies. If a smoothed spectrum is heavily unbalanced in this way, suspect nonstationarity. On the other hand, this is not a totally reliable indicator. Series with significant positive autocorrelation at short lags will also exhibit this same spectral characteristic, though usually to a lesser degree. For example, see Figure 6.18 on page 231. The use of the power spectrum for assessing inequitable power distribution is discussed in Chapter 3. The discussion of the cumulative power spectrum, starting on page 83, is particularly useful.

Stationarity

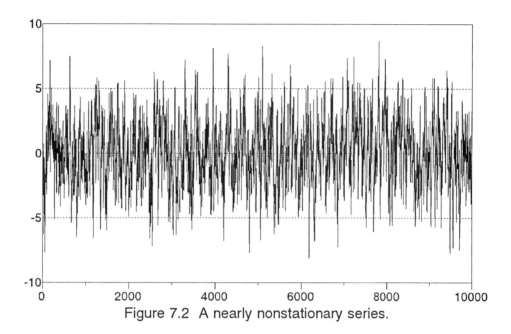

Figure 7.2 A nearly nonstationary series.

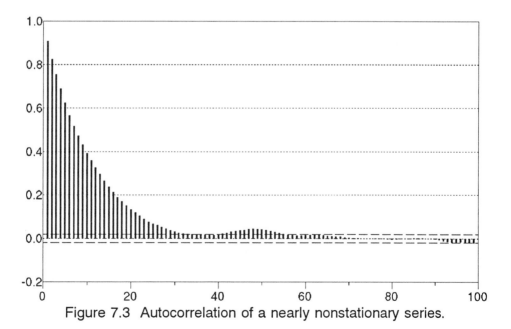

Figure 7.3 Autocorrelation of a nearly nonstationary series.

258   Differencing

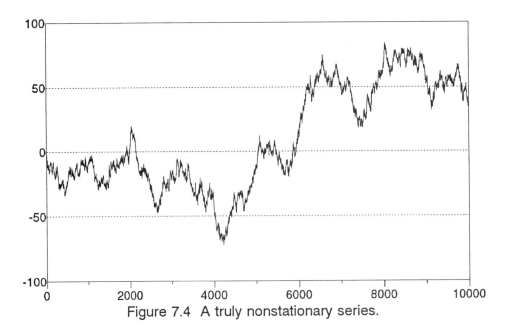
Figure 7.4 A truly nonstationary series.

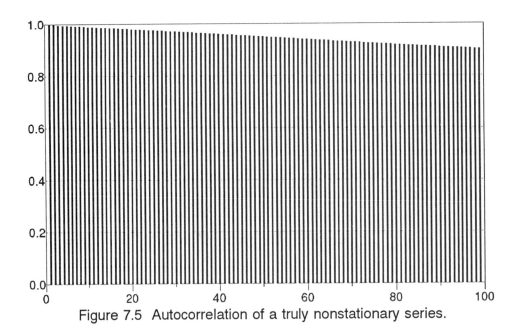
Figure 7.5 Autocorrelation of a truly nonstationary series.

Removing Nonstationarity

The most straightforward method for dealing with a nonstationary series is to compute adjacent differences. Instead of using the raw data (or, in addition to the raw data), use the change from the previous value at each time slot. Most types of nonstationarity encountered in practice can be rendered stationary by a single differencing. Sometimes a series requires two stages of differencing, but that is not common. Series that require more than two differencing operations are almost unknown.

When programming subroutines for differencing, a subtle point should be kept in mind. Although the length of the series decreases by one, the information content should be preserved. *Save the value of the first point of the original series.* The NPREDICT program does this by dedicating a member variable of the Signal class to safeguarding the first point of the original series. The reason for doing this is to enable return to the undifferenced domain after prediction. Only rarely do people care about future values of a differenced series. They want predicted values of the original series. Summing to reverse the differencing works correctly only if we start out with the original first point. See the module SIGNAL.CPP on the accompanying code disk for an example of reversible differencing.

It is enlightening to look at differencing as it appears in the frequency domain. Differencing is a potent high-pass filter. The response curve of this filter (expressed as amplitude attenuation versus frequency) is approximately graphed in Figure 7.6.

Readers who have had experience designing optimal filters will immediately see that the filter shown in Figure 7.6 is not very optimal in any usual sense of the word. It fully stops any DC component (constant offset), but then immediately starts allowing other very low frequency components to pass. Worse still is the fact that presumably valuable high-frequency components are significantly cut right up to the Nyquist frequency. Most people would agree that what is needed is more of an *S*-shaped response. The curve should be as flat as possible at the low and high-frequency ends, and rise rapidly somewhere in the middle.

The immediate conclusion is that instead of differencing, we should handle nonstationarity with some sort of sophisticated digital filter. This conclusion is only partially correct. If our ultimate goal is classification or prediction of other signals, the band-splitting techniques described in Chapter 1 are almost always superior to differencing when it comes to dealing with problems of nonstationarity. But

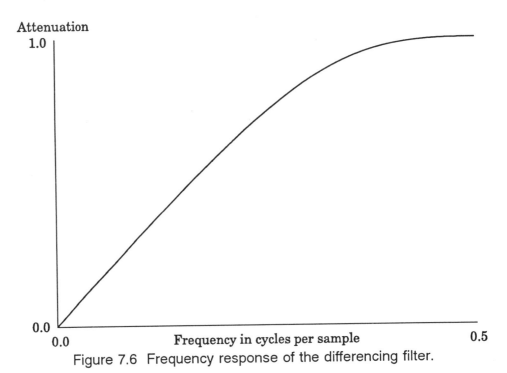
Figure 7.6 Frequency response of the differencing filter.

when our ultimate goal is recursive prediction of a nonstationary signal, differencing is the most practical approach. The reason is that it is trivially easy to invert the operation. To undo differencing, just sum. On the other hand, it is extremely difficult to find an analogue among frequency-domain techniques. Reliably undoing the effects of a high-pass filter is nearly impossible under the conditions likely to be met in practice. This leads to the following rule of thumb:

- *If a problematic nonstationary series is to be used for anything other than recursive prediction, consider band-splitting as an alternative to differencing.*

Finally, this discussion of removing nonstationarity concludes with a warning. We must not inadvertently discard important information. This is especially true when neural networks are involved. Nonstationarity can seriously impede the ability of a neural network to perform well, so such signals should not be the only input. Differences, even for stationary series, can be immensely useful. However, always consider the possibility that the wandering path of a nonstationary series can also convey useful information. If in doubt, feed the network with *both* the original and the differenced data.

ARIMA Models

No book on time series prediction would be complete without at least a token mention of ARIMA models, so here it is. When Box and Jenkins developed their ARMA time series model, they recognized that differencing falls into place in a most natural way. Trivial extension of their notation and theory allowed for the series to be implicitly differenced, treated as an ordinary ARMA model, then implicitly summed to return to the original domain. Their work is beautiful and elegant.

Since *integration* is (in a sense) another way of saying *summing*, they incorporated that term in the model's acronym. A time series that is described by differencing, ARMA modeling, then summing, is said to follow an ARIMA model. We have already described their notation ARMA(p, q) as meaning a model with p autoregressive and q moving average parameters. They extend this notation to include differencing by means of the acronym ARIMA(p, d, q), where d is the number of differencing operations that take place before the ARMA model is invoked.

This text takes a different approach to the same problem. Differencing is useful for many models other than the ARMA. Therefore, neither NPREDICT nor the developments in this text explicitly include ARIMA capability. Instead, the series is differenced as needed using explicit operations. Afterward, an ordinary ARMA model is fit. If forecasting is needed, it is done in the differenced domain, and then the series is summed to return to the original domain. The effect is exactly the same as if an explicit ARIMA model were used. Nothing more will be said on the topic.

Seasonal Differencing

How many times have we seen a series that resembles the one graphed in Figure 7.7? Periodic behavior is obvious. At the same time, the cyclic variation does not look like a simple sine wave. Also, the series is certainly not stationary. How do we cope with data like this? In

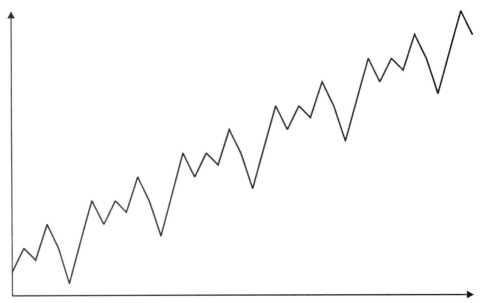

Figure 7.7 Data with a strong seasonal component.

many practical applications, *seasonal differencing* is the best way to handle this sort of data. This section explores the *whys* and *hows* of the operation.

Let's start by briefly discussing some alternatives to seasonal differencing when faced with data resembling that in Figure 7.7. One approach often considered by beginners is to remove the trend, then fit the series by means of a small set of sines and cosines. Superficially, this makes sense. The data is periodic, and sines and cosines are the penultimate periodic functions. Furthermore, a famous theorem states that we can exactly fit the data if we use enough of these functions. And therein lies the rub. It turns out that *enough* is usually a very large number. Always remember that parsimony is the golden word in times series analysis and prediction. In nearly all cases of practical interest, it is impossibly to fit observed data with a small number of sines and cosines. And even if we do, predictions based on this approach are almost always abominable. Don't even think about it.

Another approach, one that often works well, is based on the assumption that the series can be represented as the sum of two components. One of these is slow global variation, and the other is rapid local oscillation. A low-pass filter is used to isolate these two components, and they are predicted separately. Then the predictions are summed at the end. Special methods can be used to handle the periodic components in the local component. This method has passed the test of time, and it cannot be condemned. On the other hand, it is complicated and often involves decisions that are disturbingly subjective. Seasonal differencing can often be just as effective, yet vastly more straightforward. The beauty of seasonal differencing is that it is a generic preprocessing method that can be a precursor to nearly any modeling technique. Furthermore, reversal of seasonal differencing is almost trivially easy. This potent tool should be near the top of everyone's toolbox.

Even though the explanation of seasonal differencing given here does not depend on treating the series as the sum of slow and fast oscillations, it can help to visualize it this way. Consider this series: 1, 2, 3, 5, 2, 3, 5, 6, 7, 9, 6, 7, 9, 10, 11, 13, 10, 11, 7, 8, 9, 11, 8, 9. It is graphed in Figure 7.8.

For the sake of clarity, this figure has been distorted in one small but significant way. The goal is to illustrate a series consisting of a fast-moving repetitive component superimposed on a slowly varying baseline. This effect is most clearly seen when the baseline in each time period is horizontal. A more realistic representation would have the baseline sloping to follow the slow trend. In any case, this distinction is not critical to the discussion. In fact, it may be advantageous for one reason. It happens that in many real applications, the slow trend is strongly nonstationary. The baseline moves to one locality and stays there for a while. Then it moves on to a new location where it stays a while. This motion is not terribly unlike the horizontal baselines shown in the figure.

Motivation for seasonal differencing can be obtained by thinking of a periodic series as an AR process with a lag at the season length. Fireworks sales probably peak in the early summer. If we wanted to predict these sales in any given month, it would be reasonable to assume that a major component of a successful model would be the sales figure 12 months prior to the month being predicted. In other words, the prediction would be a function of several values, such as sales for recent months, perhaps recent ARMA shocks, and so on. One of the most important terms would almost undoubtedly be sales 12 months ago, since history repeats itself. We could fit an ARMA or neural model to the raw data, explicitly including the lag-12 term. That would probably work well. But why force the model to estimate the weight for the lag-12 term? We already know that it will, in all probability, be essentially 1.0. That weight at the seasonal lag can be implicitly included in the model by the simple expedient of differencing

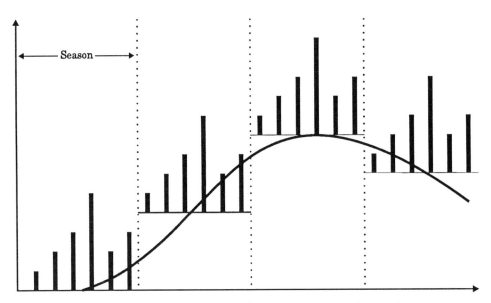

Figure 7.8 Slow and periodic components of a series.

at that lag. (A useful reader exercise is to prove this fact.) We thereby have one less term to worry about.

Another motivation for seasonal differencing is obtained by looking at the operation in the frequency domain. It should not be surprising that differencing a series at a fixed lag totally removes all frequency components at that period as well as all of its integer harmonics. In particular, the frequency response of a seasonal difference filter resembles Figure 7.9.

The strict removal of periodic components by seasonal differencing falls right into line with a cardinal rule of time series analysis: Whenever possible, use simple and reversible operations to remove potentially troublesome components before proceeding. This one primitive operation is able to remove a great deal of complexity from the variation in a time series by eliminating periodic behavior. The resulting series can almost definitely be handled more easily by additional models. Finally, seasonal summing is able to reverse the effect of seasonal differencing, providing easy return to the original problem domain.

It should be pointed out that if both adjacent-point and seasonal differencing is done to a series, the order of operation is immaterial. Exactly the same result is obtained, regardless of which type of differencing comes first.

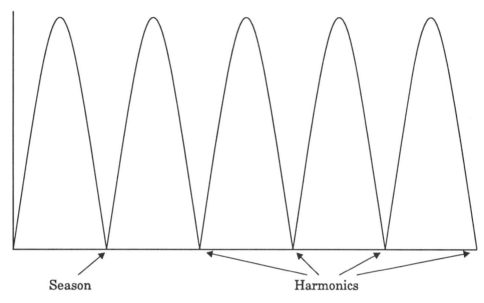

Figure 7.9 Frequency response of seasonal differencing.

Spotting the Need for Seasonal Differencing

In many cases, seasonal behavior that calls for seasonal differencing is obvious in a graph of the series. However, sometimes it is not. There are two potent tools for detecting seasonal behavior. The first is autocorrelation. A seasonal component will create a string of large correlations at integer multiples of the season's length. A strong MA component at a long lag (rare) generates just one large correlation. A strong AR component at a long lag generates a series of large correlations at integer multiples of the lag, just like a seasonal component, but they drop to zero relatively quickly. *The regularly spaced series of large correlations due to a seasonal component tends to stay large through very distant lags.*

A second tool for detecting seasonal components is the power spectrum. In most practical situations, a seasonal component will generate a spike in the power spectrum corresponding to the frequency of repetition. Additional spikes at integer multiples (harmonics) of the frequency are also common. However, using a power spectrum to detect and identify seasonal components can be risky. Look back at Figure 6.17 on page 231. This is the spectrum of a sample of white noise. It would be very easy to conclude that there is at least one seasonal component! Be warned that unsmoothed power spectrum values have a surprisingly high variance due to random sampling. Carefully study the material in Chapter 3 before making too much use of power spectra for assessing seasonality. A true seasonal component will usually stand out.

The power spectrum is also sensitive to phenomena that are related to seasonality but that may not quite fit the mold. For example, it sometimes happens that the current value of a time series is somewhat influenced by the value of the series at a relatively distant lag. This is not the same thing as true seasonality. This situation may sometimes be handled by seasonal differencing. However, it is more likely that an AR or MA component at that lag is the best approach. Figure 7.10 illustrates the power spectrum of a sampled AR series with a moderate weight at a lag of ten. We could possibly be fooled into thinking that there is a seasonal component in this series. Figure 7.11 is the spectrum of an AR series having a *negative* weight at a lag of ten. Note that troughs appear here at the places where peaks appeared in the case of a positive weight. Once again, the presence of strong local peaks might lead us to believe that there is a seasonal component. In this situation, erroneously acting on that conclusion would be disastrous!

What can we say in summary? First of all, lagged correlations are usually the most reliable seasonality indicators, at least for beginners. The very slowly diminishing correlations at very long lags are a dead giveaway. When the power spectrum is examined, do not be mislead by moderately large local peaks. These occur regularly as a product of random sampling error. They may also occur as a result of ARMA components at large lags. Only extremely prominent peaks should be taken seriously.

For advanced users, two tricks are often helpful. The first is to use the maximum entropy spectrum rather than the DFT spectrum. This is called a trick because choosing the order of the ME spectrum is difficult for beginners. But experience helps. The other trick is to consider the width of the first spectral peak, along with possible harmonics. A seasonal component will usually have a much narrower peak than an AR or MA term at a distant lag. Also, seasonal components typically have larger and more extensive harmonics. In fact, seasonal components often have harmonics that are stronger than the fundamental! These are not absolutely reliable indicators, but they are usually quite good.

Seasonal Differencing 267

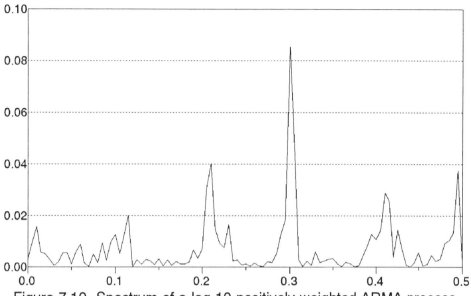

Figure 7.10 Spectrum of a lag-10 positively weighted ARMA process.

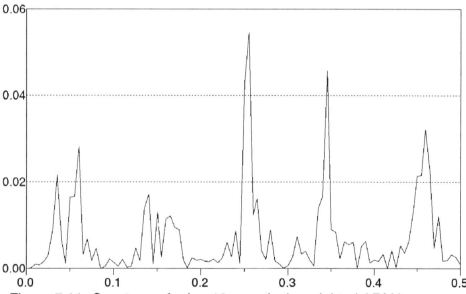

Figure 7.11 Spectrum of a lag-10 negatively weighted ARMA process.

Computational Considerations

Seasonal differencing is so straightforward that little needs to be said regarding implementation. However, for the sake of completeness, the method will be presented. Computer code can be found in the module SIGNAL.CPP on the accompanying disk.

The series resulting from seasonal differencing is shorter than the original series by as many points as the season length. If the season is 12 samples long, no difference can be computed until the 13th point from the beginning. What happens to the points in the first season? Responsible programmers keep them. The NPREDICT program uses a private vector in the Signal class. This way, the operation can be undone. To undo seasonal differencing, copy the preserved first season to the beginning of the output series. Compute subsequent values by adding the differenced value to the output value one season earlier. This exactly reverses the differencing operation. Moreover, it can be continued into the future for as long as desired. Thus, if future values of the differenced series were computed, these future values are easily transformed back into the original data domain.

An Example of Differencing

We conclude this chapter with a step-by-step example of differencing in action. Look at the series graphed in Figure 7.12. Two problems are immediately obvious: It is nonstationary, and it contains a strong periodic component. Usually, it is best (but not mandatory) to start by dealing with nonstationarity, as this problem often masks other problems. The adjacent-point differenced series is shown in Figure 7.13. The single differencing has apparently rendered the series stationary, so we need not consider a second adjacent-point differencing. This conclusion is supported by the autocorrelation of the differenced series, shown in Figure 7.14. Note the absence of the massive autocorrelations exemplified in Figure 7.5 on page 258. However, Figure 7.14 does bear the unmistakable signature of a seasonal component with a period of ten samples. We are thus led to perform seasonal differencing at a lag of ten.

At this point, we may be tempted to avoid the adjacent-point differencing and attack the original series with a seasonal difference only. Sometimes this works, but not often. The result of seasonally differencing the original series is shown in Figure 7.15. Note that it is moderately nonstationary, as also evidenced by its autocorrelation shown in Figure 7.16. The correct response is to perform both types of differencing.

The result of both adjacent-point and seasonal differencing is shown in Figure 7.17. This looks good. Its autocorrelation, shown in Figure 7.18, is a little suspicious. There are significant peaks at lags of 3, 7, and 9, as well as others that diminish slowly. In a real application, it would definitely be worthwhile to try to determine the cause of these peaks. For now, we will be content with having removed the most serious problems, trusting whatever model follows to handle the rest.

The first 175 points of the series graphed in Figure 7.17 are used as a training set for a probabilistic neural network. This network is then used to predict 15 future samples of the series. The training set, with the 15 predictions appended, is shown in Figure 7.19. Note the characteristic behavior of the predictions: They diminish in magnitude, and wild points do not appear. While this behavior is not absolutely universal, nearly all neural and ARMA models generate predictions having this property. It is a simple reflection of the principle that when information is scarce, the central tendency is the best guess for future values.

The last step is to take the predicted series shown in Figure 7.19 and undo the two differencing operations. The original series, with the final predictions appended, is graphed in Figure 7.20. Compare this with Figure 7.12. Not bad!

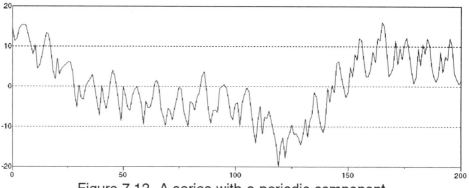
Figure 7.12 A series with a periodic component.

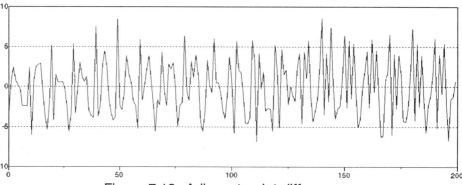
Figure 7.13 Adjacent point differences.

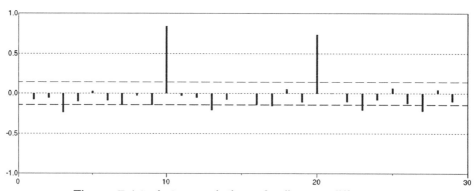
Figure 7.14 Autocorrelation of adjacent differences.

An Example of Differencing

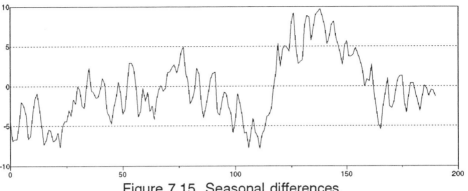

Figure 7.15 Seasonal differences.

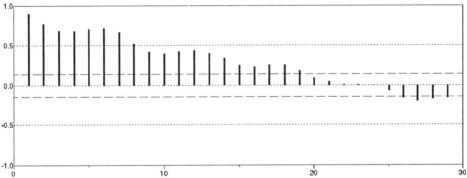

Figure 7.16 Autocorrelation of seasonal differences.

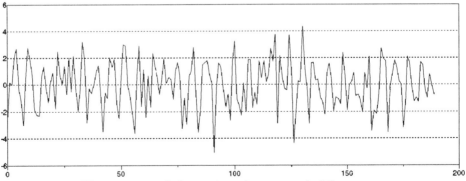

Figure 7.17 Adjacent and seasonal differences.

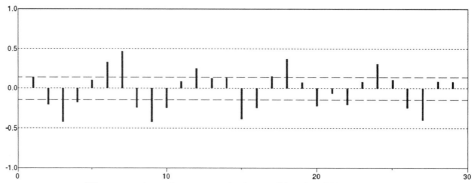
Figure 7.18 Autocorrelation of both differences.

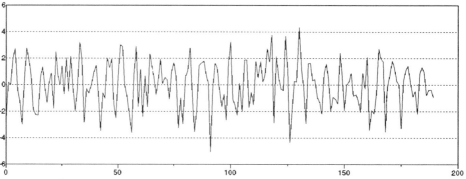
Figure 7.19 Both differences, extended by prediction.

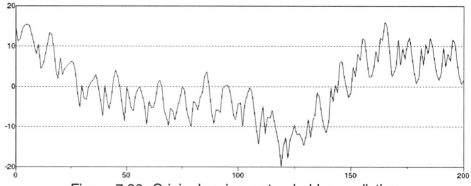
Figure 7.20 Original series, extended by prediction.

8

Robust Confidence Intervals

- Overview

- Collecting errors for neural networks

- Collecting errors for ARMA models

- Compensating for differencing

- Compensating for transformations

- Confidence in the confidence

One of the most insidious crimes committed by professionals is making predictions without simultaneously providing an idea of the reliability of these predictions. This crime is not entirely unpardonable, as there are many impediments to computing confidence figures for predictions. The neural network literature is almost totally devoid of information on the subject. Box and Jenkins have extensively studied the problem for ARMA and ARIMA models, and algorithms for univariate and bivariate confidence calculation are readily available. However, these algorithms rely heavily on assumptions of normality and independence that are not often justified in actual applications. Also, multivariate generalizations of their methods are difficult to find. A totally new approach is needed.

Overview

The confidence computations in this chapter are derived from information collected from a set of predictions within a known dataset. Ideally, this dataset should be different from the dataset on which the model was trained. However, experience indicates that reuse of the training set may be acceptable if care is taken to avoid overfitting. The prediction errors in the dataset are studied, and the assumption is made that the distribution of errors observed in the known dataset will be similar to the distribution expected when the model is put to use. Standard nonparametric statistical algorithms (which are, by definition, valid regardless of the error distribution) are used to formalize the generation of confidence intervals for subsequent predictions.

This method is not perfect. There are several problems. A fundamental limitation is availability of data for exemplifying prediction errors. Time-series applications are notoriously short on data. We usually want to use every bit that is available to train the model. If we do grudgingly hold back some data for testing, it is rarely enough to provide a clear picture of the situation. The jackknife and the bootstrap, our traditional refuges in times of scarcity, are nearly impossible to apply to time-series data. The implication is that it may be necessary to use the training set to simulate the error distribution. This is always a scary action, as the resulting defect in the computed confidence is the worst possible: It is excessively optimistic. The confidence intervals obtained by using the training data instead of an independent dataset are narrower than they should be, leading to undue trust in the predictions. On the other hand, the situation may not be too bad. If the number of trainable parameters in the model is small compared to the number of observations in the training set, the optimistic bias will be small. If the only alternative is holding back a few observations at the end of the series, the ensuing loss will not be worth the gain. As will be seen later, the error variance of confidence intervals computed from a small dataset can be so large that they are useless. Naturally, the best situation is when it is easy to collect two independent datasets, one for training and one for testing. But when this is impossible, the best bet is usually to pool the data into one set that is as large as possible and use it for both training and testing. As long as care is taken to keep the model small enough to avoid overfitting, the bias will probably be acceptable.

Another problem with the method given here is mostly theoretical, but it should be mentioned. A fundamental assumption of these techniques is that the error samples are mutually independent. Time series data, by its very nature, is clearly prone to violation of that assumption. However, the violation may not be as serious as might be thought at first. Recall that the whole purpose of modeling a series is to account for all (or for as much as possible) of its serial dependencies. A successful model will, by definition, transform a structured input series into an output error series that is (we hope) random noise. As long as the model is effective at this task, the prediction errors will be mutually independent. If they are not independent, we had better take another look at the model!

One more potential problem with the confidence intervals espoused here is that their robustness comes at a high price in power. If we are willing to assume a known distribution (such as the normal distribution) for the errors, confidence intervals computed under that assumption are optimal in the sense that they are as stable as any that could possibly be computed. Once we compute a confidence interval, we can be sure that it is good, at least relative to any confidence interval computed by other means. Many people fail to realize that the upper and lower limits of a confidence interval are random variables, subject to the vagaries of sampling error. Limits based on assumptions about the true distribution have minimal error variance themselves.

Alas, when we are not willing to make any assumptions about the error distribution (which should be a general rule!), the price we pay is relatively high error variance in the computed confidence limits. Therefore, an essential part of confidence computations is careful consideration of the possible error in the limits themselves. This will be the subject of the final section of this chapter.

Sampling the Prediction Errors

There is a straightforward way of sampling prediction errors at various distances into the future. Pretend that the known series ends at a point earlier than it really does end. Recursively predict from that point onward, using the known values to assess the error of each prediction. This provides one error measurement for each distance into the future. To collect more measurements for each distance, simply move the hypothetical end of the series. This is illustrated in Figure 8.1.

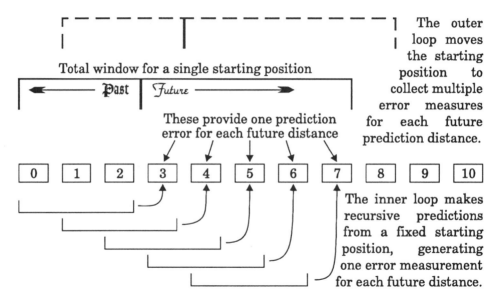

Figure 8.1 Generic collection of error measurements.

This figure focuses on a single collection sequence. Suppose the model employs three past values of the series to predict the current value. The collection process starts by assuming that only the first three values of the series are known. These are used to predict the fourth value (at offset 3). By comparing this predicted value with the known value, we find the error of this one prediction at a distance of one time slot into the future.

The next step is to use the known values at offsets of 1 and 2, along with the *just predicted* value at offset 3, to predict the value at offset 4. Comparing this to the true value there gives an error measurement for a distance of two time slots into the future. This recursive prediction is repeated as many times as desired, stopping when the maximum desired future prediction distance is reached. The key is that all model inputs beyond offset 2 are the predicted values, as opposed to the known values. This fairly simulates true lack of knowledge of these values, which will be the situation when the model is put to use.

Once the maximum desired prediction distance is reached, one error measurement for each distance has been collected. One sample is hardly enough to infer anything useful about the error distribution at a particular prediction distance, so more are needed. These are obtained by moving the prediction window. This is illustrated by the dotted lines at the top of Figure 8.1. By advancing the window one time slot at a time, repeating the set of recursive predictions within each position, multiple samples of the errors are obtained. The number that can be collected is limited by the length of the series, the number of historical values that serve as model inputs, and the maximum future distance whose error is to be assessed. The prediction window placement starts at the time slot determined by the maximum input lag, and it ends when the future distance reaches the end of the known series.

Collecting Errors for Neural Network Models

When a neural network is used for prediction, the algorithm shown in Figure 8.1 is inadequate in two ways. First, it does not clearly illustrate the fact that neural networks are able to make nonrecursive predictions at large lead-time jumps into the future. Figure 8.1 only deals with using historical values to predict the current value. Second, it does not address the issue of multiple input and output series, any of which may recursively affect future predictions at arbitrary leads. These complications are taken into account in Figure 8.2.

This figure illustrates the use of two signals to predict each other. The top signal (referred to as *Signal 1*) provides its three most recent values to the neural network as inputs. The bottom signal (*Signal 2*) provides only its most recent value as the network's fourth input. The maximum historical distance needed to make a prediction, referred to as maxlag in the code that appears later, is 3.

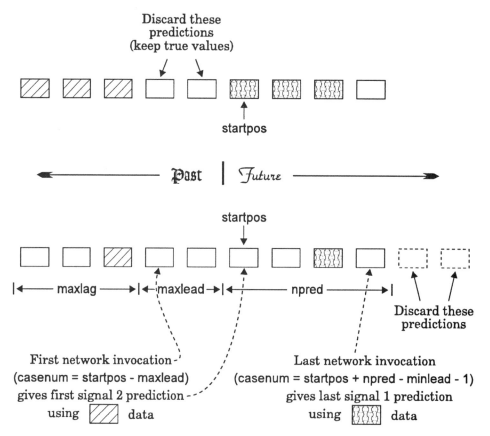

Figure 8.2 Collecting neural network error measurements.

The network has two outputs. The first output is the predicted current value of Signal 1. The second output is the predicted value of Signal 2 at a lead of two time slots into the future. The maximum lead time, called maxlead in the code that appears later, is thus equal to 2. The minimum lead time, minlead, is 0 because the Signal 1 prediction is of its current value.

Suppose our application will require us to make a total of four predictions into the future. These would be the current and three successive values of Signal 1, and leads of two through five for Signal 2. Therefore, we need to do this many recursive predictions for each window position (plus a few more at the start and end of the process). This quantity is called npred in the code that follows.

Two different time reference points are needed in this discussion. The first, startpos, is the point separating the future from the past. It is this point that moves in the outer loop, thereby generating multiple error measurements for each prediction distance. All values prior to startpos are treated as known, while values from startpos onward are treated as unknown.

The second time reference point is casenum. This is home base for the prediction currently being made. All input lags and output leads are relative to this point. In the code that appears later, both casenum and startpos are absolute subscripts (origin zero) in the series.

Figure 8.2 illustrates the error collection process for the startpos location shown. The first prediction will be for that time slot, since it contains (by definition) the first unknown values for both series. Since the network predicts Signal 2 at a lead of two, it is necessary to make this prediction from a reference point two time slots prior to startpos. In other words, casenum for the first prediction is equal to startpos − maxlead. This invocation of the neural network also produces a prediction for the value of Signal 1 in the casenum time slot. However, when the network is ultimately put to practical use, there will be no need for this prediction. This time slot's value, being prior to startpos (our defined start of the future), is still known. We acknowledge this fact by discarding the prediction. Only the startpos prediction of Signal 2 is used to compute an error measurement. The four inputs to the neural network for this prediction are identified in the figure by being filled with slanted lines.

We are now ready for the second prediction. Move casenum one slot to the right. This provides the second error measurement for Signal 2. As before, the prediction of the current value of Signal 1 is discarded, since this value is still in the known area. Note, by the way, that we have not yet had to resort to recursion. The second prediction of Signal 2 is based on known values. This is correct operation, since the same effect will be had when the network is put to real use. This is the beauty of predicting at a lead. As long as the user is willing to forgo the use of immediately prior information, long-distance predictions can be made without suffering the distortion of recursion.

The third invocation of the neural network is with casenum = startpos. This provides the third of four errors for Signal 2. It also provides the first of the four errors for Signal 1, since it is that signal's first foray into the future.

Three more predictions are needed in order to complete all four error measurements for Signal 1. Since the lead of Signal 2 gave it a head start on error collection, the last two predictions of that signal are discarded. As a general rule, the casenum of the last necessary prediction is startpos + npred − minlead − 1. In this example, minlead (the minimum lead across all outputs) is zero, since the current value (lead 0) of Signal 1 is predicted.

How many error measurements can be collected for each prediction distance? Figure 8.2 should make that clear. The smallest startpos is limited in two ways. Its first casenum is set back maxlead points. That value must, in turn, be preceded by maxlag slots in order for all network inputs to be defined. Thus, the first startpos is maxlag + maxlead. We then need npred future time slots, starting with startpos, for comparing predictions with known values. Thus, the total number of window positions is the length of the series, minus the sum of maxlag, maxlead, and npred, plus one.

It's time for some code. The following listing is a heavily edited version of the code in module NET_CONF.CPP on the accompanying disk. A line-by-line explanation follows.

```
offset = maxlag + maxlead ;             // First fully valid test is here
ncases = shortest - offset - npred + 1 ; // Number of fully valid tests

if (ncases < 20)                        // Arbitrary limit
    return 2 ;                          // Error flag

/*
    This array will be used to flag input sources.  In the case that a signal is an input
    only, its entry will be NULL.  But if the user is using an input as an output also, the
    output pointer will go here.
*/

inlist = (double **) malloc ( n_input_signals * sizeof(double*) ) ;

/*
    This is used for the network's input vector
*/

in_vector = (double *) malloc ( n_inputs * sizeof(double) ) ;

/*
    Allocate memory for the signals that will be predicted.
    For each startpos, we keep npred predictions.
*/
```

Sampling the Prediction Errors

```
        outputs = (double **) malloc ( n_outputs * sizeof(double *) ) ;

        for (i=0 ; i<n_outputs ; i++)                  // For each predicted signal
           outputs[i] = (double *) malloc ( npred * sizeof(double) ) ;

/*
   We usually want to predict (recursively) signals that also serve as input.
   This necessitates a little preparation.  Run through the input signals.
   For each that is also an output, flag that fact by storing its output pointer.
   Otherwise store a NULL as a flag to get the actual signal.
*/

        for (i=0 ; i<n_input_signals ; i++) {          // Check all input signals
           inlist[i] = NULL ;                          // Assume this is not recursive
           sigptr = Pointer to signal i                // This is the input signal
           for (j=0 ; j<n_outputs ; j++) {             // Check every output name
              if (! strcmp ( output_names[j] , sigptr->name )) { // This one?
                 inlist[i] = outputs[j] ;              // This input is here among outputs
                 break ;                               // So flag its recursive nature
                 }                                     // No need to keep looking
              }
           }

/*
   This is the main outer loop.  Each 'outputs' vector (npred long) starts at 'startpos'.
   For each new starting position, the first step is to set the recursive output vectors to
   their true values.  As the predictions are done, these will be the reference for
   keeping track of the errors.  Also, the true values will be overwritten with the
   predicted values for subsequent recursive use as inputs.  It is assumed that correct
   values are known up to (but not including) startpos.  Everything from startpos
   onward is a prediction.
*/

        for (startpos=offset ; startpos<ncases+offset ; startpos++) {

           for (j=0 ; j<n_outputs ; j++)               // Copy output signals to local vectors
              memcpy ( outputs[j] , pointer to signal j + startpos , npred * sizeof(double) ) ;

/*
   We are ready to do the predictions.  Everything before startpos is known.
   Start predicting at startpos-maxlead so we bag the longest lead.
```

Go as far as we need to get the last of the npred predictions for the signal with the shortest lead.
*/

```
endcase = startpos + npred - minlead ;

for (casenum=startpos-maxlead ; casenum<endcase ; casenum++) { // Predictions
    inptr = in_vector ;                          // Will build input vector here
    for (i=0 ; i<n_input_signals ; i++) {        // Pass through all inputs
        for (lag=min lag ; lag<=max lag ; lag++) { // All needed lags of this signal
            k = casenum - lag ;                  // This ordinal case is an input
            if (k < startpos)                    // If still in known cases
                *inptr++ = this signal [k] ;     // Use true value
            else if (inlist[i] != NULL)          // Beyond.  If recursive
                *inptr++ = (inlist[i])[k-startpos] ; // Use this prediction
            else                                 // Rare.  Dup final point.
                *inptr++ = this signal [startpos-1] ; // Usually causes distortion
        }
    }

    net->trial ( in_vector ) ;                   // Evaluate network for input

    for (j=0 ; j<n_outputs ; j++) {              // Check every output name
        lead = lead of this prediction ;         // Lead was recorded when trained
        k = casenum + lead - startpos ;          // It goes in this outvars slot
        if ((k >= 0)  &&  (k < npred)) {         // In the window of npred tests?
            err = (outputs[j])[k] - net->out[j] ; // Err of this prediction
            (outputs[j])[k] = net->out[j] ;      // Keep this prediction
        }
    }

} // For all npred cases
} // For all startpos starting positions
```

Let us now work through this code from beginning to end. Remember that although this listing is based on actual code from the module NET_CONF.CPP, it has been heavily edited to clarify operation.

The first step is to find the location of the first legal startpos. This offset from the beginning of the series is called offset, and it is computed according to the discussion following Figure 8.2. The number of cases is also computed according to that discussion. An arbitrary (and admittedly small) limit of 20 cases is imposed as a primitive sort of quality control.

There are two count variables associated with inputs. The first, n_input_signals, is the number of different signals used as inputs (and perhaps outputs also). The second, n_inputs, is the number of neural network inputs. This will be at least as large as n_input_signals, and it will be greater if more than one lag is used for any signal.

Two scratch arrays are allocated. One of these, inlist, is used to keep track of recursion. If input signal i is also an output, then element i of this array will contain a pointer to the dual-use input/output vector. If this input is not recursive, element i will be set to NULL. The other scratch vector, in_vector, is used to hold the input vector for the neural network.

Scratch storage is needed to hold the npred predictions for each output signal. It would probably be best to allocate this as a single n_outputs by npred matrix. However, we use a different method here. A vector of n_outputs pointers is allocated. Then, each output gets its own npred vector allocated. The only reason for this seemingly unusual choice is consistency. The program on which this code fragment is based does this operation many different places, and the method shown here is appropriate for those uses. Readers who wish to study this code in its original context have an easier job when output allocation follows a consistent pattern.

We now fill in the elements of inlist. For every input signal, initially assume that it is not recursive. But then pass through the names of all output signals. If any output name matches this input signal name, copy the pointer to that output's prediction array into the corresponding spot in inlist. This facilitates recursive use of predicted outputs as succeeding inputs.

The outer loop, which moves startpos across time, now begins. Recall that we allocated only enough output memory for the npred predictions. Therefore, the first position in each output array will correspond to position startpos in the actual series.

The first step in the outer loop is to copy the npred true values of the series, starting at position startpos, into the output vectors. These vectors serve a dual use. When each prediction is made, it will be compared to the corresponding value in the output vector, giving the error. Then, the predicted value will overwrite the true value in this array, where it will be used as an input for subsequent predictions.

The inner loop, which does recursive predictions for each fixed startpos, now begins. The range of this loop was discussed in conjunction with Figure 8.2. The variable inptr will be used to access successive elements of the neural network input vector as it is built.

The input signals are processed one at a time. Each may contain a range of user-specified lags. For a given lag, the actual time position of that case is casenum − lag. If this position is less than startpos, the defined boundary between the known past and the unknown future, this network input is fetched from the known input signal. Otherwise, the true value of this input is not known (at least for our purposes). Generally, it will be available as an output that has already been predicted. These recursive inputs were flagged by putting a pointer to the output array in inlist. Recall that the output arrays start at time startpos, not time 0. Therefore, the subscript in the array is k − startpos. It may be that a brave (and foolish) user has included one or more input signals that are not recursive. In this case, we arbitrarily choose to duplicate the last known point in the signal. Some readers may prefer to use the mean instead. In any event, the results will almost always be less than excellent, so it doesn't matter which choice is made.

The input vector for this casenum prediction has been set up. The neural network is invoked with this input, and its predicted outputs are now to be processed. For each output, its lead is retrieved. (A good place to store this information is in the trained network object.) The actual time corresponding to this output is casenum + lead. However, we must subtract startpos when this subscript is used in the output vectors.

Recall from Figure 8.2 that the first and last few predictions may need to be ignored due to discrepancies in the output leads. This is accomplished with the if statement. The error is computed by subtracting the predicted value (in net->out[j]) from the true value. Finally, the predicted value replaces the true value so that it will be used in all subsequent recursive predictions at this value of startpos.

Collecting Errors for ARMA Models

Error collection for ARMA models is considerably easier than for neural networks because there can be no prediction leads. All predictions are for the current value. There are some small complications, though. If the model contains any MA terms, a set of shocks must be dealt with. This set is the recursively computed shocks that may be needed for predictions into each hypothetical future. This is in addition to recursive input terms! And the situation is complicated even more by the lack of knowledge of the infinitely long history of the process, which induces an annoying disturbance into the beginning of each set of predictions. This appears to be an insurmountable theoretical problem, at least when we are limited to reasonably straightforward solutions to multivariate problems. Fortunately, unless the largest MA lag is a significant fraction of the length of the shortest series, this does not seem to induce too much error into the computations.

This discussion will focus on prediction using a single series. Extension to the multivariate case is trivial, so there is no reason to complicate the illustrations by introducing any more terms than the minimum needed to demonstrate the procedure. With this in mind, look at Figure 8.3, which shows a single series being predicted. The model contains AR terms up to a maximum lag of three, and MA terms up to a maximum lag of two. We want to collect error measurements for up to three future time slots.

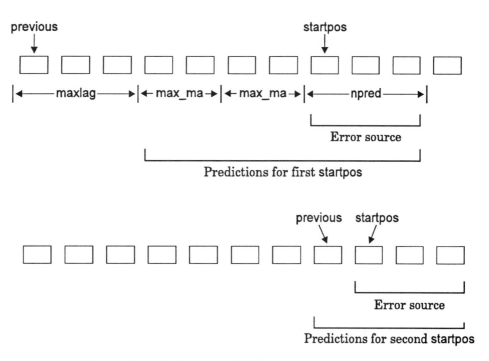

Figure 8.3 Collecting ARMA error measurements.

The first thing we need to do is to make an arbitrary decision. When should we start comparing predictions with their true values to generate error measurements? In other words, where should we place the leftmost position of *startpos*, the point at which the future starts and the past is left behind? Following the tradition set in Chapter 6, we know that we could do so as early as *maxlag*. At this point, all AR terms are defined. However, this is where the lack-of-history jolt hits full force, as the MA terms here are totally unknown and must be assumed zero. So collecting errors this soon would induce excess pessimism. If we skip as many more slots as there are MA terms, we could probably justify beginning the collection there. Once we hit that slot, all shocks are available, so the model is complete. Still, the shocks that enter into the model there are only the very first set, themselves based on incomplete versions of the model. If we skip this many slots again, the predictions from there on out will be based on MA terms derived from full predictions. This is admittedly an extremely heuristic argument. In fact, the distortion due to lack of history does not suddenly drop off at any magic spot. It smoothly tapers toward zero. Nonetheless, this is the choice made here. We skip the mandatory *maxlag* slots, then skip an additional 2 * *max_ma* slots, where *max_ma* is the maximum MA lag in the model. Conservative readers may want to skip a few more, while brave users facing a data shortage may want to start sooner. It's probably not that crucial anyway. This is illustrated in the top half of Figure 8.3.

We have decided to begin with *startpos* = *maxlag* + 2 * *max_ma*. Now we need to work up to that spot. The first prediction is at offset *maxlag*. This also provides a shock value for that time slot. Predictions continue until *startpos* is reached, with shocks being saved all along the way for MA use.

When the prediction at *startpos* is eventually made, things change in several ways. Up until this time, the predictions were used to compute the shocks, then discarded. Time slots prior to *startpos* are defined to be in the known past, so the true values are used for predictions. But from *startpos* onward, we are in the unknown future. Thus, the predictions are saved and used recursively for AR terms. Also, beginning with *startpos*, the computed shocks define the error measurements, but then they are discarded. Recall that, by definition, all shocks beyond the end of the known series are zero. Therefore, all MA terms from *startpos* onward are zero, so there is no need to keep the shocks. We continue predicting this way until we have collected error measurements for as far into the future as desired.

Sampling the Prediction Errors

At this point, we have in our possession exactly one error sample for each future prediction distance. It's time to advance *startpos* one slot to the right to collect a second sample of each. Look at the bottom half of Figure 8.3. The naive procedure would be to repeat the same steps that were done for the first value of *startpos*. That would certainly give correct results. But it would be tremendously wasteful. The only reason for starting at the beginning is to give the shocks time to stabilize. The computed shocks through the time slot just prior to the first *startpos* are still the same, so there is no need to compute them again. Only the shocks starting at the first *startpos* are invalid (having been set to zero). This inspires definition of a variable called previous that points to the leftmost time slot in need of initialization or calculation. When the algorithm starts, this is initialized to zero to start at the beginning. Thereafter, it is set to the previous value of *startpos*. In fact, if the model is pure AR, we can set previous equal to the value of *startpos* about to be used, as there are no shocks to compute.

Let's look at some code. This is an edited extract from the module ARMACONF.CPP on the accompanying disk. An explanation of this code follows.

```
offset = maxlag + 2 * max_ma ;            // First valid test here
ncases = shortest - offset - npred + 1 ;  // Number of valid tests

if (ncases < 20)                          // Arbitrary limit
    return 2 ;

/*
   Allocate memory for the signals that will be predicted.
*/

if (max_ma)
    outvars = (double *) MALLOC ( 2 * shortest * n_outputs * sizeof(double) ) ;
else
    outvars = (double *) MALLOC ( shortest * n_outputs * sizeof(double) ) ;

/*
   This is the outer loop that moves startpos across time,
   generating multiple error measurements at each prediction distance.
*/
```

```
      previous = 0 ;
      for (startpos=offset ; startpos<ncases+offset ; startpos++) {

/*
   Copy the output signals into the work area.
*/

         ivar = 0 ;                              // Index output variables
         for (i=0 ; i<nio ; i++) {               // Pass through all inputs/outputs
            if ('i' is an input)                 // If this is an input
               continue ;                        // Ignore it for now

            for (j=0 ; j<i ; j++) {              // Have we seen this output before?
               if (('j' is an output)  &&  ('i' and 'j' refer to the same signal))
                  break ;
               }

            if (i == j) { // If this is the first appearance of this output
               memcpy ( outvars + ivar * shortest + previous ,
                        This signal + previous ,
                        (startpos + npred - previous) * sizeof(double) ) ;
               ++ivar ;
               }
            }

/*
   This is the inner loop that does predictions based on a fixed startpos.
*/

         if (previous < maxlag)                  // This happens the first time only
            previous = maxlag ;                  // Avoid using undefined AR terms

         for(casenum=previous ; casenum<startpos+npred ; casenum++) {

            optr = outvars + casenum ;           // Point to this output
            shocks = optr + shortest * n_outputs ; // Shocks go here (if MA terms)

            for (iout=0 ; iout<n_outputs ; iout++) {  // For each predicted output

               wptr = Weight vector for this output ; // ARMA weights for this output
               pred = 0.0 ;                      // Will cumulate prediction here
```

```
for (i=0 ; i<nio ; i++) {                        // Check all ins and outs
   if ('i' is an input) {                        // If this is an input
      dptr = This signal ;                       // Its values are here
      for (j=min lag ; j<=max lag ; j++) {       // All lags of this input
         k = casenum - j ;                       // Position of this input case
         if (k < startpos)                       // Common: still within known
            pred += dptr[k] * *wptr++ ;          // So use actual value
         else if ((k = Index of output series) >= 0)   // Recursive?
            pred += optr[k*shortest-j] * *wptr++ ;     // Use predicted
         else  // Rare situation: beyond end of unpredicted signal
            pred += dptr[startpos-1] * *wptr++ ; // Duplicate last
      }
   }
   else {       // This is an output
      k = Index of this output vector ;          // Output vector for this MA term
      for (j=min lag ; j<=max lag ; j++) {
         if (! j)                                // Only lags are MA terms
            continue ;
         if (casenum - j >= maxlag)              // Shock available?
            pred += shocks[k*shortest-j] * *wptr++ ; // Use it
         else                                    // Shocks before the first prediction
            ++wptr ;                             // Are assumed zero
      }
   } // This is an output
} // For all inputs_outputs

if (! fixed)                                     // If constant not fixed at zero
   pred += *wptr ;                               // Last weight is the constant offset
if (pred > 1.e30)                                // Prevent runaway
   pred = 1.e30 ;
if (pred < -1.e30)
   pred = -1.e30 ;
err = optr[iout*shortest] - pred ;               // Error of this prediction
if (casenum >= startpos) {                       // If we are in the future
   optr[iout*shortest] = pred ;                  // Record for recursion
   err = 0.0 ;                                   // Future shocks are zero
}
if (max_ma)                                      // If there are MA terms
   shocks[iout*shortest] = err ;                 // Must save the shocks for MA
} // For all predicted output signals
} // For casenum
```

```
        previous = startpos ;              // Do not recompute same stuff
        if (! max_ma)                      // If MA terms, need prev shock
           ++previous ;                    // But if pure AR, no need

    } // For startpos
```

The first step is to compute offset (the first startpos) and the number of cases that will go into the error set. See Figure 8.3 on page 285 for an explanation of these formulas. If the number of cases is too small for reliable error estimation, quit with an error flag set. This limit is admittedly arbitrary and may be changed by the reader if desired. The length of the shortest input or output series is shortest.

A matrix is needed for scratch storage. Fancy coding could reduce the amount allocated by using the memory cyclically. However, this would slow computation and add horrendous complexity to the algorithm. Therefore, we allocate for the entire extent of all output signals. If there are any MA terms, we allocate twice as much. The shocks will be stored in the second half of this array.

The outer loop moves startpos across time, gathering multiple sets of prediction errors for each future distance. Remember that our definition of startpos is that it is the beginning of the unknown future. Everything before it is known, and everything from it onward is unknown. As was discussed earlier, previous is initialized to zero so the prediction loop starts at the beginning, reducing the impact of the lack of infinite history.

For each new value of startpos, the first step is to copy the known output signals to the scratch array. It is legal for the same output signal to appear among the nio input/output specifications more than once, perhaps as several disjoint MA lag ranges. We must act on only the first appearance of each output. The copying starts with the previous time slot and extends through the npred predictions that start at startpos. Note that outvars is treated as an n_outputs by shortest matrix, with the signal points changing fastest.

Normally, we begin the predictions at previous. This was discussed in conjunction with Figure 8.3. However, the first time we execute the outer loop, previous is zero. If there are any AR terms, they must be skipped in order to guarantee that all model inputs are defined. The inner loop, which computes all future predictions for the fixed value of startpos, starts at previous and extends through the npred predictions that start at startpos.

Sampling the Prediction Errors

Each prediction will be for all n_outputs output series. Initialize optr to point to this case of the first output. Similarly initialize shocks to point to the corresponding shock. Then loop through all predicted outputs.

For each output, set wptr to point to the ARMA weight set for that output. The prediction will be cumulated in pred. Loop through all nio inputs and outputs. Each is either an input (AR term) or an output (possible MA term). Act accordingly.

If it is an input, get a pointer to the start of the series and pass through all of its user-specified lags. For each individual input term, the index in the input series is the current location (casenum) minus the lag. There are three possibilities for this input term. If the index is prior to startpos, it is still in known territory and is fetched from the series. Otherwise, it is in the unknown future. In most practical applications, this series is a combined input/output used recursively. This fact is flagged by k being nonnegative, in which case k is the index of the output series simultaneously used as input. Get that previously predicted value. Recall that optr is already offset to point to the current time slot, so we need only to subtract the lag. The final possibility is that this input is not also an output. In this unusual situation, we make the arbitrary decision to replicate the last point in the series. It may be preferable to use the mean of the series instead. Either action introduces so much distortion that this should not be done unless the situation absolutely demands it.

This input/output item may refer to an output series, possibly functioning as an MA term. Pure outputs (those having no MA use) are flagged by having a lag of zero, something that is obviously impossible for an MA term. The time index of each lagged MA term is the current time (casenum) minus the lag. Recall that no shocks can be computed for the first maxlag cases in the series, so check for this situation, which happens near the start of the algorithm. If we are past this spot, incorporate the shock into the prediction.

After all AR and MA terms have been cumulated into the prediction, the last step is to add in the constant offset, if there is one. It is sometimes possible (though hopefully rare) for a weight set to be nonstationary. When this happens, the prediction may blow up. The effect compounds rapidly if it is not kept under control. Some hardware responds by doing nasty things like unceremoniously bouncing the offender back to the operating system. To avoid this indignity, the prediction is kept modestly constrained.

Finally, the error is computed by subtracting the prediction from the true value that was placed in the scratch array at the start of the startpos loop. If we are into the unknown future, the prediction overwrites the true value for possible recursive use as an AR input. Also, all future shocks are zero. Save the shock if there are any MA terms.

The last action in the startpos loop is to set previous to minimize computation during the next iteration. If there are any MA terms, the next set of predictions must start at the startpos just done to get the shock for that time slot. But if the model is pure AR, we can jump right in. Remember that we do not need to be concerned with the fact that the true values in the outvars array have been overwritten with predictions. All AR terms prior to startpos are fetched directly from the original series.

Compensating for Differencing and Transformations

So far, this chapter has only dealt with confidence intervals for predictions as they emerge from the model. In practice, we almost never care about the predicted series themselves. The reason is that we have almost certainly modified the original data before training the model. After the model has spit out its predictions of the modified data, we need to undo the modifications to take the predictions back to the original application domain. Confidence intervals for the modified data are useless. What we want is confidence intervals in the original application domain. This subject will now be discussed.

Let us begin with a generic discussion of how this problem is solved. Think about a single trial prediction set. In the language of the preceding section, this would be the set of npred future predictions based on a single startpos reference point. Don't be fooled into believing that we compute the errors, then somehow transform the errors. This does not work because modifications generally affect points in different ways depending on their value and context. Instead, what we need is to undo the modifications on both the true and the predicted values, treating each short series as an individual. Then compute the errors from the inversely modified data.

There are three main families of modifications that are popular. The easiest to handle are pointwise linear operations, in which each point in the series is individually modified by a deterministic linear operation. The most common examples of this sort of modification are centering and detrending. These modifications do not need any special treatment because they affect the true and the predicted values in exactly the same way. To undo centering, add the same constant to both series. When the series are subtracted to compute the error, the constant cancels. To undo detrending, we do exactly the same thing, except that the constant that is added depends on the time slot. Nevertheless, it is the same constant that is added to both series in each time slot, so once again it cancels when the error is computed. Compensation is trivially easy.

Multiplicative operations are almost as easy. When a constant multiplies both series, the associative law dictates that the error is multiplied by that same constant. So to compensate for multiplicative operations, simply multiply the basic confidence interval limits by the constant. There is no need to process the true and predicted series separately. No more will be said on the subject of pointwise linear modifications.

Pointwise *nonlinear* transformations are a little trickier in that separate processing is needed for both series. But the operation itself is trivial. Just apply the inverse transformation to the series and compute the errors. This is illustrated in Figure 8.4, which also demonstrates the vital need for this approach.

This figure shows three future predictions of data that was transformed by taking its log before the model was trained. The top graph shows the true values at each of the three future times as plain rectangles, and the predicted values as filled rectangles. Notice that at all three times, the magnitude of the prediction error is the same: 2.

The middle graph shows these same three pairs of points after exponentiation to invert the prior log transform. This data is now in the original application domain. Observe how nonlinear the effect is. The new values are extremely impacted by the level of the old values.

The bottom graph is the error in the application domain: the difference between the exponentiated true and predicted values. The most important point to get out of this discussion is the fact that although the directly predicted series had exactly the same error for all three time slots, the errors after compensation for the log transform are dramatically different. This should clearly indicate why we cannot simply compute the raw error and transform it. The true and predicted series must each be individually transported back to the application domain before errors are computed.

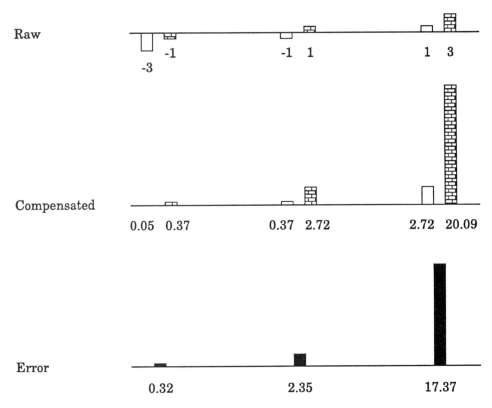

Figure 8.4 Compensating for a log transform.

The most complicated transformations are those in which the context of the data affects the results. In other words, the transformation is not pointwise. The most common such transformation is differencing, so that will be our focus. Look at Figure 8.5.

Once again, the top graph shows the true (plain rectangle) and predicted (filled rectangle) values for each of three predictions. These are the outputs of a model trained on differenced data.

The middle graph shows the three pairs of values after compensating for differencing. This compensation is done by working from left to right, adding each point to the previous point. To simplify the graph, assume that the baseline is equal to the value of the last known point in the series. Thus, to invert the differencing operation for the first point, simply add its value to the previous point (which is the baseline). As can be seen, inverting a differencing operation has no effect on the first future prediction. The same baseline is used for both the true and

Compensating for Differencing and Transformations

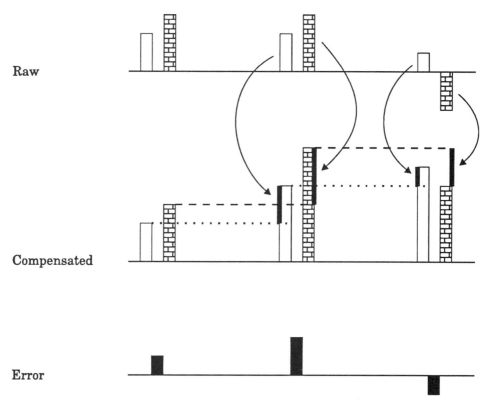

Figure 8.5 Compensating for differencing.

the predicted series, so the baselines cancel when they are subtracted to compute the error.

The second pair of points is computed by adding their values to the previous pair. This is illustrated in the figure by stacking the rectangles as shown. The third point is handled in a similar fashion. The error, graphed at the bottom of the figure, is the difference between the summed series.

To compensate for multiple differencing, repeat the summing operation the same number of times for both series. Compute the error after all repetitions have been done. To compensate for seasonal differencing, do exactly the same type of summing, except that instead of summing adjacent points, sum points across a lag equal to the period of the season.

Computer code for these operations is very simple on a superficial level, yet relatively difficult in practice. By way of introduction, here is a heavily edited section of the module NP_CONF.CPP on the accompanying disk. However, please note that *this code is incorrect as it stands*. This short fragment is meant to illustrate the basic operations while omitting the details for the sake of clarity. The serious problem with this exact code will be discussed shortly, along with a more complex listing that corrects the problem.

```
kptr = Pointer to the known values ;
pptr = Pointer to the predicted values ;
ncc = Number of modification compensations
ccs = Pointer to them

for (icc=ncc-1 ; icc>=0 ; icc--) {   // Process all confidence compensations
   ccptr = ccs+icc ;                  // Point to this one
   switch (ccptr->which) {            // Which operation?
      case CCdifference:              // Difference
         for (ipass=0 ; ipass<Number of differences to undo ; ipass++) {
            for (ipred=1 ; ipred<npred ; ipred++) {
               kptr[ipred] += kptr[ipred-1] ;
               pptr[ipred] += pptr[ipred-1] ;
               }
            }
         break ;
      case CCseasonal:                // Seasonal difference
         period = Period of the season
         for (ipred=period ; ipred<npred ; ipred++) {
            kptr[ipred] += kptr[ipred-period] ;
            pptr[ipred] += pptr[ipred-period] ;
            }
         break ;
      case CClog:                     // Log transformation
         for (ipred=0 ; ipred<npred ; ipred++) {
            kptr[ipred] = exp ( kptr[ipred] ) ;
            pptr[ipred] = exp ( pptr[ipred] ) ;
            }
         break ;
      } // Switch on operation
   } // For all compensations
for (ipred=0 ; ipred<npred ; ipred++)
   x [ ncases * (npred * ivar + ipred) + icase ] = *kptr++ - *pptr++ ;
}
```

The first thing to notice about this code is that it processes the ncc compensations in the *reverse* order from which they are stored in the ccs array. This must be done. For example, economists will often take the log of a series to move it from a multiplicative domain to an additive domain, then difference the series to induce stationarity. When the confidence intervals are compensated for this pair of operations, it is vital that the series be summed first, then exponentiated. Reversing that order produces an entirely different result!

Also notice that the seasonal difference compensation is practically identical to the adjacent-point difference compensation. In fact, the latter is a special case of the former in which the period is one.

Finally, look at the last line which shows the error being computed as the difference between the final known and predicted values. The fierce-looking subscript on the x (error) array illustrates an effective storage method. Treat it as a three-dimensional array in which the ncases cases change fastest, the npred future predictions change next, and the output series (represented by ivar) changes slowest. This facilitates handling each output series separately by passing a single pointer to its entire error collection. It also facilitates sorting of each prediction-distance error set by being able to pass to the sorting subroutine a pointer to a contiguous array of all ncases cases.

What's wrong with this code? Throughout this discussion it has been implied that each set of npred predictions is an entity unto itself. Figure 8.5 uses a baseline equal to the previous value of the series, implying that any value (including zero) is acceptable. The code just listed holds the first value constant during difference compensation, assuming a baseline of zero. Modification starts with ipred=1. This is fine as long as only one modification is done, or if all modifications are linear. However, this isolationist policy falls apart as soon as a nonlinear operation follows a differencing compensation. Consider this simple example: Suppose the user originally took logs, then did a single differencing. To reverse these modifications, we integrate, then exponentiate. If we arbitrarily take the integration baseline to be zero, which we have been doing, the first totally modified value will be equal to the exponential of the first raw value. But if we use the true value of the baseline, as determined by exactly undoing the original differencing, the first totally modified value will be the exponential of the sum of the first raw point plus the baseline. Look back at Figure 8.4. It is obvious that these two methods of compensating for the user's pair of modifications will, in general, produce completely different results.

What do we infer from this? The simple expedient of treating each set of npred predictions in isolation, assuming a baseline of zero, is fine as long as we are doing just one operation, or if all operations are linear. Otherwise, we must exactly undo the user's operations right through the point prior to the first prediction. For differencing, this implies starting at the beginning of the series. Naturally, it would be a terrible waste to redo all that computation each time we process another set of npred predictions. The correct method is to undo each operation across the entire series, saving all intermediate results. Then, when the sets of predictions are computed, the prior data can be referenced as needed. The logic gets a little hairy, but it is feasible. Here is an edited fragment of the code in NP_CONF.CPP. This code computes each stage of compensation, saving all intermediate results in a gigantic array.

```
n = Length of original signal ;
prevdata = The original signal ;
dptr = Work array where results will be saved

for (icc=ncc-1 ; icc>=0 ; icc--) {    // Check all confidence compensations
   ccptr = ccs+icc ;                  // Point to this one
   switch (ccptr->which) {

      case CCcenter:
         memcpy ( dptr , prevdata , n * sizeof(double) ) ;
         for (i=0 ; i<n ; i++)
            dptr[i] += Original center ;
         prevdata = dptr ;
         dptr += n ;
         break ;

      case CCdifference:
         passes = Number of differences previously done ;
         for (pass=0 ; pass<passes ; pass++) {
            memcpy ( dptr , prevdata , n * sizeof(double) ) ;
            sum = Original first point for this pass ;
            for (i=0 ; i<n ; i++) {
               temp = dptr[i] ;
               dptr[i] = sum ;
               sum += temp ;
            }
```

```
        dptr[i] = sum ;
        ++n ;
        prevdata = dptr ;
        dptr += n ;
        }
      break ;

    case CCseasonal:
      period = Period of this seasonal differencing ;
      memcpy ( dptr , prevdata , n * sizeof(double) ) ;
      for (p=0 ; p<period ; p++) {
        sum = Original first point at this offset into period ;
        for (i=p ; i<n ; i+=period ) {
          temp = dptr[i] ;
          dptr[i] = sum ;
          sum += temp ;
          }
        dptr[i] = sum ;
        }
      n += period ;
      prevdata = dptr ;
      dptr += n ;
      break ;

    case CClog:
      memcpy ( dptr , prevdata , n * sizeof(double) ) ;
      for (i=0 ; i<n ; i++)
        dptr[i] = exp ( dptr[i] ) ;
      prevdata = dptr ;
      dptr += n ;
      break ;
    }
  }
```

For the sake of saving space and increasing clarity, only a few representative operations are covered in the code above. The most important aspect for the reader is the straightforward method of storing all intermediate results. The pointer prevdata is initialized to the original signal, and n is its length. As each modification is about to be undone, the previous series is copied into what will be the new current series. The operation is undone in place, and prevdata is updated to point to the result. The length of the new series is computed, and the pointer to the storage area is advanced accordingly.

Once all results have been saved across the entire signal, we can process individual sets of npred predictions. Here is an example:

```
n = 0 ;                                              // Offset due to undoings
dptr = The previously computed intermediate results
pptr = The set of npred predictions start here
for (icc=ncc-1 ; icc>=0 ; icc--) {  // Process all confidence compensations
  ccptr = ccs+icc ;                 // Point to this one
  switch (ccptr->which) {

    case CCcenter:
      for (ipred=0 ; ipred<npred ; ipred++)
        pptr[ipred] += Original center ;
      kptr = dptr ;                         // Keep track of final known values
      dptr += Original signal length + n ;  // Next undoing record
      break ;

    case CCdifference:
      npasses = Number of previous differences ;
      for (ipass=0 ; ipass<npasses ; ipass++) {
        ++n ;
        pptr[0] += dptr[casenum+n-1] ;
        for (ipred=1 ; ipred<npred ; ipred++)
          pptr[ipred] += pptr[ipred-1] ;
        kptr = dptr ;                         // Keep track of known values
        dptr += Original signal length + n ;  // Next undoing record
      }
      break ;

    case CCseasonal:
      period = Period of this seasonal difference ;
      n += period ;
      for (ipred=0 ; (ipred<period) && (ipred<npred) ; ipred++)
        pptr[ipred] += dptr[casenum+n+ipred-period] ;
      for (ipred=period ; ipred<np ; ipred++)
        pptr[ipred] += pptr[ipred-period] ;
      kptr = dptr ;                         // Keep track of known values
      dptr += Original signal length + n ;  // Next undoing record
      break ;
```

```
      case CClog:
        for (ipred=0 ; ipred<npred ; ipred++)
          pptr[ipred] = exp ( pptr[ipred] ) ;
        kptr = dptr ;                            // Keep track of known values
        dptr += Original signal length + n ;     // Next undoing record
        break ;
      } // Switch on operations
    } // For all confidence compensations

  for (ipred=0 ; ipred<npred ; ipred++)
    x [ ncases * (npred * ivar + ipred) + casenum - offset ] =
        kptr[casenum+n+ipred] - pptr[ipred] ;
```

Again, only a few representative operations are shown here. However, they should suffice to illustrate the basic principles. We start with dptr pointing to the array of intermediate results computed with the code fragment shown several pages ago. The npred predictions start at pptr. We use n to keep track of expansion of the signal length due to these operations.

Centering is straightforward, so this discussion starts with differencing. The baseline for this operation, the true value just prior to the first prediction, is accessed with the subscript casenum+n-1. Let's analyze this. We are at an offset of casenum relative to the signal seen by the ARMA model or neural network. The undoings that have been done so far, including the undifferencing in progress, have increased the length of this signal by a total of n points. These new points have been added to the *front* of the signal, thus pushing everything ahead. (Review Chapter 7 if this is not clear.) So we need to add n to keep the time consistent. Finally, we subtract one to refer to the previous point. Note that the code could be made trivially more efficient here by slightly rearranging these lines. The price would be less clarity.

Seasonal differencing is only a bit more complex. For the predicted points within the first period, we need to get the true values for use as baselines. These are accessed with the subscript casenum+n+ipred-period, which should make sense if the reader understands the previous paragraph. For points past the initial period, which may or may not be present, we can integrate in place.

The final step, computing the error of each prediction, involves the modified predictions in pptr and the known values in kptr. Observe that as operations were performed, kptr was kept pointing at the most recent true values. The same subscripting principles apply here as in the individual operations, with one small exception. The offset term compensates for the fact that we started *startpos* at *offset* cases into the signal set. Thus, there is no need to reserve storage space for those unused time slots.

From Errors to Confidence Intervals

Now that we have the means of collecting a set of error measurements for each prediction distance, what do we do with them? There are an infinite number of ways they could be used to help judge the quality of predictions. The most straightforward approach is to use them to compute the mean squared error at each prediction distance. Unless the error distribution is very heavy-tailed, this knowledge alone is quite informative. If we are willing to really go out on a limb and assume that the errors follow a normal distribution, we may use the mean squared error to compute actual confidence intervals. Or we may take an intermediate approach. We may assume the errors follow some more general distribution defined by several parameters, then estimate the parameters to compute confidences. These methods are great if we are willing to accept whatever baggage they bring along in the form of assumptions. But if we wisely want to avoid building our confidence assessments on unstable ground, we can still compute decent intervals.

There is a price to pay: The robust confidence intervals presented in this chapter exhibit greater variation than intervals based on assumptions about the error distribution. If we were to collect a second dataset and use it to compute a new set of confidence intervals, this new set would generally be different from the first set. How different? We don't know. It is at the mercy of chance. But what we do know is that on average, the difference will be greater than for intervals based on assumptions. Is this a price worth paying? In the rare event that we are sure the necessary assumptions are satisfied, the price is probably excessive. But in the vast majority of practical situations, there is no contest. Go with the robust intervals.

Note that these confidence intervals have been described using the deliberately vague term *robust*. The terms *distribution-free* and *nonparametric* have been avoided. The reason is that, in the strictest mathematical sense, the sampling distribution of these intervals does depend to some degree on the error distribution. There does not appear to be any effective method for computing totally distribution-free confidence intervals for prediction errors. On the other hand, the intervals given here come close. As the sample size grows, these computed intervals converge to the true intervals, regardless of the error distribution. Furthermore, we will demonstrate later that fully distribution-free confidence intervals *for the confidence intervals* can be computed. If this all sounds confusing, don't worry. It is really very simple, and all will become clear soon.

Let's start with intuition. Consider a particular prediction distance, say the first prediction after the known values end. We have a set of n cases at this distance. This collection of n error measurements was computed by doing (hopefully) realistic predictions across a known dataset, comparing each predicted value with the known correct value. What can we infer about upcoming prediction errors from this set of observed historical errors? The most obvious inference is that errors of a given size will tend to occur with about the same frequency now and later. If in our historical collection we observe that about one-quarter of the errors exceed 2.3, it would not be unreasonable to infer that when the model is eventually put to practical use, its errors will exceed 2.3 about one-quarter of the time. Naturally, this inference is subject to the vagaries of random sampling error, with the quality dependent on the size of the historical dataset. This will be quantified later. But for now, accept this as the motivation for what follows.

It's time for some notation. Sort the collection of n error measurements in ascending order. Let $x_{(1)}$ denote the smallest, $x_{(2)}$ the second smallest, and so forth, with $x_{(n)}$ being the largest. (Preserve the signs of the errors. Do not use just the magnitude.) These quantities are often called the *order statistics* of the sample. Define the *sample distribution function* of the errors as shown in Equation (8.1). In other words, $S_n(x)$ is the fraction of the collection less than or equal to x.

$$S_n(x) = \begin{cases} 0, & x < x_{(1)} \\ r/n, & x_{(r)} \leq x < x_{(r+1)} \\ 1, & x_{(n)} \leq x \end{cases} \quad (8.1)$$

If we are satisfied that the dataset from which the errors were computed is representative of the data that will be encountered in practice, and if n is large enough that $S_n(x)$ can be taken to be close to $F(x)$, the true error distribution, it is trivial to compute confidence intervals for upcoming prediction errors. Simply keep as much of the interior of $S_n(x)$ as is desired, and use the limits of this truncated collection to define the confidence interval. In practice, we usually want the interval to be symmetric in probability, even though this implies that it will not generally be symmetric in x. For example, suppose we want a 90 percent confidence interval. We would usually want to choose the lower and upper limits, x_l and x_u respectively, to be such that $S_n(x_l) = 0.05$ and $S_n(x_u) = 0.95$, splitting the 0.1 exterior probability equally below and above the confidence limits.

There is a subtle complication. $S_n(x)$ does not vary smoothly. It takes discrete jumps at each order statistic. As a result, there is an infinite number of x values for which the sample distribution function attains any fixed level. Also, the definition of the sample distribution function in Equation (8.1) has an annoying asymmetry. The lower end of the function employs strict inequality, while the upper end is not strict. This is necessary for the function to be well defined, and it poses no problem when dealing with asymptotic results. But when we must deal with algorithmic details and finite sample sizes, this asymmetry is a nuisance. The traditional approach is to achieve uniqueness by choosing an exact order statistic, and to achieve balance by discarding an equal number of samples from both ends. This violates the strict definition, but it provides results that are more appealing on both an intuitive and a practical level.

How many extreme cases do we discard? The easy answer is np, where p is the fraction of probability in each tail. For very large samples and moderate values of p, this is fine. But for small samples, or for small values of p, a little more thought is needed. First of all, np will rarely be an integer. Do we round it? Truncate it? And into which group do we toss the selected order statistic, the outliers or the interior confidence interval? Consider this example: Suppose we have $n = 10$ cases, and we want an 80 percent confidence interval. The upper limit must be such that 10 percent of the errors exceed it, and the lower limit must be such that 10 percent of the errors are less than it (keeping signs, remember). Think about the upper limit. Suppose the largest error is 2.3, and the second-largest is 2.1. Do we compute $np = 1$ and discard the largest? This gives us an upper limit of 2.1, which is somewhat justifiable. We can say that 90 percent of the errors are 2.1 or less. Only 10 percent of the errors exceed 2.1. But wait. We could just as well say that 10 percent of the errors are 2.3 or more, with 90 percent of the errors being less than 2.3! Which is it? It all depends on how much risk we like in our life. Statistical conservatives will definitely choose the latter course. It provides wider confidence intervals than the former, but if one must commit an error, this sort of error is preferable to computing excessively narrow (optimistic) intervals. The method used in the NPREDICT program supplied with this text is to compute $np-1$, truncating any fraction. This determines the number of extreme cases to discard.

Figure 8.6 illustrates this procedure. It uses the unrealistically small sample size of $n = 10$ and provides confidence intervals for five future predictions. This is a 60 percent confidence interval, so $p = 0.2$ and $np-1 = 1$. For the first future prediction, the errors range from -0.3 to 0.4. Each of these extremes is discarded. We see that 20 percent (two cases) of the collection equals or exceeds 0.2, and another 20 percent of the collection is less than or equal to -0.2. In other words, 60 percent of the collection lies strictly within these bounds.

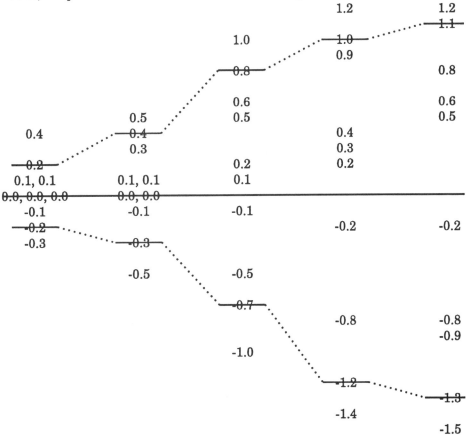

Figure 8.6 Confidence intervals from observed fractiles.

The issue of which bin gets the boundary value is important only when np is so small that we are operating far out in the tails of the sample. As will be seen in the next section, confidence intervals based on $S_n(x)$ are reliable only when a significant number of extreme cases are discarded. When this is the situation, moving one way or the other by one slot doesn't really affect the outcome very much. However, since we will sometimes need to work with small samples and take our chances, the conservative approach is recommended. The only thing worse than computing confidence intervals that are too wide is computing intervals that are too narrow.

Confidence in the Confidence

Anyone who is conscientious enough to compute confidence intervals for predictions will want to go one step further and investigate the reliability of those confidence intervals. They don't do much good if they are wildly wrong! This section discusses the important topic of confidence interval quality.

Before digging into this topic, it is important to review some fundamental assumptions that are germane to *all* confidence computations. It must be emphasized that these assumptions are not peculiar to the robust methods of this chapter. They apply to all methods. We can get away with relaxing most of the dangerous assumptions that plague many traditional methods, but there are some that always apply. They are the following:

- The test set (the dataset used to compute the error samples) must thoroughly and fairly represent the population that will be encountered in practice. The computed confidence intervals will apply only to the population implied by the test set.

- The test set should ideally be different from the dataset used to train the model. The reason is that the training process capitalizes to some degree on idiosyncrasies of the data, giving the model an advantage on this dataset that it will not have on later data. The result is excess optimism (overly narrow confidence intervals). However, if the number of free parameters is small compared to the number of cases, this may not be a severe problem. Sometimes you have no choice.

- The error samples are assumed to be independent and identically distributed. If, for example, the degree of error of a prediction depends on the value of that prediction, results will be skewed in unpredictable ways. We will return to this at the end of the chapter. If there is a tendency for groups of contiguous errors to run high or low, the effective number of cases will be reduced. Although on average the computed confidence intervals will not be directly affected by this problem, they will be less stable. The assessment methods that appear in this section may overestimate the confidence in the confidence. If the prediction model is well designed, this will rarely be a serious problem.

That's the bad news. Now for some good news. Although the robust confidence intervals of this chapter are not distribution-free, we are able to make statements about their reliability that *are* distribution free. This is a remarkable fact, and it largely compensates for the annoyance of having the distributions of the intervals dependent on the error distribution.

Foundations

The mathematics of this and the next section is a little deeper than it is in other parts of this text. Nevertheless, any reader who wants to really understand the concepts must wade through it. Anyone who made it through a single semester of statistics should have no trouble following the material. Give it a try.

Look back at Equation (8.1) on page 303. This equation defines the sample distribution function of a particular collection of errors. The errors really have the unknown distribution function $F(x)$ that we are approximating with this observed $S_n(x)$. We also introduced the concept of order statistics, the sorted values of the errors (including their signs). In particular, $x_{(i)}$ is the ith smallest error, otherwise known as the ith order statistic of the collection.

Our ultimate goal is to estimate to the best of our ability the pth fractile of F. Let us call it X_p. By the definition of fractile, there is probability p that any error sampled from F will be less than or equal to X_p. In other words, X_p is the lower confidence bound for future errors. If we let X_{1-p} be the upper confidence bound, we see that there is a probability of $2p$ that any future error will lie outside the confidence band, with the probability being equally split between the two possibilities. Equivalently, we have a 1–2p confidence interval for future errors. So to find a 90 percent confidence interval, we need $p = 0.05$.

Now we come to a result that never fails to surprise and amaze students the first time they encounter it. Pay close attention, as this is the foundation on which the remainder of this section is built. We want to think about the number of errors in a collection that are less than or equal to a specified X_p. This quantity is $nS_n(X_p)$. Recall that, by definition, the probability of any observed error being less than or equal to X_p is p. Therefore, in a collection of size n from distribution F, the total number of cases less than or equal to X_p follows a binomial distribution with parameter p. The fact that $nS_n(X_p)$ has a binomial distribution, *regardless of the shape of the underlying distribution F*, allows us to make a number of useful statements that are free of troublesome assumptions about F. But first, it is time to review the salient features of the binomial distribution.

The binomial distribution comes into play when we have an event that occurs with probability p in any single trial, and we are concerned with the number of times that event occurs out of a total of n trials. The most common example is flipping a coin. In this case we may say that the event is the coin landing heads-up, and $p = 0.5$. But there is no reason why the probability must be limited to 0.5. Any value from zero to one is legal. The probability of the event occurring exactly m times out of a total of n trials is given by Equation (8.2).

$$B(m; n, p) = \frac{n!}{m!\,(n-m)!} p^m (1-p)^{n-m} \qquad (8.2)$$

Let's put some actual numbers in this equation. Say we perform $n = 20$ trials, and the probability of the event occurring is $p = 0.1$. The probabilities associated with various values of m are shown in Figure 8.7.

Since there are 20 trials, and the probability of the event is 0.1 on each trial, intuition tells us that the most likely number of occurrences in a collection is $np = 2$. This is supported by the pictured distribution, which shows the probability that $m = 2$ is 0.285, the maximum of all possible outcomes. Note that the figure does not bother with outcomes from 10 through 20, as these have almost zero probability. If there is only a probability of 0.1 that the event occurs in any single trial, we do not expect it to happen many times in 20 trials.

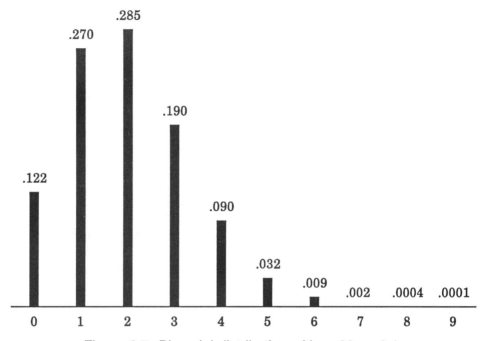

Figure 8.7 Binomial distribution with n=20, p=0.1.

The binomial distribution is related to confidence intervals by the following fact: $B(m; n, p)$ is the probability that, for a randomly sampled collection of n cases, the pair of order statistics $x_{(m)}$ and $x_{(m+1)}$ for that collection surround the unknown but oh-so-important X_p. This fact is proved in many intermediate-level statistics books. We will be satisfied with a brief intuitive approach in which we also sidestep the troublesome but zero-probability issue of exact equality. Think about what it means for this pair of order statistics to enclose X_p. This situation happens if and only if exactly m of the cases, $x_{(1)}$ through $x_{(m)}$ to be precise, are less than or equal to X_p. But the probability of this occurring is $B(m; n, p)$, as already discussed.

What about the extremes of the distribution? $B(0; n, p)$ is the probability that every case in the entire collection exceeds X_p, and $B(n; n, p)$ is the probability the X_p equals or exceeds the entire collection. If these facts are not clear, stop now and ponder them until they become clear. Recall that the binomial event here is a case being less than or equal to X_p, and this event occurs with probability p.

For a good understanding of this section, one subtle point should be kept in mind. It is tempting to think of X_p, the pth fractile of F, as a random variable, and to think of our sample as a known fact. We may then go on to make probabilistic statements concerning the possible values of X_p. Actually, just the opposite is true. X_p is a fixed number. It is a parameter of F, and it has a definite value. We are simply ignorant of that value, to our dismay. Our collection of error measurements is the random quantity here. Any probabilistic statements that we make must concern outcomes related to the error collection, given the unknown but fixed value of X_p. This is a small distinction, but serious readers will constantly keep it in mind.

Bounding the Confidence Bounds

The foundation is in place. We come now to the meat (or tofu, as the reader prefers) of this material. A collection of error measurements is in hand, and we wish to make some statements about X_p for at least two values of p (the upper and lower confidence bounds). Early in this discussion, it was decided that an intuitively good way to estimate X_p for use as a lower confidence bound is to discard the $np-1$ smallest cases, using $x_{(m)}$ as a point estimate of X_p, where $m = np$. Let's put some numbers in place and look at an example.

Suppose we have $n = 20$ cases and we want an 80 percent confidence interval, giving $p = 0.1$. We can use Figure 8.7 for probabilities. For starters, it is obvious that our choice of $x_{(2)}$ as a point estimate of $X_{0.1}$ is reasonable. Of all possible adjacent pairs of order statistics, the pair $(x_{(2)}, x_{(3)})$ has the highest probability (0.285) of containing that elusive $X_{0.1}$ when the sample size is 20. Since we are being statistically conservative, choosing the lower bound of this interval is wise.

This choice may be good, but just how good is it? Observe that there are many other intervals, including the awful bottomless pit at zero, that have a significant probability of enclosing $X_{0.1}$. In particular, there is a probability of 0.27 that $(x_{(1)}, x_{(2)})$ encloses $X_{0.1}$. This is a high probability and a very close second to the chosen interval. Worse still is the fact that there is a probability of 0.12 that $X_{0.1}$ is somewhere below the smallest case in the sample. We don't even have a lower bound. It could be *anywhere* out there! This should be a sobering thought.

In practice, things are not nearly so bad. The reason this example proved so pessimistic is that the sample size is small. If we increase the sample size to $n = 50$, the probability at $m = 0$ drops to 0.005, a far cry from 0.12. Once we get into the realm of several hundred cases, things become very nice. The width of the binomial distribution shrinks, and $x_{(np)}$ becomes a perfectly respectable estimator of X_p.

The preceding discussion can lead to at least three ways of formalizing confidences in the confidence limits: First, we could supplement the point estimate of $X_{0.1}$ with a confidence interval. We already know that there is a probability of 0.29 that $X_{0.1}$ lies inside the observed interval $(x_{(2)}, x_{(3)})$. We could combine this interval with the one below it and the one above, stating that there is a probability of 0.27 + 0.29 + 0.19 = 0.75 that $X_{0.1}$ lies within $(x_{(1)}, x_{(4)})$. However, careful consideration tells us that this particular action would generally be of little practical use. We don't usually care if our estimate of $X_{0.1}$ is smaller than the true value. An error of this sort, which is entirely possible, would be in our favor, for it would mean that our computed confidence interval for the prediction errors is wider than it should be. In other words, our predictions will be better than what is indicated. This may be a problem if the unwarranted pessimism causes us to give up, but that error is surely better than its converse! Therefore, there is almost never any reason to bring more interior intervals, such as the one with probability 0.19, into the picture.

This is a good stopping point for the reader. Do not continue until the preceding concept is clear. Here is an exercise. Suppose the first four order statistics (the smallest errors in the collection) are –6, –4, –2, and –1. We know that there is a probability of 0.29 that $X_{0.1}$, the true lower bound of an 80 percent confidence interval, lies within (–4, –2). Our conservative point estimate of $X_{0.1}$ is –4. The previous paragraph tells us that there is a probability of 0.75 that $X_{0.1}$ lies within (–6, –1). But do we really care if it's as high as –1? It would certainly be nice having a true lower bound on the error of –1 instead of our computed –4, but that sort of mistake is better than the true lower bound being –6, unbeknownst to us. Ponder this.

The preceding discussion may inspire us to use wider confidence intervals than the ones espoused so far. For example, Figure 8.7 informs us that there is a probability of only 0.12 that $X_{0.1}$ is less than $x_{(1)}$, the smallest observed error. So if we use this for the lower confidence bound instead of $x_{(2)}$, we can be 88 percent sure that our lower bound is low enough. In fact, ultraconservative fanatics may wish to always use the extreme values of the observed errors as the confidence bounds. However, there are many disadvantages to this approach. There is much to lose and little to gain. If the error distribution has tails that are even moderately heavy, these confidence intervals will be so wide that they are nearly useless. It is strongly to our advantage to discard the wild outliers in order to represent the majority best. (This is not a political statement, only a statistical observation.) Moreover, we have been dealing with an extreme example. Basing confidence intervals on just 20 cases is unrealistic. When, as is usual, we have at least several hundred cases, the computed confidence bounds will be much more stable than those discussed in this example.

We have been examining the topic of confidence in the confidence intervals from one viewpoint: p is specified and we attempted to bound X_p. This is sometimes useful. But it contains many complicating factors. Its dependence on observed order statistics makes implementation difficult. Also, interpretation can be tricky. This direction will not be pursued any further, and it is not included in the NPREDICT program that accompanies this text. We now move on to a generally preferable approach.

This approach is exactly the opposite of the preceding one. Here, we have chosen an estimate of X_p, and we ask whether our observed error collection is consistent with possible values of p. In other words, we previously fixed p and bounded possible values of X_p. Now we fix a presumed X_p and bound the possible values of p. It turns out that this is usually a more practical viewpoint.

We'll start this discussion on an intuitive level. Suppose that once again we ultimately want an 80 percent confidence interval for future errors, implying $p = 0.1$. Following our tradition, we use the observed $x_{(2)}$ as a point estimate of $X_{0.1}$, the lower bound of the confidence interval. Formerly, we agreed to hold fast to $p = 0.1$ and use the observed cases to bound the estimate of $X_{0.1}$. Now we stubbornly hold fast to using $x_{(2)}$ as our lower confidence limit and ask how far off we might be in terms of the true value of p.

We almost never care about committing one type of error. If our chosen estimate of $X_{0.1}$ is really the fractile of a smaller probability, we should be happy. For example, if in truth the observed $x_{(2)}$ happens to be equal to $X_{0.07}$, the lower limit of our confidence interval will really be better than the 80 percent interval we think we have. Its tail contains only 7 percent of the error distribution, not the 10 percent implied by an 80 percent interval. This is good.

Our concern from now on is the opposite error. What if our chosen lower bound actually corresponds to a larger fractile of F? This implies that our confidence interval for future errors has smaller probability of enclosing the prediction errors than we think. This is bad.

The issue to be addressed in this specific example is the following: We have chosen a lower confidence bound. The fact that we happened to use $x_{(2)}$ is not directly relevant at this point. We could just as well have visually examined the error collection and picked a lower limit out of the air. We now, with fear and trembling, hypothesize that our chosen lower limit is some dangerously large fractile of F, something greater than the hoped-for 0.1. We then examine the error collection and ask if it is consistent with this hypothetical value of p. In particular, we count the number of cases in the sample of 20 that are less than or equal to our chosen lower limit. We ask whether, given the hypothetical p, it is reasonable to expect to find as few small cases as we do find. If p is large, we expect a lot of cases to be less than the presumed X_p. If we find very few, then it is unlikely that p is so large, and we breathe a sigh of relief.

The use of $x_{(2)}$ as the lower confidence bound comes into play here. We do not have to count the number of smaller (or equal) cases because we already know how many there are: two. Hence, the question reduces to this: If the observed $x_{(2)}$ is really X_p (for a specified p larger than we hope), what is the probability that two or fewer cases in the observed sample of 20 would be less than or equal to this X_p? If this probability is small, it is unlikely that p is so large.

It's time for a picture. Figure 8.8 shows a binomial distribution with $n = 20$ and $p = 0.2$. Compare it with Figure 8.7 on page 309, which has the same sample size but smaller event probability.

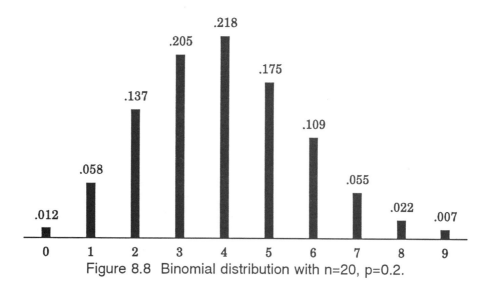

Figure 8.8 Binomial distribution with n=20, p=0.2.

Continuing this same example, we choose the observed $x_{(2)}$ as a point estimate of $X_{0.1}$, our lower confidence bound. But what if, due to bad luck in sampling, $x_{(2)}$ is really equal to $X_{0.2}$, a disastrous possibility that gives us a confidence interval much smaller than the 80 percent we requested. If this were true, the distribution of the number of cases in the sample of 20 that are less than or equal to the point estimate would be as shown in Figure 8.8. Not surprisingly, the most frequent count would be four cases. But we had only two (by virtue of using $x_{(2)}$). The probability of getting so few cases smaller than the chosen value is $0.012 + 0.058 + 0.137 = 0.207$. So we see there is about a 21 percent chance that $p = 0.2$ instead of the 0.1 that we hope for. This is uncomfortably large. But remember that the number of cases, 20, is unrealistically small. In practice, things will be much better.

The intuitive introduction is finished, so we advance to the general situation. Although the binomial distribution is discrete, it helps to visualize it as continuous. Look at Figure 8.9, which shows the binomial distributions associated with the presumed true p and a trial value that is larger than we hope.

Our traditional choice of using the npth order statistic to estimate the pth fractile of F implies that the lower confidence bound is determined by the center of the binomial distribution under the assumed p. This is shown by the vertical line in the figure. In the example just presented, this chosen limit corresponds to $np = 2$, implying the use of $x_{(2)}$ as the estimate of $X_{0.1}$, the lower bound. But what if we are wrong? What if our chosen limit is actually a larger fractile of F? This possibility is shown as a dotted line depicting the binomial distribution associated with the (scary) larger p. It is obvious that, as p increases, the probability of getting so few cases under the limit decreases. This probability is shown as the shaded region. So we do two things. First, we arbitrarily choose a probability that is comfortably small, say 0.05. Then we increase p until the shaded area drops to the comfort zone. This provides a confidence limit for the true value of the probability corresponding to our chosen limit. For example, when some increased value of p causes the shaded area to drop to 0.05, we know there is only a 5 percent chance of getting a sample this extreme if the chosen limit really is a p-fractile. We will be comforted if that p is not too much larger than the one we hope for.

How do we compute the shaded area? Remember that, despite the continuous appearance of Figure 8.9, the binomial distribution is discrete. Therefore, the shaded area is the sum of the binomial probabilities from zero through the number of cases less than or equal

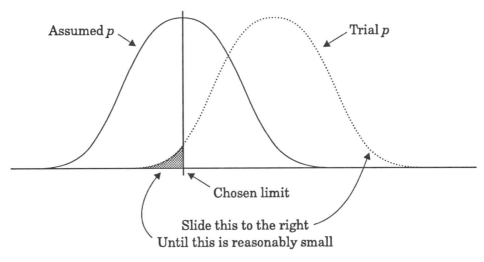

Figure 8.9 Bounding the fractile.

to the limit. This is expressed in Equation (8.3), where n is the sample size, m is the number of cases less than or equal to the limit, and p is the trial probability.

$$\text{SHADED AREA} = \sum_{k=0}^{m} \frac{n!}{k!\,(n-k)!}\, p^k (1-p)^{n-k} \qquad (8.3)$$

This entire discussion so far has focused on the lower confidence bound. We used an order statistic near the bottom of the sorted list to estimate a low fractile of F for use as the lower confidence bound, then we investigated the possibility that the probability associated with that bound is actually higher. Exactly the same arguments apply to the upper confidence bound. We use an order statistic near the top of the sorted list to estimate a high fractile (typically $1-p$) of F, then worry about the true probability being less than the assumed value. In practice, we typically discard the same number of sorted cases at the top as at the bottom. This provides a symmetric confidence interval in the sense that the probability of a prediction error being less than the lower bound is equal to the probability of it exceeding the upper bound. The only rare exception might be if one type of error is more damaging than the other. As an extreme example, it might be that errors of one type are considered totally harmless. In this situation we might use just one confidence limit instead of an interval.

This discussion will end with a small numeric example. Suppose we want a 90 percent symmetric confidence interval for future prediction errors. This implies $p = 0.05$. Assume we have a collection of 300 representative errors. We discard $np-1 = 14$ cases from each extreme of the sorted error list, using $x_{(15)}$ for the lower confidence bound and $x_{(286)}$ for the upper bound. Suppose the probability under F of an error less than the $x_{(15)}$ we observed really is greater than the 0.05 we are hoping for. The probability of a sample of 300 cases containing 15 or fewer cases that are less than or equal to the pth fractile is given by Equation (8.3) with $n = 300$ and $m = 15$. It turns out that Equation (8.3) yields the value 0.01 (a very conservative choice) when $p = 0.087$. This fact can be expressed as the following confidence statement: If our chosen lower confidence limit, $x_{(15)}$, really is the 0.087 fractile of the error distribution (as opposed to the 0.05 we wanted), there is only a 1 percent chance that our collection of 300 errors would contain so few (15 or less) cases less than or equal to the limit. This is comforting.

Keep in mind that we have dealt only with the lower confidence limit. There is also a 1 percent chance that the upper limit defined by $x_{(286)}$ really corresponds to a probability of $1-0.087 = 0.913$ instead of 0.95. If we put all of these facts together, and trade in some ostentatious statistical rigor for plain language, we can make the following useful (if trivially flawed) statements:

- For a collection of 300 cases, the 15th and 286th order statistics are a good guess for a symmetric 90 percent confidence interval.

- There is a 1 percent chance that each tail probability could be as high as 0.087 (instead of 0.05).

The NPREDICT program makes a statement similar to the second one in the audit log file after computing confidence intervals. The probability is fixed at 0.05 instead of the 0.01 used in this example.

There is one last way of evaluating the quality of our confidence intervals. For motivation, reread the second bulleted point just given. It is tempting to warp that statement into the following safe, but incorrect statement:

- There is a 98 percent chance (1 percent error in each tail) that the overall confidence interval covers at least $1 - 2 * 0.087 = 82.6$ percent of the error distribution.

The flawed reasoning goes like this: There is a 1 percent chance that the upper confidence limit has a tail probability as large as 0.087, and the same is true of the lower tail. Thus, there is a 2 percent probability that the total tail area is as large 0.174, giving an interior area possibly as small as 0.826. The flaw is that *these are not independent events*, so we cannot simply add them. Crudely stated, when one limit is bad, the other is probably not bad. This works in our favor, since we are not likely to suffer the indignity of having *both* the lower and upper limits be bad. In fact, the true fraction of the error distribution encompassed by $(x_{(15)}, x_{(286)})$ with 2 percent probability is approximately 86 percent, not 83 percent. So if we do make the preceding erroneous statement, at least we have the comfort of knowing that we are being conservative.

If we are interested in computing the true figures, there is a way. Here is how to compute a distribution-free *tolerance interval* based on order statistics. A tolerance interval is a pair of points that, with specified probability β, enclose at least the fraction γ of the distribution. For simplicity, we will assume that the interval is symmetric in that the same number of observations are discarded from each extreme. As usual, let n be the number of cases in the collection, and let $m-1$ observations be discarded from each end. Thus, $x_{(m)}$ is chosen as the lower confidence limit, and $x_{(n-m+1)}$ is the upper limit. The probability associated with this tolerance interval is given by the Incomplete Beta Function as shown in Equation (8.4).

$$\beta = 1 - I_\gamma(n-2m+1, 2m)$$
$$= 1 - \sum_{k=n-2m+1}^{n} \frac{n!}{k!\,(n-k)!} \gamma^k (1-\gamma)^{n-k} \quad (8.4)$$

This equation is presented without explanation. The reason is that, even though it looks straightforward, its derivation is actually quite complex, requiring relatively advanced mathematics beyond the scope of this text. Many graduate-level statistics books present the complete proof.

How can we make effective use of this equation? The first answer that might occur is to use it to compute the probability that our requested confidence interval really does cover the area of the error distribution that we require. In other words, suppose we have asked for a 90 percent confidence interval. We could use this equation to compute the probability that we really have at least a 90 percent interval. But closer examination shows this to be silly. Remember that we defined the interval to be a best guess. It might be a little too big, or a little too small. The method of this chapter has computed it so that both possibilities are approximately equally likely. The implication is that if we use Equation (8.4) to compute the probability that the interval really does cover at least 90 percent of the error distribution, we will get an answer in the vicinity of 50 percent. There is a 50–50 chance (approximately) that our computed interval is too big or too small. So this particular computation is useless.

A good technique is to ask how likely we are to have a confidence interval that is at least as good as one that is somewhat smaller than the one we requested. For example, suppose we requested a 90 percent interval. What is the probability that the computed interval is at least an 80 percent confidence interval for future errors? It is surprising how high this probability usually is, even with small sample sizes.

This is done in the NPREDICT program that accomanpies this text. The program considers a confidence interval whose tails are twice the size of the tails in the user's requested interval. A statement similar to the following appears in the audit log file after confidence computation is complete:

There is a 98.3 % chance that the interval exceeds 80 %.

The first number is the value of Equation (8.4) when γ is equal to the width of the tentative interval. In other words, this example assumes the user has requested a 90 percent interval. Each tail area is 0.05. Doubling the tail area and subtracting from 1.0 gives $\gamma = 0.8$ here.

Code for Bounds on the Confidence Limits

Here are a few subroutines that are useful for performing the computations described in the preceding section. These are extracted from the module NP_CONF.CPP on the accompanying code disk. We start with a simple little workhorse. This computes the binomial coefficient: the number of combinations of n items taken m at a time. This is the term involving three factorials in Equation (8.2) on page 308. Beware that when n gets large, typically over 1,000 or so, most floating-point hardware will overflow. We will take care of that later.

```
double binom ( int n , int m )
{
  double x = 1.0 ;

  while (m) {
     x *= n-- ;
     x /= m-- ;
     }

  return x ;
}
```

Equation (8.2) is the foundation of the binomial distribution. Here is a subroutine that evaluates its sum over a range of values of m so that it may be used in applications like that shown in Equation (8.3). Note the safety precaution to avoid illegal floating-point operations. Also note that some readers may wish to replace this simple routine with a much more complex but elegant incomplete beta routine. The choice is yours.

```
double binom_sum ( int start , int stop , int n , double p )
{
  int i ;
  double sum, log_p, log_1mp ;

  if (p <= 0.0)
     p = 1.e-14 ;

  if (p >= 1.0)
     p = 1.0 - 1.e-14 ;

  log_p = log ( p ) ;
  log_1mp = log ( 1.0 - p ) ;

  sum = 0.0 ;
  for (i=start ; i<=stop ; i++)
     sum += binom ( n , i ) * exp ( i * log_p ) * exp ( (n-i) * log_1mp ) ;

  return sum ;
}
```

Here is the subroutine that uses the preceding workers to compute the probability corresponding to a specified value of the shaded area in Figure 8.9. Its operation is a little tricky, though basically straightforward. An explanation follows the listing. As usual, n is the size of the collection, m is the number of cases less than or equal to the chosen lower confidence limit, and limit is the desired shaded area, typically 0.05 or so.

```
#define P_LIMIT_INC 0.1
#define P_LIMIT_ACC 1.e-7

double p_limit ( int n , int m , double limit )
{
  int i ;
  double p1, p2, p3, y1, y2, y3, denom, trial, y ;

/*
   The first point is known
*/

  p1 = 0.0 ;
  y1 = limit - 1.0 ;

/*
   This scans upward until it changes sign
*/

     for (p3=P_LIMIT_INC ; p3<1.0 ; p3+=P_LIMIT_INC) {
       y3 = limit - binom_sum ( 0 , m , n , p3 ) ;
       if (fabs ( y3 ) < P_LIMIT_ACC)           // Very unlikey
         return p3 ;                            // But check anyway
       if (y3 > 0.0)                            // When we change sign
         break ;                                // The root is bracketed
       p1 = p3 ;                                // Move on to next trial interval
       y1 = y3 ;
       }

/*
   The root has been bracketed between (p1, y1) and (p3, y3).  Refine.
*/
```

```
for (i=0 ; i<100 ; i++) {              // This limit is insurance that is never used
    p2 = 0.5 * (p1 + p3) ;             // Midpoint of current bounding interval
    if (p3 - p1 < P_LIMIT_ACC)         // This is a secondary convergence test
        return p2 ;
    y2 = limit - binom_sum ( 0 , m , n , p2 ) ;   // Compute criterion
    if (fabs ( y2 ) < P_LIMIT_ACC)                // This is a main convergence test
        return p2 ;

    denom = sqrt ( y2 * y2 - y1 * y3 ) ;          // Remember y1, y3 opposite sign
    trial = p2 + (p1 - p2) * y2 / denom ;         // New test point for root
    y = limit - binom_sum ( 0 , m , n , trial ) ; // Compute criterion
    if (fabs ( y ) < P_LIMIT_ACC)                 // This is another main converge test
        return trial ;

    if ((y2 < 0.0)  &&  (y > 0.0)) {              // Root between mid and test point?
        p1 = p2 ;
        y1 = y2 ;
        p3 = trial ;
        y3 = y ;
    }

    else if ((y < 0.0)  &&  (y2 > 0.0)) {         // Ditto, but other way?
        p1 = trial ;
        y1 = y ;
        p3 = p2 ;
        y3 = y2 ;
    }

    else if (y < 0.0) {                           // Must keep one of the current bounds
        p1 = trial ;
        y1 = y ;
    }

    else {                                        // Ditto, but other bound
        p3 = trial ;
        y3 = y ;
    }
}

return -1.0 ; // Only get here if convergence failure
}
```

The basic algorithm goes like this: We need to advance p until the sum expressed in Equation (8.3) drops to limit. This is done in two steps. First, p is advanced in relatively large discrete jumps equal to P_LIMIT_INC until the threshold is crossed. Then the exact value of p is iteratively refined to an accuracy of P_LIMIT_ACC.

These steps are accomplished by finding the root of the function defined by limit minus Equation (8.3). When p is zero, Equation (8.3) evaluates to 1.0 (which is all due to the very first term in the sum). In this case, the function is negative: limit−1. We keep incrementing p until the function changes sign. As soon as this happens, we know that the value of p we seek lies between the two most recent points. From now on, (p1, y1) will lie to the left of (p3, y3). Also, y1 will be negative and y3 positive, thus bracketing the root.

We now go to iterative refinement. The algorithm is a simple adaptation of Ridders' method, documented in Press *et al.* (1992). A limit of 100 iterations is imposed, although, in practice, needing more than six iterations is rare. The first step is to compute the midpoint of the bracketing interval and return if the interval is small enough. Otherwise, evaluate the function there and return if the root is found to satisfaction. A new trial point is computed where the root should theoretically lie, and the function is evaluated there. Again, if the root has been satisfactorily located, return. Otherwise, replace one or both of the bounding points. This logic looks more complicated than it really is. If we are lucky, the midpoint and the trial point will be on opposite sides of zero, in which case we replace both endpoints. If we are not so lucky, just replace whichever endpoint is opposite the trial point.

If the sample size is large (beyond about $n = 1000$), the subroutines just listed will suffer floating-point overflow. However, we are saved by the fact that when n is this large, the binomial distribution is almost identical to a normal distribution. The following code computes the value of p that gives a shaded area equal to 0.05, which is the value used in the NPREDICT program:

```
p = (m + 0.5) / (double) n ;
std = sqrt ( n * p * (1.0 - p) ) / n ;
for (i=0 ; i<12 ; i++) {
   pp = p + 1.64 * std ;
   std = sqrt ( n * pp * (1.0 - pp) ) / n ;
   }
p += 1.64 * std ;
```

This code uses a trivial normal approximation. It computes p as m/n, plus the standard correction for approximating a discrete distribution with a continuous one. The standard deviation of the distribution of this fraction is given by the second line of code. Skip the middle lines for now. The normal distribution has 5 percent of its area beyond 1.64, a fact that can be found in any statistics text. Thus, we increment the mean by 1.64 times the standard deviation to get the confidence limit. But there's a slight catch. We just shifted p, so the standard deviation already computed is no longer correct. The middle section of this code iterates a few times to allow the process to stabilize. This code may be modified to compute p for any value of the shaded area. Simply replace 1.64 with the corresponding point in the normal distribution. For example, use 1.96 for an area of 0.025, and use 2.33 for 0.01.

Finally, here is a subroutine to compute the probability that the interval covers at least a specified minimum fraction (tolerance interval) of the distribution. This is a trivial implementation of Equation (8.4). If n exceeds 1,000 or so, putting computations in jeopardy of overflow, do not worry. As long as limit is at least moderately smaller than the confidence interval implied by m, which is what this text recommends, the computed probability would be essentially 1.0, so just going with that value is fairly safe. As an alternative, standard library routines may be used to directly evaluate the incomplete beta function.

```
double t_limit ( int n , int m , double limit )
{
   return 1.0 - binom_sum ( n-2*m+1 , n , n , limit ) ;
}
```

Multiplicative Confidence Intervals

This entire chapter has relied on a major assumption that has not received any attention beyond a single sentence on page 307. Violation of this assumption invalidates the computed confidence intervals. It must be emphasized that precisely the same assumption applies to all other standard confidence computation methods known to the author, so the method of this chapter does not have a special weakness. Nevertheless, a brief discussion is in order.

The error distribution is computed by cumulating (for each fixed prediction distance) a single grand error vector. This vector provides a range of errors that is used to bound future predictions. When we use this approach, we are assuming that the error distribution is exactly the same for each prediction. In particular, we are assuming that the nature of the errors does not depend on the values of the series. By adding the chosen representatives of the error to each and every prediction, we are acting as if the possible errors are the same, regardless of whether a given prediction is large or small. If the size of the prediction affects the error distribution, we are in trouble.

A fully general solution to this problem is practically hopeless. We would have to segregate the known data into bins according to size, processing each bin separately. Even when we lump all errors into one bin, as is done in this text, we almost always wish we had more samples on which to base the confidence intervals. If we deplete our supply even more by using multiple output bins, we would almost never have enough data to achieve satisfactory results.

Luckily, things are not usually so bad. First of all, a great quantity of real-life data does satisfy this independence assumption, at least to enough of a degree that reasonably reliable confidences can be obtained. And in those situations in which the data does violate the assumption, it nearly always does so in one particular, predictable way. This common violation happens when the data is multiplicative. Suppose we measure a stock price. Prediction errors when the price is around 40 are probably double what they are when the price is around 20. How do we deal with this? It's not difficult. If we have set up our model intelligently, 90 percent of the battle is already won.

Back in Chapter 1, we learned that the correct way to handle multiplicative data is to take its log before doing anything else. This transforms it into the additive data that is so dear to most prediction models (even neural networks). When we do this, the errors in the log domain satisfy the independence assumption that has been the source of our grief. So the answer to the problem is to use the methods of this chapter to compute confidence intervals *in the log domain*, then transform these intervals back to the original application domain by exponentiating them along with the predictions.

Another way of looking at this solution is to see that the correct confidence intervals for multiplicative data are themselves multiplicative. In other words, instead of bounding predictions by adding and subtracting fixed constants, we bound the predictions by multiplying and dividing them by fixed constants. These constants are the exponentials of the bounds in the log domain.

The occurrence of multiplicative data is so common that the NPREDICT program supplied with this text takes a daring but virtually always safe course. When confidences are computed, the list of confidence compensations is examined. If the *first* compensated operation on a signal is a logarithm, the program decides that the data is multiplicative and proceeds accordingly. When the compensations are performed, the final act of exponentiating the predicted and true data is omitted. As a result, the error distribution is computed in the log domain where the independence assumption is (presumably) valid. A flag is set so that when confidence intervals are displayed, they are computed by multiplication rather than by addition. This is correct operation in the vast majority of applications.

Note that there is an infinitude of other possible violations of the independence assumption. It is impossible to incorporate facilities for handling all of them. But the fact of the matter is that multiplicative data is the only common violation. So the NPREDICT program takes one of two courses. If the user does not specify a log as the first confidence compensation, it is assumed that independence is satisfactory, and normal additive intervals are computed. If a log is the first confidence compensation, multiplicative intervals are computed. Other possibilities must be programmed as needed.

Confidence Intervals in Action

This chapter ends with a few examples of confidence intervals. These examples are created with contrived data to illustrate specific points. Several examples of confidence intervals based on actual data can be found elsewhere in this text.

What happens if the prediction model is terrible? It may be that the model was trained using a dataset that is not at all representative of the population to which the model will ultimately be applied. If we are lucky (judicious?) enough to compute the confidence intervals from data that *is* representative of the true population, the method of this chapter will provide intervals that are correct, even if the predictions are not! As a rather extreme example, Figure 8.10 on page 329 illustrates a model trained on data whose mean is approximately 4. This model is then used to compute confidence intervals and predict future values from data that has a mean of –5. As soon as the future begins, the predictions take right off according to the response dictated by the faulty training. Despite this bizarre occurrence, the computed confidence intervals correctly bracket the values that are clearly expected.

An MA(1) model (moving average with a single weight at a lag of one) is too weak for many uses, since it degenerates to a constant value after the first prediction. Nevertheless, it can be appropriate when only a single prediction is needed. Also, it provides a good illustration of confidence intervals. Figure 8.11 shows the use of an MA(1) model to predict a fairly noisy series. Observe that after the first prediction (the positive peak), all future predictions are fixed at the mean of the series as demanded by the model. The confidence intervals are also flat from there on out. This is common, but not guaranteed. Random events in the dataset may cause the confidence intervals to fluctuate slightly. However, there is a strong tendency for their width to remain constant, especially if the series is long. It is an excellent reader exercise to think about why this is so. Look back at the beginning of this chapter where the general philosophy was discussed, and consider the sample cases that go into the collection for each future prediction distance.

A somewhat more common prediction model is the IMA(1), which is an MA(1) model applied to a differenced series. This is shown in Figure 8.12. In the IMA(1) model, the *slope* of the prediction line remains fixed, instead of the predictions being fixed. The important concept to be obtained from this example is the tremendous rate at which the confidence intervals widen as the predictions move into the future. This is a direct result of the summing that must be done to return the predicted differences to the original problem domain. Errors compound rapidly. The small dropoff of the upper confidence interval at the far right of the figure is a typical artifact of random sampling error. Such effects are common and generally small, being no cause for alarm.

The serious nature of differencing is best illustrated by taking it to an extreme. Only in the rarest of circumstances does one need to difference a series three times to achieve stationarity. But suppose we do this, just to see what can happen. Figure 8.13 shows a series whose predictions are made by differencing three times, predicting the triple-differenced values, then undoing the differencings. Not surprisingly, the model heads for the wild blue yonder, with the confidence intervals following along and getting wider all the time. Always be wary of differencing more than needed. It is risky business.

Predictions from the simple AR(1) model are slightly more interesting than those from the equally simple MA(1) model. AR(1) predictions do not flatten out immediately. Instead, they taper off gradually, asymptotically approaching the mean. This is illustrated in Figure 8.14. Observe that the width of the confidence interval increases quickly at first, but then remains fairly constant.

Also observe that the slow wandering of the series suggests that it may be nonstationary. With this as inspiration, the series is differenced before prediction. The result of predicting this way is shown in Figure 8.15. Several noteworthy items appear. First, the predictions are obviously different from those made with the original series. This is not unusual. Differencing a series causes the prediction model to look at the data from an entirely different perspective. There is also a strong possibility that a model that is appropriate for the raw data is inappropriate for the differenced data. *Always consider this possibility, as this effect is common.*

Despite the dramatically different predictions made using the two approaches (raw versus differenced), the confidence intervals for the first prediction are nearly identical for both methods. However, their behavior diverges after that. The differencing operation causes the intervals in Figure 8.15 to widen rapidly as they move into the future. The turndown of the upper limit at the far right is again an artifact of random sampling, exaggerated by the fact that relatively few cases went into the computation. If the prediction distance were extended, the limit would turn back up again very soon. This degree of instability is common when short series are used to compute confidence intervals. But it is really small in the grand scheme, and larger samples work wonders at ensuring stability.

Seasonal differencing is a potentially useful method for dealing with periodic phenomena. Figure 8.16 shows predictions made by seasonal differencing, predicting the differenced series, then undoing the differencing. Since the prediction model chosen for this example is one whose predictions go rapidly to the mean in the future, the final predictions are an almost perfect repeat of the last known points. It is interesting to compare these results with those obtained using a general regression neural network to predict the raw data. This is shown in Figure 8.17. The predictions are similar but not identical. Most importantly, note that by avoiding the error-compounding effect of seasonal differencing followed by undoing the differences, the neural-network confidence intervals are significantly narrower than those in Figure 8.16. Of course, this is also somewhat due to the fact that the model in Figure 8.16 is deliberately weak, while the neural network is powerful. It is difficult to separate these effects. However, there is an important general principle here: A good way to assess the quality of a prediction model is to look at the width of the confidence intervals.

These examples may be a little discouraging to the reader. In most cases, the confidence intervals rapidly expand, quickly becoming so wide that the predictions appear to be almost useless. Do not worry. Most practical applications do better than these examples for two reasons. Most importantly, in order to illustrate clearly the nature of the computed confidence intervals, they have been extended into the future further than any sane user would attempt. Nobody would use an IMA(1) model to predict eight observations ahead. In fact, as has been repeatedly emphasized throughout this text, recursive prediction more than a few samples into the future is always dangerous due to the compounding effect of accumulated errors. If you look closely at these figures, the confidence intervals for the first one or two predictions into the future are actually quite narrow compared to the overall range of the observed series. As long as we are reasonable in what we ask of our model, we will obtain reasonable confidence in our predictions.

The second reason for encouragement is that the models employed for this set of examples have been deliberately chosen to be weak relative to their task. This allows clear portrayal of the confidence intervals. In most situations, more effective models would be employed, which, in turn, leads to narrower intervals.

Finally, keep in mind the warning that appeared near the beginning of this chapter. These confidences are rigorous in that they are nearly independent of the distribution of the data. The price paid for this rigor is that the computed intervals are somewhat wider than intervals based on distributional assumptions. If any reader compares these intervals to those computed by, say, direct application of Box and Jenkins' formulas, he or she may be surprised by the difference. But do you really want to be constrained by those assumptions? Remember that the intervals in this chapter are based on actually observed errors!

Confidence Intervals in Action 329

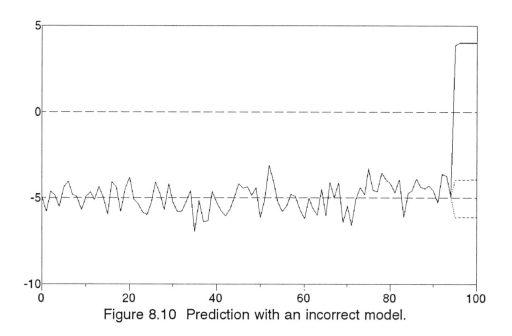

Figure 8.10 Prediction with an incorrect model.

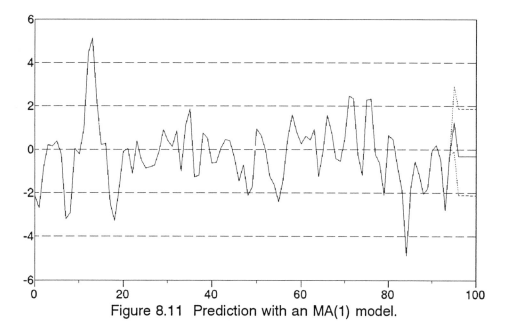

Figure 8.11 Prediction with an MA(1) model.

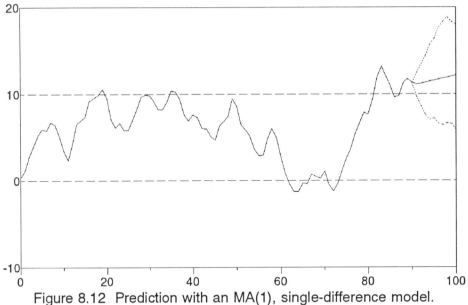
Figure 8.12 Prediction with an MA(1), single-difference model.

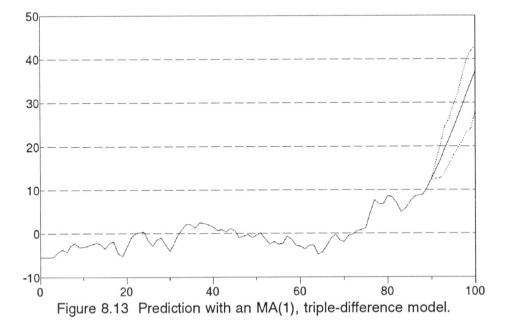
Figure 8.13 Prediction with an MA(1), triple-difference model.

Confidence Intervals in Action

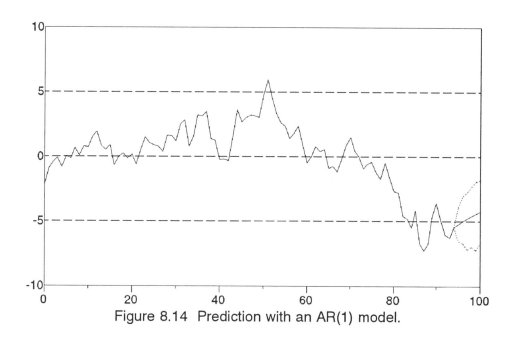

Figure 8.14 Prediction with an AR(1) model.

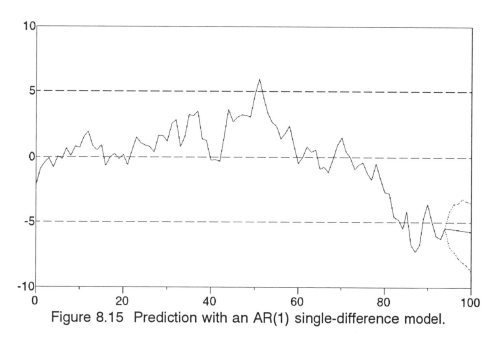

Figure 8.15 Prediction with an AR(1) single-difference model.

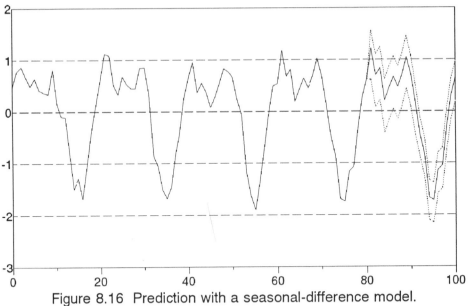
Figure 8.16 Prediction with a seasonal-difference model.

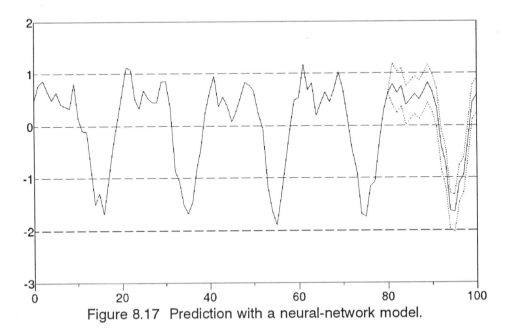
Figure 8.17 Prediction with a neural-network model.

9

Numerical and Statistical Tools

- Random number generation

- Equation solution

- Deterministic optimization

- Stochastic optimization

- Eigenvalues and eigenvectors

- Fourier transform

- Data reduction and orthogonalization

- Autoregression by Burg's algorithm

Even the most sophisticated programs rely on the mundane numerical subroutines that abide in their core. These include random number generators, equation solvers, function optimizers, and any number of matrix operation routines. This chapter presents a toolbox full of the most important such subroutines. No detailed discussion of the theory behind the algorithms will be given. The reader will instead be referred to other sources. Only concise guidance on the proper use of these subroutines will appear. Complete source code for all of the routines is supplied on the disk that accompanies this text.

Random Numbers

There is an easy way to judge the experience and competence of someone who claims to be an expert in numerical methods. Just ask that person how easy it is to generate random numbers. If the expert claims that it is simple, find another expert. There are probably more subtle pitfalls in random number generation than in any other area of numerical methods. What makes these errors particularly dangerous is that problems capable of totally ruining an experiment may be virtually invisible without both the knowledge of what to look for and the tools to detect the problem. Therefore, readers are strongly advised to avoid blindly experimenting with random number generation schemes that only look good. Find something that's known to be good and stick with it. The author is of the opinion that the random number generators supplied with this text meet the highest quality standards. However, it must be emphasized that nothing short of counting emissions from radioactive decay is truly random. The numbers generated by these subroutines are deterministic. A rigorous rule defines their behavior. They only give the appearance of being random. Therefore, it is always possible that they will be totally unsuitable for some task. No guarantee of their universal quality is stated or implied. *Caveat Emptor.* Let the buyer beware.

Uniform Random Numbers

The foundation on which all other random numbers are built is almost invariably a source of uniform random numbers. In the world of computers, a uniform random number generator generally is a source (or is closely based on a source) of integers spanning a fixed range, typically zero to one less than a power of two. Every integer in that range is equally likely. The generator provided here spans the range 0 to $2^{32}-1$. In other words, it is a 32-bit generator, in which every possible 32-bit integer is equally likely.

A common rule of thumb for random number generators is that the high-order bits are more random than the low-order bits. The implication is that if fewer than the full 32 bits are needed, it is better to use the high bits than the low bits. If a range from zero through a number not one less than a power of two is needed, there is a standard generation method. Add one to the desired upper limit, then find a multiple of that quantity that is as large as possible (for efficiency) but that does not exceed the upper limit of the source generator. Invoke the source generator repeatedly until it produces a random number that is strictly less than the upper limit of the multiplied range. Divide that random number by the multiplier to produce a new random number that lies within the desired range. If this is not clear, see the subroutine rand16_1 in the file FLRAND.CPP for an example of using this method to produce a 16-bit random number from one that is not limited to an exact power of two.

Two subroutines are needed in order to handle generation of uniform random numbers. The first, sflrand, sets the random seed. The second, flrand, actually generates a random integer. Both of these subroutines may be found in FLRAND.CPP. Each 32-bit random number is created by combining two separately generated 16-bit random numbers, one for the upper half and one for the lower half. These 16-bit components are produced by shuffled linear congruential generators having different parameters. The 32-bit result is also shuffled, perhaps a bit of overkill.

```
void sflrand ( int seed )
int flrand ()
```

A closely related problem is to produce a floating-point number in the range [0–1]. Most applications require that a value of exactly 0.0 can be obtained, but exactly 1.0 cannot. This is easily done by dividing a uniform random integer by one more than its maximum possible value. More sophisticated algorithms will go one better by generating two random integers to increase the resolution. Be warned that with most hardware, this high-resolution approach is capable of generating a value of exactly 1.0 due to internal rounding in the floating-point processor. Appropriate precautions should be taken. The routine unifrand () in FLRAND.CPP uses this method.

> unifrand ()

Gaussian (Normal) Random Numbers

Another commonly needed random number distribution is the normal distribution. The method used here is the popular Box-Muller algorithm, which generates standard normal deviates (zero mean, unit standard deviation) in pairs. Early versions of this algorithm used iterative methods to avoid computation of transcendental functions, but the implementation given here directly computes the sine and cosine. Modern coprocessors perform these operations so quickly that there is no longer any need to avoid them.

Two normal random number subroutines are provided in RANDOM.CPP. If more than one random number is needed, the usual case, the normal_pair routine should be used. The Box-Muller algorithm generates the normal random deviates in pairs, so this is the most efficient use of computer time. The normal subroutine generates a pair internally, then simply discards one of them.

> double normal ()
> void normal_pair (double *rand1 , double *rand2)

Multivariate Cauchy Deviates

Harold Szu (1986) demonstrated that, under some conditions, a Cauchy distribution is superior to a normal distribution for simulated annealing. As a result, this formerly obscure statistical anomaly has become popular. It is very easy to generate a univariate Cauchy random variable. Just compute the tangent of a uniform random deviate in the range zero to pi. But the multivariate Cauchy distribution is another story. The problem is that the marginal distribution of each component is very complicated, and the marginals are correlated. The easiest method is to compute the direction of the vector in one step, and its length in another step. For full details, see Masters (1995).

The subroutine that generates Cauchy random vectors takes three parameters. The first is the number of variables. The second is the scale parameter, typically 1.0. The random vector is returned via the third parameter. Note that the term *scale parameter* is used because this distribution has such heavy tails that its variance is infinite. Code for this subroutine may be found in RANDOM.CPP.

```
void cauchy (
    int nvars ,
    double scale ,
    double *rand )
```

Other Random Number Generators

The RANDOM.CPP module contains several other random number generators. These include a source of points randomly located on the surface of a unit sphere, gamma random deviates, and beta random deviates.

```
void rand_sphere ( int nvars , double *rand )
double gamma ( int v1 )
double beta ( int v1 , int v2 )
```

The rand_sphere subroutine returns random vectors whose length is 1.0 and whose direction is uniformly distributed. Subroutine gamma returns gamma random deviates having parameter v1 / 2. Finally, beta returns beta random deviates having parameters v1 / 2 and v2 / 2.

Singular Value Decomposition

We frequently need to solve systems of linear equations similar to that shown in Equation (9.1).

$$\mathbf{A}x = \mathbf{b} \tag{9.1}$$

When the **A** matrix is square and nonsingular, there are many ways of computing x. But when **A** contains more rows than columns, a least-squares approximate solution is generally called for. Worse, when the columns of **A** are linearly dependent, or nearly so, simple methods fail. Since these problems are common in prediction applications, a robust method for solving linear systems is required. Singular value decomposition is the standard approach. This is described in Press *et al.* (1992). The particular implementation provided here, whose source is in the file SVDCMP.CPP, is described in Masters (1993).

The user creates a SingularValueDecomp object, fills in its a member array, then calls svdcmp. If the constructor was called with save nonzero, the a matrix is preserved. In order to solve a linear system, place the right-hand side in the member array b and call backsub, passing the singular-value limit (perhaps 1.e-8) and a pointer to the array where the solution is to be placed.

```
class SingularValueDecomp {
public:
    SingularValueDecomp( int rows, int cols, int save );
    ~SingularValueDecomp () ;
    void svdcmp () ;
    void backsub ( double limit , double *soln ) ;
    int ok ;         // Memory allocation successful?
    double *a ;      // Rows by cols inputs A, outputs U
    double *u ;      // If save != 0, U output here
    double *w ;      // Cols vector output of svalues
    double *v ;      // Cols by cols output of 'V' matrix
    double *b ;      // Rows vector of RHS to backsub
};
```

The user who wants to solve one or more linear systems starts by creating a new SingularValueDecomp object. The constructor allocates all necessary memory (a, w, v, b, and optionally u). The third parameter in the constructor, save, should be input as any nonzero number if the input a matrix is to be preserved. Otherwise, this matrix is overwritten with **U** when the decomposition is computed.

After the object is created, the user should check the value of ok. If it is zero, there was insufficient memory to create the object completely. Otherwise, the **A** matrix should be placed in the public vector a and the member function svdcmp() should be called. If save was input as zero in the constructor call, a will be replaced with the **U** matrix. Otherwise, **U** will be available in u. The w and v arrays are also computed. Note that most users have no need of these three quantities. See Masters (1993) for an explanation of their meaning.

The user is now ready to solve one or more linear systems having the same **A** matrix. To solve a system, place the right-hand side vector in the public vector b and call backsub. The solution will be returned in the soln vector supplied by the user as the second parameter of backsub. The first parameter, limit, should be set to some small value. The author typically uses 1.e–8. Smaller values produce higher accuracy in the solution but decreased stability of the solution. Most standard texts on numerical methods discuss this tradeoff. Press *et al.* (1992) does a good job.

Deterministic Optimization

Six subroutines are provided for deterministic minimization of a continuous function. Two of them, glob_min and brentmin, bound and refine (respectively) the minimum of a univariate function. The other four subroutines minimize a multivariate function. The basic conjugate gradient algorithm, implemented in CONJGRAD.CPP, requires storage only on the order of the number of variables. Also, it requires only the gradient of the criterion function. This makes it suitable for a very wide variety of tasks. The subroutine in DERMIN.CPP is also a conjugate gradient algorithm, but in addition to the gradient, it requires the second derivative of the criterion function with respect to each variable (pure partials only). This information is used to optimize the linear search component of the algorithm. The Levenberg-Marquardt algorithm is implemented in the file LEV_MARQ.CPP. This algorithm has two restrictions. First, the criterion being minimized must be a sum of squared errors of a computed value relative to fixed correct values. Second, it requires scratch storage on the order of the square of the number of variables. This makes it infeasible for large problems. But for small problems, it is often tremendously faster than any competitor. Finally, for situations in which the gradient cannot be computed, Powell's algorithm is available in POWELL.CPP.

Bounding a Univariate Minimum

The most basic minimization operation is locating an interval that contains a minimum of a univariate function. There are many sophisticated approaches to this problem. One excellent general method involves golden sections and quadratic extrapolation. Press *et al.* (1992) treats this subject well. On the other hand, it has been the author's experience that a high degree of sophistication in solving this problem is often wasted, or perhaps even counterproductive. The approach taken here is simply to divide a specified interval into equally spaced divisions, evaluate the function at each point, and choose the subinterval containing the minimum. Provision is made for extending the search beyond either endpoint if the function is decreasing there. Many readers will initially cringe at this seemingly crude approach. On the other hand, it actually has broad applicability and reasonable efficiency if used intelligently. Unlike most traditional methods, it has the ability to locate the global minimum among local minima. This is valuable in many prediction problems. Also, if the user has prior knowledge about the probable location of the minimum, this knowledge can be incorporated into the choice of the search interval and the number of search points. This is actually a powerful and useful utility. Mathematical sophistication is best saved for refinement, discussed in the next section.

Operation of the glob_min subroutine is mostly straightforward. The source code is in the file GLOB_MIN.CPP, and a detailed discussion of the algorithm can be found in Masters (1995). Normally, this subroutine returns zero. If the user pressed ESCape before a minimum was bounded, one is returned. The subroutine user_pressed_escape(), which must be defined by the user, is called to detect this event and to return a nonzero value when ESCape is pressed. Also, the user supplied criterion function, criter, may abort and return the negative function value (or any negative number) when the user presses ESCape.

Deterministic Optimization

```
int glob_min (
    double low ,          // Lower limit for search
    double high ,         // Upper limit
    int npts ,            // Number of points to try
    int log_space ,       // Space by log?
    double critlim ,      // Quit if crit drops this low
    double (*criter) (double) , // Criterion function
    double *x1 ,          // Lower X value
    double *y1 ,          // and function there
    double *x2 ,
    double *y2 ,          // Middle (best)
    double *x3 ,
    double *y3 ,          // And upper
    int progress )        // Print progress?
```

The first three parameters specify the endpoints of the interval to be searched for a minimum and the number of points to evaluate in that interval. The minimum number of points is three, as that many are required to define an interval that contains a minimum. If there is any significant possibility of multiple local minima, a larger number of points should be specified.

If log_space is input zero, the points are linearly spaced in the search interval. Otherwise, they are logarithmically spaced. This is appropriate for parameters that have a multiplicative effect.

The critlim parameter allows the user to specify a satisfactory minimum. If the function value drops this low, glob_min will return as soon as the minimum is bounded. Generally, this parameter would be input as zero to force a complete search of the interval.

The criterion function must normally return nonnegative values. If it returns a negative number, glob_min will interpret that as a signal that the user wants to abort. No more function evaluations will be performed. If a minimum has been bounded, zero will be returned. Otherwise, one will be returned.

The next six parameters output three points. The function value, y2, at the center point, x2, is guaranteed to be less than or equal to the function value at the other two points.

There is one special feature of this subroutine that may be useful. If npts is input as the negative of the number of points desired, this flags the fact that the user is supplying in *y2 the value of the function at the leftmost point in the interval, low. This is an efficiency aid, since in many applications that function value will be known at the time glob_min is called, thus saving an expensive function evaluation.

Every time the function is evaluated, the evaluation point and the function there will be formatted to a text string. If the user inputs progress as any nonnegative integer, the external subroutine write_progress will be called with that text string as its only parameter. Otherwise, the external subroutine write_non_progress will be called with that text string as its only parameter. The user may define those two subroutines to take whatever action is desired with the text.

Refining a Univariate Minimum

Once a minimum has been bounded, it usually must be refined to higher resolution. Brent's algorithm, originally described in Brent (1973), is generally agreed to be the most efficient of all competitors. An excellent discussion of this algorithm can be found in Press *et al.* (1992). A detailed description of the particular implementation supplied here in BRENTMIN.CPP can be found in Masters (1995). Suffice it to say that second-order interpolation is used whenever possible, and golden-section refinement is kept in reserve for pathological situations. It doesn't get much better.

This subroutine returns the best (minimum) value of the criterion function. The first parameter imposes a limit on the number of iterations. This is rarely needed, as the convergence tests built into the algorithm are excellent. However, if the user happens to ask for more precision than is numerically attainable, an iteration limit is nice to have in place.

The critlim parameter allows the user to specify a degree of satisfaction. If the function drops this low or lower, refinement ceases. In most cases, this parameter will be set to zero so that the next two parameters control convergence.

There are two different methods for ascertaining convergence. Refinement ceases if either of them is satisfied. The eps parameter specifies accuracy in the function value, while tol specifies accuracy in the x value. The tests are absolute or relative, whichever is more lenient. Since refinement stops when either is satisfied, it is acceptable to let one of these parameters be zero to force full use of the other parameter. As a general rule of thumb, eps should strictly exceed the machine's floating-point precision, and tol should not be much less than the square root of that quantity. If both of these rules are violated, it is quite possible that refinement will uselessly continue until itmax mercifully terminates the process.

The user must input a bounding interval. This is defined by the lower endpoint, the location of an interior minimum, and the upper endpoint. The quantities computed by glob_min are the usual source of these numbers. Also, the value of the function at the interior point, x2, must also be provided via y. The final best point is returned in x2, and the function value there is returned by brentmin.

```
double brentmin (
    int itmax ,            // Iteration limit
    double critlim ,       // Quit if crit drops this low
    double eps ,           // Function error tolerance
    double tol ,           // X error tolerance
    double (*criter) (double) , // Criterion function
    double *x1 ,           // Lower X value, in and out
    double *x2 ,           // Middle (best), in and out
    double *x3 ,           // And upper, in and out
    double y ,             // Function value at x2
    int progress )         // Print progress?
```

During each iteration, this subroutine formats six numbers to a text string. These are the lower endpoint of the current interval, the best point, the upper endpoint, and the best, second best, and third best function values. If the user inputs progress as any nonzero number, the external subroutine write_progress will be called with this text string as its only parameter. Otherwise, the subroutine write_non_progress will be called with the text string as its parameter. The user may define these two subroutines to use the text string in whatever way is desired.

The external subroutine user_pressed_escape() is called after each iteration. If that subroutine returns any nonzero value, brentmin immediately terminates, and it returns the negative of the best function value obtained up to that point. Also, it is expected that the user's criterion function, criter, normally returns a nonnegative value. If it returns a negative value, brentmin takes that as a flag that the user wants to abort.

Conjugate Gradient Minimization

The conjugate gradient algorithm is probably the most effective broadly applicable minimization algorithm available. It only requires computation and storage of first-order derivative information, yet it usually converges to a minimum at practically the speed of full second-order methods. In the author's opinion, the conjugate gradient algorithm's nearly miraculous behavior makes it one of the seven wonders of modern mathematics.

```
double conjgrad (
    int maxits ,            // Iteration limit
    double critlim ,        // Quit if crit drops this low
    double tol ,            // Convergence tolerance
    double (*criter) (double * , int , double * ) ,
    int n ,                 // Number of variables
    double *x ,             // In/out of independent vars
    double ystart ,         // Input of first function value
    double *base ,          // Work vector n long
    double *direc ,         // Work vector n long
    double *g ,             // Work vector n long
    double *h ,             // Work vector n long
    int progress )          // Print progress?
```

Source code for the conjgrad subroutine may be found in CONJGRAD.CPP. It returns the attained function value, or the negative of that value if the user pressed ESCape during computation. In order to determine if the user has pressed ESCape, conjgrad regularly calls the external subroutine user_pressed_escape(), which must return a nonzero value if ESCape was pressed.

The conjugate algorithm as implemented here is very similar to the traditional algorithm. The principal difference has to do with convergence issues. If things are going badly, limits on gamma may be imposed. Also, the accuracy of line minimization is dynamically adjusted. For full details of the algorithm, see Masters (1995).

The first parameter is for insurance only. It limits the number of iterations that will be performed. In practice, it would usually be set to zero, in which case it is ignored. The second parameter is also unimportant in most situations. If the function being minimized ever drops this low, the search for the minimum ceases. Most of the time this parameter would be set to zero so that the search is exhaustive.

The tol parameter is the normal means by which convergence is measured. It is set to a small number, perhaps something in the range of 1.e-5 (very crude) to 1.e-12 (extremely strict). When several iterations in a row fail to reduce the function value by this relative amount (or absolute if the function is tiny), convergence is assumed.

The criterion function criter takes three parameters. The first is the vector of n variables at which the function is to be evaluated. The second is set by conjgrad to either zero or one. If the user's criter function is called with this parameter equal to zero, only the function should be evaluated, and the last parameter must be ignored. If the parameter is nonzero, then criter must also evaluate the gradient and return it in the array pointed to by the last parameter.

The remaining parameters in the conjgrad call list are straightforward. The user specifies the number of variables to be optimized and provides a vector that, on input, contains their starting values and that, on output, will contain their final optimized values. The value of the criterion function at the starting point must also be input. This is an efficiency issue. In most cases this value is already known, so providing it to the subroutine avoids wasting time evaluating it again. The next four parameters are work vectors, each of which is n doubles long.

The last parameter has to do with keeping the user informed of progress. At frequent intervals during the optimization process, intermediate results are formatted to a single text line. If progress is nonzero, the external subroutine write_progress is called with this string as its only parameter. Otherwise, the subroutine write_non_progress is called with this string as its parameter. The user is free to define these two subroutines to use the text line in whatever way is desired. At the beginning of each iteration, glob_min is called for rough line minimization, so its progress output is generated. Then brentmin adds its own output as the minimum is refined. Finally, conjgrad itself generates a single line containing the scale value for search optimization, the attained function value, the length of the search vector, and the percent improvement relative to the previous iteration.

If it is possible to compute the second partial derivative of the criterion function with respect to each variable, this information can be used to make the initial phase of the line search more efficient. The dermin subroutine is almost identical to conjgrad, so the reader should refer to the description of that subroutine. The only significant difference is that the criterion subroutine criter has one additional parameter. The last parameter points to a vector where the criter subroutine will place the n second derivatives whenever the gradient is computed. Like the gradient, this information should be computed only when the second parameter is nonzero.

```
double dermin (
int maxits ,            // Iteration limit
double critlim ,        // Quit if crit drops this low
double tol ,            // Convergence tolerance
double (*criter) (double * , int , double * , double * ) ,
int n ,                 // Number of variables
double *x ,             // In/out of independent vars
double ystart ,         // Input of starting function value
double *base ,          // Work vector n long
double *direc ,         // Work vector n long
double *g ,             // Work vector n long
double *h ,             // Work vector n long
double *deriv2 ,        // Work vector n long
int progress )          // Print progress?
```

Progress reports are identical to those of conjgrad except that the scale factor is more interesting in that values close to 1.0 indicate successful use of the second-order information for search optimization.

Full details of the algorithm can be found in Masters (1995). Source code is in the file DERMIN.CPP.

Levenberg-Marquardt Minimization

When the criterion function being minimized is the sum of squared errors of a model relative to fixed correct values, the Levenberg-Marquardt (LM) algorithm may be the minimization method of choice. In most applications seen by the author, this algorithm converges to a minimum significantly faster than the conjugate gradient algorithm. The principal limitation of the LM algorithm is that it requires scratch storage on the order of the square of the number of independent variables. Its overhead also goes up at approximately that rate, or perhaps even faster. As a result, the LM algorithm is impractical when there is a large number of variables to be optimized.

This subroutine requires a criterion function that is somewhat more complicated than those for the conjugate gradient routines. It not only must compute the function and gradient on demand, but it also must compute an estimate of the Hessian matrix of second-order partial derivatives. But the key is that the second-order information does not need to be explicitly evaluated. There is a slick technique for approximating it during gradient computation. To see how this is done, assume that we have a model $f(x; w_1, w_2, ...)$ that predicts a value y when given an observed x vector. The w_i parameters are to be chosen so that the sum of squared errors of the predicted values relative to the actual values is minimized. The total squared error across a training set is shown in Equation (9.2). The derivative of this error with respect to a model parameter is shown in Equation (9.3), and the approximate second derivative with respect to a pair of parameters is shown in Equation (9.4). Justification for this approximation is given in Masters (1995).

$$e = \sum_k \left[f(x_k; w_1, w_2, ...) - y_k \right]^2 \tag{9.2}$$

$$\frac{\partial e}{\partial w_i} = 2 \sum_k \left[f(x_k; w_1, w_2, ...) - y_k \right] \frac{\partial f(x_k; w_1, w_2, ...)}{\partial w_i} \tag{9.3}$$

$$\frac{\partial^2 e}{\partial w_i \, \partial w_j} \approx 2 \sum_k \left[\frac{\partial f(x_k; w_1, w_2, \ldots)}{\partial w_i} \cdot \frac{\partial f(x_k; w_1, w_2, \ldots)}{\partial w_j} \right] \quad (9.4)$$

In order to compute the gradient and approximate Hessian, one usually starts by zeroing their storage areas, then one sums across the training set. Note that since the Hessian is symmetric, only half of it needs to be computed. The other half is obtained by symmetry. In particular, the code usually looks something like this:

```
error = 0.0 ;

for (i=0 ; i<n ; i++) {
  gradient[i] = 0.0 ;
  for (j=0 ; j<=i ; j++)            // Symmetric, so only compute half
    hessian[i*n+j] = 0.0 ;
  }

for (case=0 ; case<ntrain ; case++) { // Do all cases
  // Process this case by summing error, gradient, and Hessian
  } // for all cases

for (i=1 ; i<n ; i++) {              // Use symmetry to get other half
  for (j=0 ; j<i ; j++)
    hessian[j*n+i] = hessian[i*n+j] ;
  }
```

Actually, we usually want to compute the negative of the gradient, since most minimization algorithms (including those in this text) expect this. Also, the factors of two in Equations (9.3) and (9.4) can be ignored due to the fact that they cancel in the LM algorithm. With these facts in mind, the code for processing an individual case usually resembles the following example. In this example, out is the value predicted by the model, target is the actual observed value, and deriv is the vector of partial derivatives of the model's output with respect to each parameter.

```
diff = target - out ;
error += diff * diff ;               // Equation (9.2)
```

```
for (i=0 ; i<n ; i++) {
  gradient[i] += diff * deriv[i] ;          // Equation (9.3)
  hptr = hessian + i*n ;
  for (j=0 ; j<=i ; j++)
    hptr[j] += deriv[i] * deriv[j] ;        // Equation (9.4)
  }
```

The lev_marq subroutine, whose source code is in the file LEV_MARQ.CPP, is similar in many ways to the conjgrad subroutine. Thus, the reader should refer to the description on page 344 for the explanation of most parameters and external subroutine calls. Full details of the algorithm can be found in Masters (1995).

```
double lev_marq (
    int maxits ,              // Iteration limit
    double critlim ,          // Quit if crit drops this low
    double tol ,              // Convergence tolerance
    double (*criter) (double * , double * , double * ) ,
    int nvars ,               // Number of variables
    double *x ,               // In/out of independent vars
    SingularValueDecomp *sptr ,  // Work object
    double *grad ,            // Work vector n long
    double *delta ,           // Work vector n long
    double *hessian ,         // Work vector n*n long
    int progress )            // Print progress?
```

The criterion function here is different from that for conjgrad in two ways. First, there is no gradient computation flag. This computation is always performed. Second, the approximate Hessian must also be computed as described earlier in this section. The parameters in the criterion function are the point being evaluated, the gradient, and the Hessian, respectively.

Another consideration is that the lev_marq subroutine requires a SingularValueDecomp object for scratch use. This object is described on page 338. When it is allocated, the user must set the last parameter in its constructor (save) equal to any nonzero number so that the original matrix is preserved.

At the end of each iteration, lev_marq creates a text line containing the current function value and the value of lambda. This line is made available just as in conjgrad. No other progress reports are issued.

Powell's Algorithm

When it is possible to compute the gradient of a multivariate function, it is nearly always best to do so and use one of the previous algorithms for minimization. But sometimes there is simply no efficient method for computing the gradient. *It is always wasteful to numerically estimate the gradient by individually perturbing the variables*. The increased efficiency brought about by having the gradient is never worth the high price of numerical estimation. When the gradient is simply not available by any direct means, the best approach to minimization is usually Powell's algorithm. This is discussed in Press *et al.* (1992) as well as Acton (1970) and many other standard texts.

```
double powell (
    int maxits ,              // Iteration limit
    double critlim ,          // Quit if crit drops this low
    double tol ,              // Convergence tolerance
    double (*criter) ( double * ) , // Criterion func
    int n ,                   // Number of variables
    double *x ,               // In/out of independent var
    double ystart ,           // Input of starting fval
    double *base ,            // Work vector n long
    double *p0 ,              // Work vector n long
    double *direc ,           // Work vector n*n long
    int progress )            // Print progress?
```

The algorithm is simple yet surprisingly effective, attaining nearly quadratic convergence in good conditions. It passes through all of the variables, calling glob_min and brentmin to minimize along each direction. Then it looks at the average direction moved over the course of those individual minimizations. If several tests are passed, one of the original univariate search directions is replaced by the average direction of improvement. This process is repeated until convergence is obtained.

The parameters are nearly the same as those in the previous minimization subroutines. The user can specify a limit on the number of iterations, or zero if no limit is desired. A satisfactory value for the criterion function allows early escape. Normal escape is by means of the usual convergence tolerance, typically around 1.e–6 to 1.e–9 or so. For efficiency, the user inputs in ystart the function value at the starting point (input in x). Three work vectors are required.

At the end of each iteration, powell creates a text line containing the current function value and the degree of improvement. This line is made available just as in conjgrad. Also, whenever a search direction vector is replaced, a message to that effect is generated. No other progress reports are issued.

Stochastic Optimization

Sometimes the function being minimized is well behaved. It is shaped like a bowl. Drop a marble anywhere in it, and the marble will roll to the bottom. But more often, the function to be minimized is filled with hills and valleys. A randomly dropped marble will most likely roll to the bottom of a valley that is much more elevated than the lowest valley. In this situation we must trade away the speed of deterministic optimization for the thoroughness of stochastic optimization. This latter approach is based on random numbers. The function is evaluated at random locations as the lowest valley is sought. These are not just any random locations, though. They must be intelligently chosen. In this section we will present three subroutines that intelligently perform a random search to locate the global minimum of a function. One of them is a primitive, but broadly effective version of simulated annealing. Another is a generic annealing algorithm that can be used to implement a number of variations on the traditional approach. The third algorithm is stochastic smoothing, in which a smoothed version of the objective function is minimized, with the degree of smoothing being gradually reduced.

Primitive Simulated Annealing

Consider the following highly intuitive method of function minimization. Start by randomly trying a lot of points centered around a good starting point such as the origin. As long as new records for function smallness continue being set, keep searching in that vicinity. When a long string of tries fails to improve the function any more, move to the point at which the smallest (best) function value was found. Now use that point as the center of a random search for improvement, but limit the search to a slightly narrower domain. Repeat this procedure as needed. This is the algorithm implemented in ANNEAL1.CPP. It was originally presented in Masters (1993), and it is reviewed in Masters (1995).

```
double anneal1 (
    int n ,                  // This many parameters
    double *x ,              // They are input/output here
    double *work ,           // Work vector n long
    double (*criter) ( double * ) , // Criterion func
    double besttval ,        // Starting f value, huge forces
    int itry ,               // For more randomization
    int ntemps ,             // Number of temperatures
    int niters ,             // Iterations at each temp
    int setback ,            // Set back iter counter
    double starttemp,        // Starting temperature
    double stoptemp,         // Stopping temperature
    enum RandomDensity density , // Density
    double fquit ,           // Quit if criterion drops this low
    int progress )           // Write progress?
```

The first parameter is the number of variables to be optimized. The x parameter serves as both the input of a starting point (often the origin) and as the output of the location of the best point. The next parameter supplies a work vector. The criterion function is expected to return nonnegative values always. This allows the user to interrupt criterion computation if this is a slow process.

The bestfval parameter facilitates double duty for this subroutine. Only improvements on this value will be accepted. If the starting vector is a point that is already known to be good, and improvement is sought, then set bestfval equal to the function value at that starting point. But if we are looking for an initial estimate of the best point, in which case the starting point has no meaning other than being a reasonable search location, then set bestfval to some gigantic number. That will force acceptance of *some* point. This is handy when the origin is used as the starting point, but the final point must not be all zeros. (Some neural network training algorithms fail to operate if the starting weights are identically zero.)

The itry parameter is useful when anneal1 may be called repeatedly with the same starting point to do parallel annealing. By setting itry to a different integer each time, randomization is enhanced.

The basic annealing parameters follow. These are the number of temperatures, the number of iterations at each temperature, the amount to set back the iteration counter after improvement, the starting temperature, and the stopping temperature. The latter two quantities are the standard deviation of the random perturbation if the Gaussian distribution is used, and the scale factor if the Cauchy distribution is used. The density parameter must be set to either NormalDensity or CauchyDensity.

The fquit parameter allows the user to specify a degree of satisfaction. If the criterion function value drops to this value, annealing is immediately terminated. In most cases this would be set to zero.

The last parameter controls output of progress information. After each temperature is complete, a text line is formatted to contain the temperature and the best error that was obtained at that temperature. If progress was input as any nonzero value, the external subroutine write_progress is called with that text line as its parameter. Otherwise, the external subroutine write_non_progress is called with that text line as its parameter. The user should define these two subroutines to use the line appropriately.

Normally, the best function value is returned. However, anneal1 regularly calls the routine user_pressed_escape() to see if the user wants to abort. If that routine returns any nonzero integer, anneal1 immediately quits and returns the negative of the best attained function value.

Generic Traditional Simulated Annealing

The algorithm presented in the previous section bears only slight resemblance to the original form of simulated annealing. It uses random search based on decreasing temperatures, but that's about it. However, the fact of the matter is that at least in neural network applications, it usually performs significantly better than the traditional algorithm. Nevertheless, for the sake of completeness and fairness, traditional simulated annealing is also supplied in this text. The algorithm whose source is in ANNEAL2.CPP is fully described in Masters (1995). It is very versatile in that more than just the usual annealing parameters may be specified. In addition to the parameters described in conjunction with anneal1, the user can choose either of two different temperature reduction schedules and two different acceptance probability functions. A parameter for controlling the initial acceptance rate is also available.

```
double anneal2 (
    int n ,                   // This many vars to optimize
    double *x ,               // They are input/output here
    double *current ,         // Work vector n long
    double *best ,            // Work vector n long
    double (*criter) ( double * ) , // Criterion func
    double bestfval ,         // Starting f value, huge forces
    int ntemps ,              // Number of temperatures
    int niters ,              // Iters at each temperature
    int setback ,             // Set back iteration counter
    double starttemp,         // Starting temperature
    double stoptemp,          // Stopping temperature
    enum RandomDensity density , // Density
    double ratio ,            // For starting probability
    int climb ,               // Hill climbing?
    int reduction ,           // Temp reduction schedule
    double fquit ,            // Quit if criterion drops this low
    int progress )            // Write progress?
```

This subroutine is similar to anneal1 in many regards, so the reader should refer to the previous section for most details. Only the differences will be discussed here.

The *ratio* parameter allows the user to set the initial acceptance rate, which should usually be about 80 percent or so. In many problems, setting *ratio* equal to something in the range 0.1 to 5.0 is appropriate. Larger values produce greater acceptance.

The *climb* parameter should be any nonzero integer to use the traditional Metropolis acceptance criterion. If it is zero, Harold Szu's (1986) alternative is used.

The *reduction* parameter should be set to one of the two constants ANNEAL_REDUCE_EXPONENTIAL or ANNEAL_REDUCE_FAST. These constants are defined in CONST.H. The user may wish to slightly modify the source code in ANNEAL2.CPP to avoid use of these constants. This parameter specifies that either traditional exponential temperature reduction is used, or Szu's fast alternative is used.

Progress output for anneal2 is identical to that for anneal1 with one exception. When this algorithm starts, the text line *Initializing* is output. When initialization is complete, the mean error and its standard deviation are output. From that point on, output is the same for both subroutines.

Stochastic Smoothing

This stochastic minimization algorithm uses a radically different approach than simulated annealing. It does not base its operation on the function values at individual points. Instead, it randomly samples neighborhoods to estimate the average value of the function, and it minimizes this smoothed version of the raw criterion. As time passes, the size of the smoothing neighborhood is gradually decreased so that the smoothed function more and more closely resembles the actual function. A complete description of this algorithm can be found in Masters (1995), and the source code is in the file SSG_CORE.CPP.

This subroutine is similar in operation to anneal1; see that description on page 352 for most details. Only the differences will be discussed here.

The principal difference is that ssg_core requires a slightly unusual criterion function. The first parameter of the criterion function is the vector containing the point at which the function is to be evaluated. The second parameter is a limit. The third parameter is a flag for gradient evaluation. The last parameter is the vector in which the gradient will be returned if it is computed. The criterion function must compute the gradient if and only if the third parameter, the gradient flag, is nonzero, *and* the function value is less than the second parameter, the limit. If either of these conditions is false, the gradient should not be evaluated. This unusual requirement is for the sake of efficiency. The relatively expensive gradient is computed only when ssg_core determines that it is definitely needed.

```
double ssg_core (
    int n ,                    // This many vars
    double *x ,                // They are input/output here
    double (*criter)( double *, double, int, double * ),
    double bestfval ,          // Starting func, huge forces
    int ntemps ,               // Number of temperatures
    int niters ,               // Iters at each temperature
    int setback ,              // Set back iteration counter
    double starttemp ,         // Starting temperature
    double stoptemp ,          // Stopping temperature
    enum RandomDensity density ,   // Density
    double fquit ,             // Quit if crit drops this low
    int use_grad ,             // Use gradient hinting?
    double *avg ,              // Work vector n long
    double *best ,             // Ditto
    double *grad ,             // Ditto if use_grad != 0
    double *avg_grad ,         // Ditto if use_grad != 0
    int progress )             // Write progress?
```

It is possible to use this subroutine to minimize a function whose gradient cannot be computed. Input the use_grad parameter as zero. This will cause the gradient flag passed to the criterion function to always be zero. If use_grad is any nonzero value, the gradient will be used to optimize the random search. Note that the last two work vectors may be NULL if use_grad is zero.

The progress output is straightforward. When the routine starts, the text *Initializing* is output. When initialization is complete, the average and the best function values are output. After processing at each temperature is complete, a relatively long line is output. It contains the temperature, the percent of trials that resulted in improvement, the average function value, and the best function value so far. If the gradient is used, the gradient length is also printed.

Eigenvalues and Eigenvectors

Eigenvalues and eigenvectors of a real symmetric matrix may be computed with subroutine jacobi. If a matrix is the product of a real symmetric matrix and the inverse of a real symmetric matrix, the eigenstructure of the product $\mathbf{B}^{-1}\mathbf{A}$ may be computed with eigen_bia. The calling sequence for jacobi is as follows:

```
void jacobi (
    int n ,              // Size of matrix
    double *mat ,        // Square symmetric real matrix
    double *evals ,      // Output of n eigenvalues
    double *evect ,      // Output of n by n eigenvectors
    double *work1 ,      // Work vector n long
    double *work2 )      // Work vector n long
```

The user inputs the size of the square matrix and the matrix itself, which is destroyed. The eigenvalues are returned as a vector, sorted from largest to smallest. The eigenvectors are returned as a matrix, each of whose columns corresponds to the eigenvalues in the same order. Two work vectors must also be supplied.

Some mathematical algorithms, such as computation of linear discriminant functions, require the eigenvalues and eigenvectors of a matrix that can be expressed as the product of a real symmetric matrix $\mathbf{A}$ and the inverse of a real symmetric matrix $\mathbf{B}$. The eigenstructure of the product $\mathbf{B}^{-1}\mathbf{A}$ is computed with eigen_bia, whose calling sequence is shown below.

```
void eigen_bia (
    int n ,              // Size of matrix
    double *a ,          // Square symmetric real matrix
    double *b ,          // Ditto (both destroyed)
    double *evals ,      // Output of n eigenvalues
    double *evect ,      // Output of n by n eigenvectors
    double *work1 ,      // Work vector n long
    double *work2 )      // Work vector n long
```

The calling parameters of eigen_bia are almost identical to those of jacobi. Both input matrices are destroyed. The eigenvalues are again sorted in decreasing order, and the columns of evect are the corresponding eigenvectors.

Both of these subroutines are in the module EIGEN.CPP on the accompanying disk.

Fourier Transforms

The fast Fourier transform (FFT) included with this text is implemented as a class rather than as a set of subroutines. The reason is that many FFT applications involve repeated invocation of the algorithm for arrays of the same size. By creating an object appropriate for a certain FFT size, allocation of scratch storage and some other size-dependent operations can be consolidated and reused. The constructor for the FFT class is as follows:

```
FFT::FFT (
    int ndim ,           // Dim of this var, N for a vector
    int spacing ,        // Spacing of pts, 1 for a vector
    int n_segments )     // N of segments, 1 for a vector
```

Fourier Transforms

Signal-processing applications will invariably call the constructor with the first parameter (ndim) set to the length of the vector to be transformed (or half that length for real series; see below). The other two parameters are set equal to one. This algorithm can also be used for multivariate transformations of an arbitrary number of variables. These uses are beyond the scope of this text. See IEEE (1979) for a brief overview of multivariate FFT applications. Singleton (1969) is the ultimate reference, as he is the designer of the algorithm on which this implementation is based. Complete source code is on the accompanying disk. See the end of this section for a description of the source files.

There are several FFT functions that can be used to compute a transform. The complex-general transform is computed with cpx:

```
void FFT::cpx (
    double *real ,      // Real parts (in and out)
    double *imag ,      // Imaginary parts (in and out)
    int isign )         // Sign of exponent (+/-1)
```

The real and imaginary vectors each have a total length equal to the product of the three parameters in the constructor call. The transformation is done in place, so these two vectors are used for both input and output. The last component determines the direction of the transform. Historically, a sign of −1 was considered a forward transform, and +1 the inverse transform. However, this definition is showing a tendency to reverse in modern usage. All programs in this text use a sign of +1 for the forward transform.

Many applications involve transformation of a real vector. In this case, a fully general transform is wasteful of both time and memory. If the length of the real vector is even, an efficient forward transformation (positive exponent) can be done with the rv function:

```
void FFT::rv (
    double *real ,      // In: 0,2,4,... Out:Real parts
    double *imag )      // In: 1,3,5,... Out: Imag parts
```

There are several tricky aspects of this function. The constructor must have been called with its ndim parameter equal to *half* the length of the real series. The other two parameters are one, as usual. The transformation is done in place, so the real and imag vectors serve as both input and output. The raw series is placed in these two series by alternating terms. All even-indexed (zero origin) raw terms go into the real input, and all odd-indexed raw terms go into the imaginary input. In other words, the first raw term goes into the first real slot. The second raw term goes into the first imaginary slot. The total length of the complex input vector is $n/2$, where n is the length of the raw series.

The output is a little tricky. Recall that the Fourier transformation of a real series contains $n/2+1$ frequency components. This is one more than the length of the input/output vectors. What do we do with the extra component? The answer lies in the fact that the imaginary parts of the first (DC) and last (Nyquist) components are always zero. So this function returns the real part of the last component in the imaginary part of the first component. The quantity that would otherwise reside in real[$n/2$], an address that does not exist, is placed in imag[0], a place that theoretically contains zero. All other output terms are as expected.

The inverse transformation (negative exponent) can be done with the member function irv. This exactly reverses the effect of rv. In particular, it divides each component by the number of terms in the original series, as is required at some point in the round trip. The real and imaginary input/output vectors must be arranged as they were output by rv, with the Nyquist component in imag[0]. The calling sequence is as follows:

```
void FFT::irv (
     double *real,      // In: Real parts, Out: 0,2,4,...
     double *imag )     // In: Imag parts, Out: 1,3,5,...
```

In case the unusual order required for the efficient DFT of a real series is still not clear, it is illustrated in Figure 9.1. The left box in this figure is the arrangement of the time-domain series before rv is called and after irv is called. The right box is the arrangement of the frequency-domain series after rv is called and before irv is called. The columns in each box are the real and imaginary input/output vectors.

Fourier Transforms

	Time domain			Frequency domain	
Position	Real	Imaginary		Real	Imaginary
0	x_0	x_1		Re_0	$Re_{n/2}$
1	x_2	x_3		Re_1	Im_1
2	x_4	x_5		Re_2	Im_2
3	x_6	x_7		Re_3	Im_3
...				...	
n/2 - 1	x_{n-2}	x_{n-1}		$Re_{n/2-1}$	$Im_{n/2-1}$
				$Im_0 = Im_{n/2} = 0$	

Figure 9.1 Layout for the even real FFT.

Source code for these algorithms is supplied on the accompanying code disk. It is split among several files. All routines accessible to the user are in MRFFT.CPP. This module calls two worker routines. One of them, permute, is in MRFFT_P.CPP. The other, kernels, is in MRFFT_K.C. Note that this module is treated as a C program rather than C++. Also, its prototype in MRFFT.CPP includes the extern "C"{kernels} declaration. The reason for using C is that an alternative source file, MRFFT_K.ASM, is also supplied. This is a Pentium optimized assembler version of the program. For large series, it does not execute much faster than the C version compiled by a high-quality optimizing compiler. However, the ASM version is significantly more accurate than the C version due to the fact that many of the critical intermediate computations are done entirely in the math coprocessor stack, retaining full 80-bit precision. This avoids the loss of precision suffered when intermediate results are stored to and retrieved from 64-bit system memory. Unfortunately, unless the series is quite short, the cache thrashing inherent in mixed-radix transforms swamps out most of the gains attainable by hand-coding in assembler. On the other hand, the ASM version certainly is faster than the C version compiled by a poor compiler! And for series that are short enough to avoid cache thrashing, the assembler version is almost twice as fast as even an efficiently compiled C version.

Data Reduction and Orthogonalization

Several methods are available for reducing the dimensionality of a dataset while simultaneously enhancing the independence of its component variables. These include computation of principal components of the raw data, principal components of class centroids, and linear discriminant functions. These topics are described on page 16 of this text. Internal operation details can be found in Masters (1995).

Two files on the accompanying disk are relevant to this technique. The core algorithms, including the eigenstructure routines jacobi and eigen_bia, as well as the PrincoData and Discrim classes, are in EIGEN.CPP. Examples of the use of these classes are in ORTHOG.CPP.

Eigenvalues and eigenvectors are computed with the subroutines jacobi and eigen_bia. These are described on pages 357 and 358, respectively.

Principal components of a data array, which may actually be an array of group centroids, is accomplished with the PrincoData class. Its constructor is as shown below.

```
PrincoData::PrincoData (
    int ncases ,    // N of cases (rows in data array)
    int nv ,        // N of vars (columns in data array)
    double *data ,  // Input (ncases by nv) matrix
    int stdize ,    // Standardize to equal variance?
    int maxfacs ,   // Max number of output factors
    double frac ,   // Fraction (0-1) of variance
    int *nfactors ) // Out: Number of factors retained
```

The user inputs a matrix, each of whose rows is a single case. Its columns correspond to the measured variables. The data is preserved. If the flag stdize is input nonzero, the principal components are based on standardized inputs. This is usually recommended, but see Masters (1995) for a discussion of this issue. The user may also input maxfacs as an upper limit on the number of factors retained. It is usually best to set this to a huge number and let the number of variables set the limit. Similarly, the user may input frac as the fraction (0–1) of the total input variance to retain. The number of factors is the minimum number that accounts for at least this fraction. Again, it is often best to set this to 1.0 or greater to keep all factors, then choose the number to use later. If both maxfacs and frac are used to limit the number of factors, the number kept will be the minimum as determined by those two limits. Remember that nv and ncases−1 are also upper limits. The number of factors retained is returned in nfactors.

Instead of filling the data array with raw cases, each row may be a class centroid. The resulting principal components optimally separate the groups *when the correlation of the variables is ignored*. Since this is nearly always a bad assumption, this technique is generally discouraged. The stdize flag should be set to zero in this situation.

If one desires to compute factors that optimally separate classes, nearly always the better approach is to compute linear discriminant functions. Even though this technique assumes that the covariances are the same for all classes, this assumption is nearly always superior to ignoring correlations entirely, which the preceding technique does. Linear discriminant functions are computed with the Discrim class, whose constructor is now shown.

```
Discrim::Discrim (
    int ncases ,      // Number of cases
    int nv ,          // Number of variables
    int nc ,          // Number of classes
    double *data ,    // Input (ncases by nv) matrix
    int *classes ,    // Class ID of each case (0 to nc-1)
    int maxfacs ,     // Max number of output factors
    double frac ,     // Fraction (0-1) of variance
    int *nfactors )   // Out: Number of factors retained
```

The calling parameters for this constructor are almost identical to those of the PrincoData constructor. There are only two differences. The user must specify the number of classes via nc. The ncases elements in the vector classes contain the class membership information. Each case's class id is specified there, with classes running from 0 through nc-1. It is the user's responsibility to make sure that each of the nc classes has at least one representative present.

After an object of either the PrincoData or Discrim type has been constructed, data may be transformed by calling the member function factors. The constructor does not actually change any data. It simply computes and stores the information that will be needed to transform data. The member function that applies the transformation for the Discrim class is shown below. The PrincoData class has exactly the same member function.

```
void Discrim::factors (
    int ncases ,        // N of cases (rows in data)
    double *data )      // Input (ncases by nvars) and
                        // Uutput (ncases by nfacs)
```

There are only two parameters in the call list. The first specifies the number of cases in the array. The number of variables does not need to be specified because it is the same as was used for the constructor (or better be).

There is one potentially confusing point of operation. The column dimension of the data array changes from input to output. When the original variables are input, there are nvars columns. Upon output of the computed factors, there are nfactors columns, which will generally be less than nvars. This rearrangement facilitates subsequent use of the array for other operations that usually assume the number of columns is equal to the number of variables.

Autoregression by Burg's Algorithm

When pure autoregressive parameters must be computed for a single series, the method of choice is usually Burg's algorithm. The original source for this algorithm is in Childers (1978). The implementation in this text follows Press *et al*. (1992). An excellent derivation of the algorithm, including a lot of philosophical justification, can be found in Elliot (1987). Unfortunately, the flowchart included in that text leaves much to be desired. The calling sequence for this subroutine, found in BURG.CPP on the accompanying disk, is as follows:

```
void burg (
    int n ,              // Length of input series, x
    double *x ,          // This is the raw data
    int maxlag ,         // Compute this many coefs
    double *pcorr ,      // Output last of each iter here
    double *coefs ,      // Output maxlag final coefs
    double *prev ,       // Work vector maxlag long
    double *alpha ,      // Work vector n long
    double *beta )       // Work vector n long
```

The first three parameters are the length of the input series, the series itself, and the maximum lag in the model. Three work vectors are also needed. Two vectors are output. The autoregressive coefficients are returned in coefs. This vector contains maxlag weights corresponding to lags of one, two, and so on, through maxlag. Since the algorithm recursively computes the weights for successively larger models, a byproduct of the procedure is the partial autocorrelations of the series. These maxlag numbers are returned in pcorr. In other words, the first element of pcorr is the single AR parameter of an AR(1) model. The second element is the lag-two parameter of an AR(2) model, otherwise known as the lag-two partial autocorrelation of the series. The third element is the lag-three parameter of an AR(3) model, and so forth.

Note that the input series is centered before any other computation is done. The x vector itself is not changed; the centering is implicit.

10
Neural Network Tools

- Probabilistic neural networks

- Multiple-layer feedforward networks

This chapter describes the neural network tools that are supplied with this text. Details concerning internal operation and properties of the models are largely omitted. The reader should see Masters (1994) for a thorough discussion of all multiple-layer feedforward (MLFN) models and Masters (1995) for all members of the probabilistic neural network (PNN) family. The focus of this chapter is limited to practical implementation issues for the reader who wishes to incorporate the supplied code into his or her program.

Training and Test Sets

All neural networks in this text share a common format for training and test data. The TrainingSet class encapsulates into a simple structure all input and output values for both mapping and classification tasks. This class contains many public and private members that are of no concern to most users. The constructor takes care of setting those parameters, and the neural network member functions use them as needed. Only the important variables will be discussed here. Full source code is in the file TRAIN.CPP.

```
class TrainingSet {
public:
    TrainingSet ( int outmode , int n_in , int n_out ) ;
    unsigned ntrain ;   // Number of samples in 'data'
    double *data ;      // Actual training data here
    unsigned *nper ;    // N of cases in each class
    double *priors ;    // Prior probability weights
} ;
```

The constructor requires three parameters. The first is the output mode. It must be set to either OUTMOD_CLASSIFICATION or to OUTMOD_MAPPING. These constants are defined in CONST.H. The other two parameters are the number of inputs and outputs, respectively. If classification output is specified, the number of outputs is equal to the number of classes (exclusive of the implicit REJECT class that may be used in MLFN models). The ntrain member variable is the number of cases in the training set. If classification output is being used, the nper and priors arrays (automatically allocated by the constructor) must contain the number of cases in each class and the prior probability of each class, respectively. This information is used only by PNN models. These two arrays do not contain any information about implicit REJECT cases. There are exactly n_out elements in each array.

The training data is strung out in a single array of doubles. This data array must be allocated by the user, although it will be freed by the destructor if it is not NULL. The first n_in elements are the inputs for the first case. If mapping mode is in use, the next n_out elements of the data array are the corresponding outputs. If classification is the output mode, the class occupies a single double following the inputs. This should be the ordinal class number plus a small fraction (to avoid floating-point problems when it is truncated to an integer). The class number origin is one, as class zero is reserved for the implicit REJECT category allowed in MLFN models. This layout is repeated for each case.

Generic Network Parameters

All neural network models share some parameters in common. This includes things like the number of inputs and outputs and the network and output models. Some parameters apply only to one family of models. The network parameters are collected into one common structure to facilitate passing them to subroutines. This structure is shown on the next page.

The net_model variable is one of the NETMOD_? constants defined in CONST.H. These constants will be discussed as each model is presented later in this chapter. The number of inputs and outputs define the basic architecture. The out_mode variable is the same flag that is used in the training set constructor and defined in CONST.H. It must be either OUTMOD_CLASSIFICATION or OUTMOD_MAPPING, according to whether the classification model (one output neuron per class) or general function mapping is to be used.

The basic PNN model gives the user the choice of two kernel functions. The variable net_kernel should be set to KERNEL_GAUSS for the usual Gaussian kernel, or to KERNEL_RECIP for the reciprocal kernel.

```
struct NetParams {
    int net_model ;     // Net model ( NETMOD_?)
    int n_inputs ;      // Number of input neurons
    int n_outputs ;     // Number of output neurons
    int out_mode ;      // Output mode (OUTMOD_?)
    int net_kernel ;    // PNN Parzen kernel (KERNEL_?)
    int net_domain ;    // MLFN domain (DOMAIN_?)
    int linear ;        // MLFN linear outputs?
    int n_hidden1 ;     // MLFN hid1 neurons
    int n_hidden2 ;     // MLFN hid2 neurons
};
```

The remaining member variables apply only to MLFN models. The domain is specified by net_domain. Linearity of the output neurons is flagged by linear. The size of each hidden layer is specified by n_hidden1 and n_hidden2.

Generic Learning Parameters

As was the case for the network parameters, there are many learning parameters in common for all neural network models. There are also some that specifically apply to certain models. All learning parameters are grouped for convenience into one structure. This facilitates passing the parameters to various subroutines. This structure is listed on the next page.

The first member variable, progress, controls output of intermediate results. Most training algorithms are relatively slow. For the sake of impatient users, these algorithms regularly format intermediate results into a text string. If progress is set to any nonzero value, the user-supplied subroutine write_progress is called with the text string as its only parameter. Otherwise, write_non_progress is called with the text string as its parameter. The user is expected to define these two subroutines to take appropriate action with the text. Normally, this would be writing it to the screen for the user's edification.

Generic Learning Parameters

```
struct LearnParams {
    // These parameters are universal
    int progress ;        // Detailed progress info dest
    double quit_err ;     // Quit if error this low
    int acc ;             // Initial digits accuracy
    int refine ;          // Additional refined digits
    // These are for PNN family only
    double siglo ;        // Min for global optimization
    double sighi ;        // And max sigma
    int nsigs ;           // Number to try in that range
    // These are for MLFN family only
    int method ;          // MLFN  learning METHOD_?
    int errtype ;         // MLFN err measure, ERRTYPE_?
    int retries ;         // Quit after this many retries
    int pretries ;        // N of tries before refinement
    struct AnnealParams *ap ; // Annealing params
} ;
```

The user can specify an acceptable error level via the quit_err variable. If training progresses to the point that the error drops this low, the training terminates. In most cases, the user will set this variable equal to zero to force thorough training.

The acc and refine variables work together to determine the degree of accuracy attained during training. For MLFN training algorithms that hybridize simulated annealing with direct descent, acc is the number of significant digits computed initially, and refine is the number of *additional* significant digits computed during refinement. See Masters (1995) for a thorough presentation. For PNN family training, the individual acc and refine values are ignored. Their *sum* is the number of significant digits optimized. The siglo, sighi, and nsigs variables specify the search range and number of trial points for initial sigma bounding when training members of the PNN family.

For MLFN training, the method variable specifies the training algorithm to use, and errtype is the error measure to optimize. These subjects are discussed on pages 438 and 439, respectively.

The retries and pretries variables apply to the hybrid MLFN training algorithms described in Masters (1995). Finally, the ap structure contains the simulated annealing parameters for both pure annealing and hybrid training. See CLASSES.CPP for a listing of this structure.

Generic Neural Networks

Neural networks form a class hierarchy. The pure virtual base class is Network. It encapsulates all of the basic neural network architecture. Most users will need only a few of its member variables and functions, so only the important ones are shown here.

```
class Network {
    int ok ;
    int errtype ;
    double *out ;
    int learn ( TrainingSet *tp , struct LearnParams *lp );
    int trial ( double *input ) ;
    double trial_error ( TrainingSet *tptr ) ;
} ;
```

After constructing a new network, the user should check the value of the member variable ok. If it is zero, there was insufficient memory to construct the object completely.

The network is trained by calling the learn member function. Two parameters are passed to this function. The first is a pointer to a training set. The second is a pointer to a structure containing the learning parameters described on page 370. Normally, this function returns zero. If the user pressed ESCape before training completed, one is returned. If there was insufficient memory to train the network, minus one is returned.

As long as the network has not been trained to any degree, the errtype member variable is zero. When the network is trained at all, even just partially, this variable is set to a nonzero value. For members of the PNN family, the value is one. For members of the MLFN family, the value is the errtype specified in the learning parameter object.

The user may press ESCape at any time during training. In some situations, if training is aborted very early in the process, insufficient information will have been gathered to support any degree of confidence in the training. In this case, errtype will remain zero and learn will return the value one. If some training progress has been made before the user aborts, learn will still return one, but errtype will be made nonzero. If training is completed to the user's specifications, learn will return zero.

The trained network is executed on a case by calling the member function trial and passing the input vector to that function. The output neuron's activations are then found in the public vector out.

The mean squared error (across all cases and output neurons) of a training or test set is computed by calling the trial_error member function.

Multiple-Layer Feedforward Networks

The MLFN class, a child of the Network class, supports models having zero, one, or two hidden layers. The hyperbolic tangent activation function is used for all hidden neurons. The output neurons may either use that same function, or they may be linear (use the identity function).

Automatic classification mode is supported, with the number of output neurons being equal to the number of classes. An implicit REJECT class may be used to indicate that all output neurons are to be turned off. This is indicated in the training set by a case being in class zero. Prior probabilities, if specified, are ignored. Naturally, fully general function mapping is supported in addition to classification.

```
MLFN ( char *name , NetParams *net_params );
```

The constructor requires a name for the network, which may be empty or NULL. It also requires that the desired network parameters be passed to it via the NetParams structure. This structure was described on page 369.

MLFN Training Considerations

There is considerable versatility in training MLFN models. The error measure that is optimized can be chosen, and the learning algorithm itself can be specified. These are both identified in the LearnParams structure (page 370) that is passed to the learn member function.

Four different error measures are available. They are mean squared error, mean absolute error, Kalman-Kwasny error, and cross entropy. These measures are briefly discussed on page 439 of this text. Full details can be found in Masters (1994). The choice is set via the errtype member variable in the learning parameters object. Legal values for this variable are the ERRTYPE_? constants defined in CONST.H.

Many training algorithms are available. They include several pure stochastic methods (primarily for experimentation) and some extremely effective hybrids. These algorithms are briefly discussed on page 438 of this text. Full details can be found in Masters (1995). The choice is made via the method member variable in the learning parameters object that is passed to learn. Legal values for the method variable are the METHOD_? constants defined in CONST.H.

Probabilistic Neural Networks

Several members of the PNN family are available. These include the basic single-sigma model; the SEPVAR model, in which each variable has its own sigma; and the SEPCLASS model, in which each class has its own sigma vector. When general function mapping is required, the generalized regression version of the PNN is invoked. These models are discussed in detail in Masters (1995). All PNN models are children of the virtual parent PNNet, which, in turn, is a child of the Network class.

When classification is used, prior probabilities can be incorporated. An implicit REJECT class is not allowed. If the training set contains any cases in class zero, serious runtime errors will occur. Users wanting a REJECT category must explicitly designate one of their classes as representing rejects.

Each constructor requires a name for the network, which may be empty or NULL. It also requires that the desired network parameters be passed to it via the NetParams structure. This structure was described on page 369.

```
PNNbasic ( char *name , NetParams *params ) ;
PNNsepvar ( char *name , NetParams *params ) ;
PNNsepclass ( char *name , NetParams *params ) ;
```

Only one training algorithm is available for each PNN model. They all use cross validation to compute the mean squared error of the training set. This technique is discussed in detail in Masters (1995). The PNNbasic model uses glob_min (page 340) and brentmin (page 342) to compute the optimal sigma. The other two models use glob_min to estimate a good common sigma, then they use dermin (page 346) to estimate the individual sigmas.

11

Using the NPREDICT Program

- Overview of program capabilities and operation

- Detailed discussion of all commands

- Alphabetical summary of all commands

- Validation suite

NPREDICT is a general-purpose program for analyzing and predicting time series. It contains a wide variety of preprocessing options, including transformations, ordinary filters, quadrature-mirror and Morlet wavelet filters, and a variety of dimension-reduction algorithms such as principal components and linear discriminant functions. Multivariate ARIMA models are available for prediction, as are probabilistic neural networks, generalized regression, and multiple-layer feedforward networks. Finally, results can be displayed on the computer screen and written to disk in assorted ways.

Using This Manual

Two versions of the NPREDICT program are provided. One runs under control of the DOS operating system, and the other runs under Microsoft Windows NT™. They are identical in their capabilities, including their use of command control files. However, interaction with the user is different. The DOS version requires the user to type in commands exactly as they would appear in a command control file. The Windows version allows the user to perform many operations through menus. This adds complexity to the manual when both techniques must be addressed. Since most users will enjoy the convenience of command control files, and since these files are identical for both versions, the primary organization of this manual will feature the text version of commands. Windows menu paths should generally be obvious when the program is used.

This chapter is conceptually divided into several parts. The first and largest part groups all topics according to their content. For example, all neural network information appears together in one section. All ARMA information appears together in another section, and so forth. The second part of this chapter is an alphabetical listing of all commands for quick reference. Of necessity, there is no organization of this part other than alphabetical. The final section contains validation suites. These are command control files that test various aspects of the program. The purpose of this material is twofold. First, it allows users to test the program on their hardware to verify its correct operation. This is particularly important if the user has recompiled the source code to produce a custom version of the program. The second purpose of this information is for education. Users would do well to study these command control files in order to gain a clearer understanding of the operation of NPREDICT. The examples that appear throughout the first part of this chapter are helpful, but most of them are fragments that focus on individual topics. The files in this final part are complete and self-contained.

General Commands

Some commands are related to general operation of the NPREDICT program. These commands are described in this section.

Exiting the Program

In order to exit NPREDICT, the user types **BYE**. This command may appear in a command control file. Before returning control to the operating system, NPREDICT frees all allocated memory and simultaneously performs an internal consistency check. This may occasionally result in a very short delay. This delay should never be a problem, and it is a valuable feature in a program that may suffer user customization. This is especially true of the Windows version, as NPREDICT would be ashamed of itself if it failed to return to the system all of the resources that it borrowed.

The Audit Log File

By default, an ASCII file named AUDIT.LOG is created in the current directory whenever NPREDICT starts. If a file having this name already exists, it is erased. This file records all operations performed in the current session, regardless of whether the operation was initiated in a command control file or manually. Results of operations, when relevant, are also written to this file. For example, when training neural networks, the attained error is written. The audit log file provides a hard copy of operations that facilitates documentation and duplication of results.

The name of the audit log file can be changed during a session. This is done by means of the following command:

AUDIT LOG = FileName

The filename may be changed as often as desired, thus allowing the user to separately document different activities. If a file having the new name already exists, it will be preserved. Operations from the current session will be appended to the end of the file.

If no audit log file is desired, simply enter the command **AUDIT LOG** with no equal sign or filename. Operations will not be recorded from that time onward unless a new audit log file is requested.

Note the different treatment of audit log files according to their use. The original AUDIT.LOG is erased when the program starts, but subsequent files used via **AUDIT LOG** are preserved, with new information appended.

Progress Indicators for Slow Operations

Many operations in the NPREDICT program can be excruciatingly slow. Some neural network training can require many hours. It is unnerving to have the computer absolutely quiet for long periods of time. Invoke the following command for relief:

PROGRESS ON

After that command has been received, the program will write to the screen voluminous output detailing intermediate results whenever particularly slow operations are being performed. This output may be incomprehensible to the average user, but at least the poor soul will know that something is happening, as opposed to waiting for three days after a power glitch caused the computer to hang. Some explanation for this output can be found in this manual when the operation producing the output is presented. Hardcore users can refer to the source code provided on disk for the ultimate explanation.

By default, this option is turned off so as to avoid confusion. After turning it on, it can be turned off again with the following command:

PROGRESS OFF

Note that this pair of commands is relevant only to the DOS version of NPREDICT. The Windows version always prints progress reports to a special window that appears when slow operations start.

Command Control Files

Interactive operation is possible. The user may enter commands that are immediately executed. However, the more usual mode of control is to prepare in advance an ASCII command file. Each line of this file is a single command, identical to those that might be entered interactively. There are at least two reasons why this method is preferable to interactive control. First, many program runs will require an extended period of time: overnight or even longer. The use of a self-contained command file allows the user to engage in other activities while the program is executing. The other reason is that a command file is self-documenting. The user will have a hard record of the parameters that went into producing the obtained results. The audit log file also provides a hard record of this information, but the command control file has the advantage that it can be edited and executed again. The command to read an ASCII command file is the following:

COMMAND FILE = filename

Usually, this command is typed by the user to initiate processing of the command file. However, it may also appear in a command file. In other words, recursive processing of command files is allowed.

Every command has one or more components. The mandatory first component is the action to be taken. Many commands will also have auxiliary information. In this case, the action is followed by an equal sign (=), which in turn is followed by the auxiliary information. Blanks may optionally appear before or after the equal sign, and case (upper/lower) is ignored.

Commands may be commented by placing a semicolon (;) after the relevant parts, then following the semicolon with any text. Lines that are only comments are implemented by using a semicolon as the first character on the line.

Working with Signals

The fundamental unit of data in NPREDICT is the *signal*. A signal is a univariate time series. In other words, it is an ordered sequence of real numbers. Some operations will not care about ordering, and any arbitrary reordering of the signal will produce identical results. Other operations assume not only ordering, but they also assume that the samples were taken at equally spaced intervals. The assumptions of each operation, if unclear, will be discussed as they are presented in this manual.

Many operations support multivariate data. A different signal is used to represent each component of a multivariate dataset. There is no such thing as a single multivariate signal. For example, suppose we measure the oxygen and carbon dioxide level in a reaction vessel every minute for an hour. We have an oxygen signal containing 60 samples, and we have a carbon dioxide signal also containing 60 samples. Using appropriate rules, these can be treated as a bivariate time series for the duration of a particular operation. This allows tremendous versatility in choosing variables.

Reading Signals from a Disk File

One or more signals may be read from a disk file. This is a two-step operation. First, the signal names to be associated with the data fields are specified. Then the file itself is named. This latter operation is what actually causes the signals to be read.

The ASCII disk file must follow a strict format. Each line in the file corresponds to one observation time slot. One or more variables appear on each line. Each of these variables is separated from the others by one or more nonnumeric characters. In other words, each number consists of the digits 0–9 with an optional decimal point or minus sign. Any character other than these, including spaces, commas, or tabs, can be used to delimit the variables.

The names of the series to be read are specified with the following command:

NAME = Name1 , Name2 , ...

The names of the series are separated by commas. Any number of spaces, including none, may appear before or after the commas. Each name may optionally contain spaces, but the program will automatically change all embedded spaces to underscores (_).

It is possible to skip fields in the dataset. This is done by using one or more commas as placeholders. If there are more variables in the dataset than signal names given, the extra data will be ignored. If the user specifies more signal names than there are variables in the dataset, errors will surely follow.

The following example illustrates reading two signals from a disk file. Each line of the file contains five numbers. We are interested in the second and the fourth. The other three variables are to be ignored. Suppose the file looks like this:

```
3.1   5.2   2.2   2.4   7.1
4.2   4.6   1.1   2.6   6.8
        .
        .
        .
```

Also, suppose the signal names are specified as follows:

NAME = , oxygen , , CO2

The leading comma in that rule causes the first variable to be skipped. The extra comma in the middle causes the third variable to be skipped. As a result, the *oxygen* series will have the values (5.2, 4.6, ...), and the *CO2* series will be (2.4, 2.6, ...).

The **NAME** command only associates signal names with data fields. In order to read the file, the following rule is used:

READ SIGNAL FILE = FileName

Only the current directory is searched for the named file. A full path may also be specified as part of the filename. If a named signal already exists, it will be destroyed and replaced by the new signal.

Saving a Signal to a Disk File

It is possible to write a signal to an ASCII disk file. The signal is written with one observation per line. Only one signal at a time can be written. It is not possible to write multiple signal files. The command to save a signal is as follows:

SAVE SIGNAL = SignalName TO FileName

By default, the file is written in the current directory. A complete pathname may be specified if desired.

Generating a Signal

It is possible to generate a signal from scratch, as opposed to reading it from a file or deriving it from another signal. The name of the signal to be generated should first be specified with the following command:

NAME = Name

Then, the number of samples in the signal and the method for generating it are specified with the following rule:

GENERATE = Samples Method [Optional parameters]

The method is one of the following:

UNIFORM Min Max — The samples are random floating-point numbers uniformly distributed in the range Min to Max.

NORMAL Mean StdDev — The samples are random floating-point numbers following a normal (Gaussian) distribution with the specified mean and standard deviation.

SINE Amplitude Period Phase — The samples are a sine wave having the specified amplitude, period (in samples), and phase (in degrees). To generate a cosine wave at zero phase, generate a sine wave with a phase of 90.

RAMP Amplitude Period — A ramp wave having the specified amplitude and period is generated. The first point is zero. The signal rises linearly, with the value of point i (zero origin) being the amplitude times $(i\%p)/p$, where p is the specified period and % stands for the *mod* function (remainder after division). Unlike most other generation functions, the period must be an integer. If the specified period is not an integer, any fractional part is truncated.

ARMA AR1 AR2 AR10 MA1 MA2 MA10 — This is used to generate test series. The series has three AR and three MA terms, each at lags ot 1, 2, and 10. The user must specify all 6 weights on this command, even those that are zero.

For example, to generate 128 samples of an inverted sine wave that repeats every 32 samples, use the following pair of rules:

```
NAME = SineWave
GENERATE = 128 SINE 1.0 32.0 180.0
```

If more than one name is given on the **NAME** command, the names must be separated by commas. Multiple *identical* copies of the generated signal are made. The fact that they are identical is significant when the generated signals are random numbers! Use multiple signal names in one command to replicate the random sequence. If different random sequences are desired, separate **NAME** and **GENERATE** commands must be used for each different signal.

Modifying Signals

There is a variety of commands for modifying a signal in place. These commands do not create a new signal. They change an existing signal. As a general rule, only one modification of each possible type can be applied to a signal. For example, if you detrend a signal, you cannot detrend that signal again, even if other modifications have been done subsequent to the original detrending. The reason for this restriction is that modifications can also be undone. This facilitates many powerful procedures, especially some hybrid prediction methods. By allowing only one modification of each type per signal, ambiguities are eliminated. This restriction is almost never an obstacle in practical applications. In those rare situations in which multiple modifications of a single type are desired, the user can simply copy the signal, creating another new signal, and modify that one.

Many modifications can be undone. This is accomplished with a command that starts with the word UNDO, followed by the name of the modification to be undone, an equal sign, and the name of the signal. It is also possible to undo for one signal a modification that was done to a different signal. For example, it may be useful to detrend signal x, train a model with that detrended signal, use the model to create signal *new_x* as predicted future values of the detrended x, then retrend *new_x* to correspond to the original x. This is done by following the name of the signal to be undone with the word PER, then the name of the signal whose modification is to be used for the undo operation.

Some of the modification commands require that a parameter be specified, while others do not. Legal modifications are:

CENTER = Signal — The mean is subtracted from the signal, giving it a new mean of zero.

MEDIAN CENTER = Signal — The median is subtracted from the signal, giving it a new median of zero.

UNDO CENTER = Signal [PER Other] — The previously computed center is added to the signal.

DETREND = Signal — The least-squares trend line is removed from the signal. The balance point is the center of the signal, so the mean is not changed.

UNDO DETREND = Signal [PER Other] — The previously computed trend line is added to the signal. The same balance point is used for retrending as was used for detrending. If the retrended signal is the same length as the detrended signal, its mean will not change. However, if it is longer, as would happen for prediction of future values, the mean will change accordingly. This is correct operation and is what is desired.

OFFSET = constant Signal — Add a constant to every point of a signal.

UNDO OFFSET = Signal [PER Other] — Subtract the previously added constant.

SCALE = constant Signal — Multiply every point of a signal by a constant.

UNDO SCALE = Signal [PER Other] — Divide every point of a signal by the constant that previously multiplied the signal.

STANDARDIZE = Signal — Linearly transform the signal to have zero mean and unit variance.

UNDO STANDARDIZE = Signal — Linearly transform the signal to undo the effect of standardization.

DIFFERENCE = degree Signal — The signal is differenced one, two, or three times as specified by the degree. Each point is decremented by the previous point. The length of the signal will decrease by the degree (which cannot exceed three). The value of the first point before each stage of differencing will be preserved internally and restored during an UNDO operation.

UNDO DIFFERENCE = Signal [PER Other] — The signal is summed as many times as it was differenced in the original difference operation. Each point is incremented by the previous point. The former first point is restored prior to each summing. The signal will increase in length by the degree of original differencing.

SEASONAL DIFFERENCE = period Signal — The signal is seasonally differenced using the specified period. The points in the first period are internally preserved in anticipation of a subsequent UNDO operation. The length of the signal decreases by the period.

UNDO SEASONAL DIFFERENCE = Signal [PER Other] — The seasonal differencing is undone by summing, incrementing each point by the point exactly one period earlier. The points in the first period of the originally differenced signal are restored prior to summing. The length of the signal increases by the period.

INTEGRATE = [period] Signal — The first *period* points remain unchanged. After that, each subsequent point is increased by the value of the point *period* samples earlier. If no period is supplied, it defaults to one. The length of the signal remains the same. This command may be repeated as many times as desired, and it cannot be undone. A **DIFFERENCE** command will reverse an **INTEGRATE** command except for the fact that the first point (or period for **SEASONAL DIFFERENCE**) is lost.

LOG = Signal — The log of every point in the signal is found.

EXP = Signal — Every point in the signal is replaced by its exponential.

Copying and Subsetting a Signal

It is possible to copy one signal to another. Optionally, only part of the signal may be copied. Also, zeros may be prepended or appended during the operation. The syntax is the following:

    NAME = Destination [, Dest2 , ...]
    COPY = [count] Source

Two commands are required. First, the name of the signal that will receive the copy is specified. It is legal to specify more than one signal name on the **NAME** command, in which case multiple identical copies will be made. It is also legal to use the same name for both the source and the destination, in which case the signal will be replaced.

The second command names the source signal. An optional point count may appear. If no count is given, a direct copy is done. Otherwise, the length of the destination signal will be equal to the absolute value of the count, which may be positive or negative. If the count is positive, the copy starts at the beginning of the source signal. In this case, if the count exceeds the length of the source signal, zeros are appended to the end of the destination signal to bring it up to the full length. If the count is negative, that many points are copied from the end of the source signal. If the absolute value of the count exceeds the length of the source signal, the entire source signal is used, and zeros are prepended to the beginning of the destination signal to bring it to the correct length.

Displaying Signals

A signal is displayed with the following command:

    DISPLAY = Name

By default, the horizontal axis will be labeled with the ordinal number of each observation, using zero as the first. The user may optionally specify any nonnegative time origin with the following command:

    DISPLAY ORIGIN = Number

Working with Signals 389

Also, the sample rate may be specified with the following command:

DISPLAY RATE = Number

The sample rate is the number of sample points per unit of time. The horizontal axis is labeled in units of time. The default rate is 1.0, making the time of each sample equivalent to its ordinal number. The sample rate affects the display of signals and spectra. Correlations are always displayed in terms of ordinal lags.

There are several ways of specifying the vertical axis scaling. When a scaling command is invoked, it remains in effect for all subsequent **DISPLAY** commands. The following scaling methods are available:

DISPLAY RANGE = OPTIMAL — This is the default. The vertical axis is scaled in such a way that the entire range of the signal is displayed.

DISPLAY RANGE = SYMMETRIC — The entire signal range is displayed with the value zero in the vertical center.

DISPLAY RANGE = Min Max — The user specifies the display range. Signal values outside that range will be truncated.

Note that in all cases, the actual display range will usually be greater than the signal or the user indicates. This is to allow sensible labeling of the axis.

It is possible to display only a subset of the signal. This is accomplished with the following command:

DISPLAY DOMAIN = ALL — This, the default, causes the entire signal to be displayed.

DISPLAY DOMAIN = Min Max — This causes only the specified subset of points to be displayed. The origin is zero, meaning that these numbers correspond to the labels on the display if the rate is one.

The specified domain is used for all subsequent **DISPLAY** commands until another **DISPLAY DOMAIN** command appears.

Displaying Correlations

When the user computes a new signal by means of one of the lagged correlation commands (page 405), an internal flag is set recording that fact. When a **DISPLAY** command is issued for that signal, the signal is displayed in a special way. Instead of connecting the cases with a continuous line, each case is represented by a vertical bar. The horizontal limits are slightly extended so that all cases are fully visible. Finally, a pair of lines is drawn at plus and minus 1.96 times the approximate *theoretical* standard deviation of the correlations under the assumption that the series is white noise. Brave (or foolish) users may wish to use these lines to judge the significance of correlations.

The horizontal axis is always labeled in terms of sample lags. If the user specified a **DISPLAY RATE**, that rate is ignored when lagged correlations are displayed.

Displaying Spectra

When a signal is created with the **SPECTRUM** or **MAXENT** command, an internal flag indicates that the signal is a spectrum. When this signal is displayed, the horizontal axis is labeled according to frequency. The default sample rate is one sample per unit time. This implies that the rightmost extent of the spectrum graph (the Nyquist frequency) is 0.5. If the user specified a different rate with the **DISPLAY RATE** command, that rate determines the horizontal labeling. The rightmost extent is equal to half the sample rate.

When a cumulative spectrum deviation is displayed, a pair of confidence lines is symmetrically drawn above and below zero. These are the approximate $\alpha = 0.1$ thresholds for a Kolmogorov-Smirnov test of the null hypothesis that the time-domain signal is white noise. If the deviation graph touches or crosses either of these lines, it is likely that the series is not white noise.

The Power Spectrum and its Relatives

Two commands are needed to compute the Fourier transform of a signal. These are the following:

NAME = Real , Imag , Power , Phase , Deviation
SPECTRUM = SignalName

The first command names one to five output signals. These are the real component of the transform, the imaginary component, the power (sum of squared real and imaginary components), the phase angle in radians (from minus pi to pi), and the deviation of the cumulative power spectrum from uniformity. The source series is divided by its length before spectrum computation so that the sum of the power terms equals the variance of the original series.

Any of these five outputs may be ignored. To ignore outputs prior to the one(s) of interest, use commas as placeholders. If fewer than five output names appear, later outputs are ignored. For example, to compute only the power, use two commas to skip the real and imaginary components:

NAME = , , PowerSignal

The second command actually performs the computation. The **SPECTRUM** command names the signal whose Fourier transform is found. There are no restrictions on the length of this signal. However, maximum speed is obtained when the length is a power of two. It should be even if at all possible. If the length can be factored into small primes, speed should still be good. In the event that the length is a large prime number, computation could be very slow.

The length of the output signal(s) is equal to half the length of the source signal (truncating the fraction if odd), plus one. The first point in the output corresponds to the DC (constant) component, and the last point is the Nyquist frequency (a period of two samples).

If a signal name appears in the fifth **NAME** slot, the slot for the cumulative spectrum deviation, supplementary information is written to the audit log file. This includes DMAX (the maximum absolute deviation), and alpha (the approximate Kolmogorov-Smirnov probability of attaining a deviation this high if the time-domain signal is white noise). Small values of alpha (perhaps less than 0.1 or so) indicate that the signal is probably not white noise.

By default, a Welch data window is applied to the source series before transforming. This may be changed by explicitly specifying a window. When a window is specified, it remains in effect for all subsequent **SPECTRUM** commands until explicitly changed again.

At this time, NPREDICT offers only one data window, the excellent Welch window. Legal commands for specifying this data window (or turning it off) are:

SPECTRUM WINDOW = NONE
SPECTRUM WINDOW = WELCH

If no data window is used, the raw data is transformed. In this (unwise) case, it is strongly recommended that the series be centered prior to the transformation. If a data window is used, the series will be automatically centered during spectrum computation using the window-weighted mean. This causes the DC component to vanish, which is nearly always desirable.

Smoothing the Power Spectrum

The raw power spectrum has tremendous variance. This randomness causes the individual components to be difficult to visually interpret. They generally provide meaningful information only when groups are taken together. On the other hand, direct grouping has problems of its own. For this reason, it is recommended that some sort of lowpass filter be applied to the spectrum before it is displayed. Any of the lowpass filters provided in NPREDICT may be used. However, members of the Savitzky-Golay family are often best because they are especially good at following sharp spikes. Two commands are necessary to effect this filter. First, the user specifies the name of the generated signal with a **NAME** command. Then the filter is invoked by a command that specifies the filter half-length, its degree, and the name of the signal to be filtered. These commands are:

NAME = OutputName
SAVGOL = HalfLength Degree SourceName

As a general rule, larger values of the half-length result in a smoother output. However, the half-length should not significantly exceed the width of the narrowest spike to be resolved. Something in the range of 10 to 40 is common. The degree should be even and generally in the range of two (a good first try) to six (rather extreme). Higher degrees are most appropriate for longer filters.

The Maximum Entropy Spectrum

There is an alternative to the discrete Fourier transform for computing a power spectrum. The maximum entropy method is recommended when it is expected that there may be one or more sharp, narrow spikes that are particularly interesting. However, be warned that this method has somewhat of a craving for such spikes. Even if none naturally exists in the data, this method will quite possibly find one or more spikes. This can easily deceive the user. Therefore, it is strongly recommended that the DFT power spectrum also be displayed and used for confirmation. The special beauty of the maximum entropy method is that it can precisely locate the center of a spike. This is especially valuable when the data may contain a low-amplitude pure periodic component that might be buried in DFT noise and smoothed out of existence by a lowpass filter.

To compute the maximum entropy spectrum, use two commands. The first command names the output signal that will receive the spectrum. The second command specifies the resolution, the order, and the source signal whose spectrum is to be computed. These commands are as follows:

NAME = OutputSignal
MAXENT = Resolution Order SourceName

Most users will want to use a resolution of several times the length of the source series, and at least several hundred, even if the source series is short. The order strongly affects results. It should usually be as small as possible, yet several times the number of spikes expected. Something around 20 is often good. Higher values cause more details to appear and are needed if the user must detect numerous spikes. However, using an excessive order invites the appearance of spurious spikes. Values beyond 150 are virtually never appropriate, and values beyond 50 are not often needed.

Data Reduction and Orthogonalization

When the user is overwhelmed with a multitude of signals, perhaps exhibiting great redundancy, it can be beneficial to reduce their quantity to a few relatively independent signals. This subject is briefly described on page 16 of this text, and it is covered in detail in Masters (1995). This section shows how data reduction and orthogonalization can be accomplished with the NPREDICT program.

In order to orthogonalize a set of signals, several steps are necessary. First, the user must define a training set on which the orthogonalization model will be based. This training set is essentially identical to the training set used for neural network training, discussed on page 423. The interrelationships of the variables in this set will determine the nature of the data reduction computations. After the training set has been created, the user must specify some parameters relating to the orthogonalization model and tell the program to define a model. Finally, this model is used as often as desired to compute new sets of signals (called *factors*) from old sets. These steps are now described.

The user must first identify the signals and their lags (if lags are desired) for the data that will comprise the training set. This is done with commands having one of the following forms:

INPUT = SignalName
INPUT = SignalName Lag
INPUT = SignalName MinLag MaxLag

If a lag range is specified, the two lags may optionally be separated by a dash. This has no effect; many users think that a dash improves readability of the control file.

Every time an **INPUT** command appears, it is appended to the end of a cumulative input list. Thus, disjoint lag ranges are attained by using multiple **INPUT** commands. To clear the input list and start a new list, use the following command:

CLEAR INPUT LIST

For example, suppose we want to orthogonalize based on four variables. These are the current value of signal x and the values of signal y at lags of 2, 3, and 8. The following commands would specify this set of variables:

    INPUT = x
    INPUT = y 2-3
    INPUT = y 8

Data Reduction and Orthogonalization

There is one more small but important issue related to cumulation of a training set. Some of the orthogonalization techniques are based on examples from several classes. *Even when the chosen technique doesn't refer to class membership*, a class must be specified. This is done with the following command:

CLASS = ClassName

Note that NPREDICT keeps a running list of all classes that have been named. This list is used for many operations in addition to orthogonalization. To avoid confusion, it is usually good to clear the class list before (and/or after) orthogonalization operations. This is done with the following command:

CLEAR CLASSES

After the input variables and a class name have been supplied, the training set is built by issuing the following command:

CUMULATE TRAINING SET

As the command indicates, this is a cumulative operation. If a training set already exists, the new data is appended to the old. In order to erase any existing training data and start fresh, issue the following command:

CLEAR TRAINING SET

Three different types of orthogonalization are available. The type that is desired is specified as follows:

ORTHOGONALIZATION TYPE = Type

Detailed explanations of the three types may be found starting on page 16. Legal types are:

PRINCIPAL COMPONENTS — The principal components of the data are computed. Class information is ignored (even though some dummy class name had to be specified during creation of the training set). The maximum number of factors is equal to the number of input variables.

CENTROIDS — The principal components of the class centroids are computed. This is usually the least desirable choice, being appropriate only when the covariance matrices of the classes are very different from one another. Standardization should not be used with this type of orthogonalization. The maximum number of factors is equal to the number of variables or to one less than the number of classes, whichever is smaller.

DISCRIMINANT — The linear discriminant functions that optimally separate the class centroids relative to the pooled within-group covariances are computed. Standardization is implicit in this type of orthogonalization, so the **STANDARDIZE** option is ignored. The maximum number of factors is equal to the number of variables or to one less than the number of classes, whichever is smaller.

By default, the maximum possible number of factors is computed. It is probably best to let this be done and determine the number of factors actually retained by means of the signal names on the **NAME** command described later. However, if the user wishes to limit the number of factors, this may be done in two different ways. First, the number may be explicitly specified with the following command:

ORTHOGONALIZATION FACTORS = Number

Alternatively, the user may wish to specify that only those most important factors that account for a given percentage of the total variance be retained. This percentage (0–100) is specified with the following command:

ORTHOGONALIZATION LIMIT = Percentage

Note that the number of factors retained is the minimum of the quantities specified by the two commands just shown, as well as the mathematical limits discussed in the **TYPE** definitions. For example, suppose we have 20 variables and 6 classes. There is an automatic limitation of 5 factors (one less than the number of classes, which is less than the number of variables). Also suppose the user specified the number of factors as 4, and limited the variance percentage to 60 percent. Assume that after including 3 factors, the variance accounted for is 61 percent of the total. Then 3 factors are retained.

There is one more option that may be specified. If the orthogonalization type is either **PRINCIPAL COMPONENTS** or **CENTROID**, the following rule determines whether standardization of the raw variables is included in the computation:

ORTHOGONALIZATION STANDARDIZE = YES / NO

Note that it is highly recommended that this option be set to **NO** if the **CENTROID** type is used. For **PRINCIPAL COMPONENTS**, this option should almost always be set to **YES**. This option is ignored for the **DISCRIMINANT** type.

After the training set has been cumulated and the options just discussed have been set, the orthogonalization model may be defined. This is done with the following command:

DEFINE ORTHOGONALIZATION = Name

The user must supply an arbitrary name in this command. This name will be referenced whenever the model is used to compute new sets of signals from old sets. Any number of orthogonalization models may exist at one time.

Information about the model is written to the audit log file. This information takes the form of a matrix. There are as many columns as there are factors retained. Each column corresponds to a factor, with the first being the most important and subsequent columns having decreasing importance. The topmost number in each column is the percentage of the total variance accounted for by that factor. The remaining rows are the weights that multiply each input. The input variables are numbered for each row, and the lag of that input is enclosed in parentheses.

Orthogonalization models may be saved to disk and restored at a later time. This is handy when neural networks have been trained on reduced data sets. In this case, the runtime data must be reduced in exactly the same way. Using a previously saved orthogonalization model avoids the clumsy alternative of recomputing the model from training data each time. The save and restore operations are effected with the following commands:

SAVE ORTHOGONALIZATION = Name TO File
RESTORE ORTHOGONALIZATION = Name FROM File

The file specified on that command may optionally contain drive and path information. If only a filename is given, it is assumed to be in the current directory.

Orthogonalization models do not take up much memory, so there is no great harm in leaving them around throughout a session. However, sometimes it might be desirable to remove them. The first of the following two commands clears a single named model, while the second clears all orthogonalization models.

CLEAR ORTHOGONALIZATION = Name
CLEAR ORTHOGONALIZATIONS

Once an orthogonalization model has been defined, it may be used as often as desired to compute a new set of signals from an old set. This involves three steps. First, the input signals (and their lags, if any) must be specified by means of **INPUT** commands, just like those used when the training set was cumulated. It is *vital* that this information be specified in exactly the same order as the training set. NPREDICT cannot read the user's mind. The variables in the model that was based on the training set are matched up with the variables specified now, one by one, in exactly the same order. It is the user's responsibility to make sure that the orders correspond.

The second step is to specify the names of the new signals that will be generated. It is legal to use any or all of the same names as the input signals, in which case the inputs will be replaced with the new data. However, this would probably not be desirable very often. The names of the created signals are specified with the following command:

NAME = Name1 , Name2 , ...

As many names as desired may be specified, each separated from the others by a comma. The number of factors can always be determined by examining the audit log file entry created by the **DEFINE ORTHOGONALIZATION** command. If fewer names than factors are given, the remaining (least important) factors are ignored. If more names than factors are given, the excess signals are set equal to zero in their entirety.

The final step in computing a new signal set is to issue the following command, which names the model that is used:

APPLY ORTHOGONALIZATION = Name

In many situations, the input data will contain lagged values. This has different implications for training the model and for using the model to compute a reduced dataset. Suppose signal x contains 100 points. Also, suppose that we wish to compute the principal components of the four variables defined by the current value and lags up to three samples back in time. The following commands are used to build the training set:

```
INPUT = x  0-3              ; The four input variables
CLASS = dummy               ; Ignored: Any name will do
CUMULATE TRAINING SET       ; Builds the training set
```

The first case in the training set will contain the values of x at time slots 3 (the current time), 2, 1, and 0 (the maximum lag). In other words, the concept of *current time* must begin three slots in from the beginning in order to avoid including in the training set data that is undefined (prior to the start of the series). The training set will contain 97 cases. It would be disastrous if invalid data found its way into the training set, where it would introduce errors into all subsequent computations.

A different situation occurs later, when we are computing a reduced set of principal component variables. Let us use the same series and input list. We still have the problem that the first three time slots are based on missing data. But do we really want to skip those slots when computing the output series? Life is a lot easier if we do not skip them, computing the full 100 output cases. The reason is that subsequent use of the signals is simplified if we do not have to keep track of time offsets. If we were to discard the first three observations, computing only 97 points for the output series, we would be forever after burdened with the responsibility of remembering that each point in the output series corresponds to a time slot three samples later than the other series that are probably being used in the application. (It is, after all, likely that other series are in use. Even if not, a consistent time reference is always nice.) At best, this is an annoying burden. At worst, careless user errors are possible. Luckily, the price of computing the first three invalid points is very small. All we need to remember is that they (and they alone) are invalid. In fact, the NPREDICT program replicates the first point in each input series for use as prior values, so the amount of error in the first three outputs is probably not too large. This is a price well worth paying in return for keeping all time references concurrent.

An Example Ignoring Class Membership

It's time for some examples. To start, suppose we are doing a mapping problem such as prediction of current values of some dependent series, or predicting future values of any series. Or perhaps we are predicting class membership, but we do not care about optimizing the orthogonalization to separate the classes. Let us say that we want the raw inputs to be the current value of variable $x1$, and values of $y1$ at lags of 0 (current value), 1, 2, and 7. This is a total of five inputs. The first step is to create the training set. Remember that we need to name a class, even though class information will be ignored. For this example, assume we are starting fresh, so that no clearing of class, input, or training set information is needed. A later example will demonstrate clearing old information. The following commands would be used:

```
INPUT = x1                   ; First variable
INPUT = y1 0-2               ; Second, third, and fourth
INPUT = y1 7                 ; Fifth
CLASS = dummy                ; Any name will do
CUMULATE TRAINING SET        ; This builds the training set
```

The next step is to use the training set to define an orthogonalization model. Since classes are to be ignored, the PRINCIPAL COMPONENTS type is used. Standardization of input variables is almost always a good choice for this type. The name *princo* seems like a good name for the model.

```
ORTHOGONALIZATION TYPE = PRINCIPAL COMPONENTS
ORTHOGONALIZATION STANDARDIZE = YES
DEFINE ORTHOGONALIZATION = princo
```

We are now ready to use the model to create a new set of variables from the old. Since the number of factors has not been limited by the user (with the **FACTORS** or **LIMIT** option), the mathematical limit is five, the number of input variables. We decide that we want only the three most important factors. So we name three signals and compute them.

```
NAME = factor1 , factor2 , factor3
APPLY ORTHOGONALIZATION = princo
```

An Example Using Class Membership

Now let's take a different situation. Suppose we have a classification problem, and we want to compute the optimal linear discriminant functions as new variables. Again, the first step is to cumulate a training set. If we are in the midst of a program session, we need to clear the decks. Whenever a new set of operations involving its own set of classes is begun, the class list should be cleared. If this is not done, and there are some unrelated classes left over from previous work, NPREDICT will complain that not all classes are represented in the training set. Speaking of training sets, we want to start a new one. Issue the two following commands:

```
CLEAR CLASSES
CLEAR TRAINING SET
```

We will use the same variable structure as in the previous example. Suppose that the samples for the first class are in signals $x1$ and $y1$. Cumulate those samples into the training set. Remember to clear the input list before specifying new members of that list.

```
CLEAR INPUT LIST        ; Start a new input list
INPUT = x1              ; First variable
INPUT = y1 0-2          ; Second, third, and fourth
INPUT = y1 7            ; Fifth
CLASS = class1          ; These represent first class
CUMULATE TRAINING SET   ; Start the training set
```

Suppose the samples for the second class are in signals $x2$ and $y2$. Append this information to the existing training set.

```
CLEAR INPUT LIST        ; Start a new input list
INPUT = x2              ; First variable
INPUT = y2 0-2          ; Second, third, and fourth
INPUT = y2 7            ; Fifth
CLASS = class2          ; These represent second class
CUMULATE TRAINING SET   ; Append to the training set
```

We can now use the training set to compute an optimal orthogonalization. Specify the type and give the model a good name.

ORTHOGONALIZATION TYPE = DISCRIMINANT
DEFINE ORTHOGONALIZATION = discrim

In most situations we now want to apply the orthogonalization to the samples in both classes so that the new variables can be used to train a neural network or some other model. There are two classes, so only one factor can be computed. The samples from the first class are used to compute a new signal, *dfact1*, by means of the following commands:

```
CLEAR INPUT LIST           ; Start a new input list
INPUT = x1                 ; First variable
INPUT = y1 0-2             ; Second, third, and fourth
INPUT = y1 7               ; Fifth
NAME = dfact1              ; Create this signal
APPLY ORTHOGONALIZATION = discrim
```

The corresponding signal for the second class may be computed as follows:

```
CLEAR INPUT LIST           ; Start a new input list
INPUT = x2                 ; First variable
INPUT = y2 0-2             ; Second, third, and fourth
INPUT = y2 7               ; Fifth
NAME = dfact2              ; Create this signal
APPLY ORTHOGONALIZATION = discrim
```

Filters

A variety of filters are available. Two commands are needed to filter a signal. The first, a **NAME** command, names the output signal(s). The input (source) signal may be replaced by naming it as an output if desired. (This would not often be done.) Otherwise, the input signal is not changed. The output signal has the same number of points as the input signal. No time shift takes place, so the absolute time of each point in the output corresponds to the same time in the input. Except for quadrature-mirror (and Morlet wavelet) filters, all responses are in phase. The individual filters are now discussed.

Lowpass, Highpass, and Bandpass Filters

These three filters have the same syntax and general operation except for the keyword used. The syntax is the following:

> **NAME** = Signal [, Signal , ...]
> **LOWPASS** = Frequency Width SourceName
> **HIGHPASS** = Frequency Width SourceName
> **BANDPASS** = Frequency Width SourceName

The **NAME** command usually names exactly one signal, which is the filtered output. If more than one signal is named, with the names separated by commas, multiple identical copies of the filtered output are made.

Lowpass filters pass all frequencies below the specified frequency, and they are usually employed for smoothing. Highpass filters pass all frequencies above the specified frequency. They are usually used to extract information on local variation while suppressing overall signal levels. Bandpass filters pass only those periodic components in the vicinity of the specified frequency.

The frequency must range from 0.0 (the DC or constant frequency) to 0.5 (the Nyquist frequency, which is the highest correctly detectable frequency). The reciprocal of the frequency is the period of the component. So, for example, to keep only those components having a period longer than 12 samples, use a lowpass filter with a frequency of $1/12 = 0.0833$.

The width parameter controls the rate at which the filter drops off. Smaller values produce a faster dropoff. However, smaller values also extend the length of the implicit time-domain filter. In other words, smaller values of the width cause each filtered output point to be influenced by more distant input points. As a rough rule of thumb, divide 0.8 by the width to estimate the distance of the furthest sample affecting each output point. A width of 0.01 causes a very sharp dropoff, but inputs as far away as 80 samples in both directions affect each output point. A width of 0.1 is quite gentle, but reasonable for many common applications other than special narrow bandpass filters. With this width, only points up to about 8 samples away influence each output point. A width of 0.05 is a good compromise for general use. See Figure 4.5 on page 113 for an illustration of the effect of various width parameters.

Quadrature-Mirror Filters and Morlet Wavelets

Quadrature-mirror filters are similar to the three filters just discussed in that they require a **NAME** command to specify the output signal(s), and they require specification of a frequency and width. However, they are different in that up to four different output signals are generated. These are the in-phase output, the in-quadrature output, the amplitude, and the phase angle in radians (from minus pi to pi). A Morlet wavelet filter is a special type of QM filter that is modified to have zero response at a frequency of zero. This is important if very wide filters are used. The syntax is as follows:

> **NAME = InPhase , InQuad , Amp , Phase**
>
> **QMF = Frequency Width SourceName**
> **MORLET = Frequency Width SourceName**

Any of these four outputs may be ignored. To ignore outputs prior to the one(s) of interest, use commas as placeholders. If fewer than four output names appear, later outputs are ignored. For example, to compute only the amplitude, use two commas to skip the in-phase and in-quadrature outputs:

> NAME = , , AmpSignal

Padding

By default, the input series is preprocessed by being centered and detrended in such a way that its first and last points are equal to zero. Zero is then used for padding. This reduces potentially serious end effects for series that contain a deterministic trend or are nonstationary. For series that are fundamentally centered around a stable mean value, it is better to avoid detrending and just pad with the observed mean. The commands that determine which action is taken (for all subsequent filters) are:

> **PADDING = DETREND**
> **PADDING = MEAN**

Moving Average

A moving-average filter with specified period is computed by naming an output with a **NAME** command, then issuing the following command:

MOVING AVERAGE = Period SourceName

Autocorrelation and Related Operations

This section discusses computation of autocorrelation, partial autocorrelation, crosscorrelation, and partial crosscorrelation. These operations refer to the generation of a new signal whose values are the correlations of one signal with lagged samples of itself or a different signal. The implementation here is one-sided in that only lags are used. This is in accord with most common uses for these operations. Two-sided operations are for specialized applications only.

Two steps are needed to compute a new signal with one of these techniques. First, the name of the signal to be created must be specified with the following rule:

NAME = SignalName

It is legal to let this name be the name of a signal on which this operation takes place, in which case the new signal will replace the old after computation is complete. However, this would rarely be useful. It is also legal to specify more than one name, with the names separated by commas. This has the effect of producing multiple copies of the result. For example, the following rule would produce three identical copies of the computed signal:

NAME = Copy1 , Copy2 , Copy3

Four related lagged correlation operations are possible. They are invoked with the following rules:

AUTOCORRELATION = N Signal
PARTIAL AUTOCORRELATION = N Signal
CROSSCORRELATION = N Main AND Lagged
PARTIAL CROSSCORRELATION = N Main AND Lagged

The user must specify the length of the new signal as an integer following the equal sign. The first lag is one, so the maximum lag is equal to the specified length. For crosscorrelation, the second named signal is lagged relative to the first. For example, consider the following commands:

```
NAME = z
CROSSCORRELATION = 10 x AND y
```

A new signal named z is created, and it has ten points. The first point is the correlation of x with the previous value of y. The second point is the correlation of x with the lag-two value of y, and so forth. The last value in z is the correlation of x with y at a lag of ten.

Three of these four operations are reasonably fast. However, be warned that **PARTIAL CROSSCORRELATION** is quite slow. Worse, its computation time increases rapidly as the number of lags increases. Avoid trying to compute more than a dozen or so lags if the input series are long. In an emergency, this operation (and only this one of the four) can be aborted by pressing the ESCape key. Even then, significant time may be required before the key press is detected. A future version of the NPREDICT program may contain a faster algorithm.

Box-Jenkins ARMA/ARIMA Prediction

ARMA models have been a mainstay of time series prediction since Box and Jenkins (1976) rigorously presented their theoretical and practical foundations. They have been traditionally used for univariate applications, although bivariate transfer-function versions and several multivariate versions also exist. This program incorporates a broad generalization of the ARMA model. Any number of lagged inputs can be used to predict any number of outputs, and moving-average (MA) terms can be derived from the implicit shocks of any outputs. Seasonality is fully generalized, allowing multiple season lengths and even different seasons for different variables. Naturally, ordinary and seasonal differencing can be applied to any series prior to fitting an ARMA model in order to implement that less general form of the seasonal ARIMA model. The two versions can even be mixed in a wide variety of combinations. Finally, a constant offset may be estimated or may be fixed at zero.

Specifying the Model and Signals

If lagged values of any signals are to be used in the model, they are specified by means of one or more **INPUT** statements. An **INPUT** statement takes one of the following forms:

INPUT = SignalName Lag
INPUT = SignalName MinLag MaxLag

The first form is used when a single lag is desired. The second form is used when a contiguous range of lags is desired. The lags must be positive. A dash may optionally be inserted between the minimum and maximum lags if the user feels that readability is enhanced. Any number of **INPUT** statements may be used to specify multiple lags that are not contiguous.

If MA terms involving output shocks are part of the model, they are specified by **OUTPUT** statements. The two legal versions of **OUTPUT** statements are not surprising, and the rules for their use are the same as for **INPUT** statements.

OUTPUT = SignalName Lag
OUTPUT = SignalName MinLag MaxLag

If any outputs are being predicted, but their shocks are not to be used in MA terms, they are specified by means of an **OUTPUT** statement with no lag:

OUTPUT = SignalName

It's time for some examples. A basic univariate AR(1) model has one input at a lag of 1, and it has one output, itself. This model, using a signal called x, would be specified with the following two commands:

```
INPUT = x 1
OUTPUT = x
```

Another very simple example is the MA(1) model. Since the prediction is based on only the output shocks, there is no input. Just one command is used.

```
OUTPUT = x 1
```

Now let's look at a more complex univariate model. We want three AR terms at lags of 1, 2, and 12 (a general seasonal term). We also want two MA terms, at lags of 1 and 2. This model would be specified as follows:

```
INPUT = x  1-2
INPUT = x  12
OUTPUT = x  1-2
```

Suppose we want to predict signal y based on the value of signal x at lags of 1 and 10. This model is specified with three statements, or four if we also want to predict x:

```
INPUT = x 1
INPUT = x 10
OUTPUT = y
OUTPUT = x    (optional)
```

Finally, let's consider a fairly complex model. We have three variables. The prediction will be based in part on the value of x at lags of 1 and 10, as well as on the value of z at lags of 1 and 2. The value of y is not considered as an input for the model. The prediction is also based on the shock of x at a lag of 5 and the shock of y at lags of 1 and 2. This model is specified as follows:

```
INPUT = x  1
INPUT = x  10
INPUT = z  1-2
OUTPUT = x  5
OUTPUT = y  1-2
OUTPUT = z    (optional)
```

The final statement in that example is optional, depending on whether we want to predict z. Since we are not using its shocks as part of the model, we do not have to include it in the output list.

The **INPUT** and **OUTPUT** commands are cumulative. Every time another appears, it is appended to whatever collection already exists. If the user wishes to define a new model, it is nearly always necessary to start new input and output lists. The existing lists are cleared with the following commands:

CLEAR INPUT LIST
CLEAR OUTPUT LIST

The signals used as training data do not need to be the same length. Longer signals are truncated so that they are the same length as the shortest signal. The length of the shortest signal, minus the longest input lag, must be at least as great as the total number of weights (including constant offset) in the model. If this is not the case, the model is overparameterized, and an error message will be issued.

In most cases, the user will want to estimate the constant offsets for the output series. But occasionally it is appropriate to assume that the offset is zero. This might happen if the outputs were computed by differencing signals that are known to have no deterministic trend. Or there might be a hard physical reason why the mean must be zero. To fix the offset at zero, use the following command:

ARMA FIXED = YES

Once that command is issued, it will remain in effect until explicitly turned off by repeating the command with a parameter of **NO**. The default is **NO**, as most users will want to compute an optimal offset.

Training the Model

After the model and the training data have been specified by means of **INPUT** and/or **OUTPUT** commands, the model is trained by issuing a single command:

TRAIN ARMA = Name

The user must provide a name for the model. This name may be used later when the trained model is invoked to do a prediction. Any number of ARMA models may be present (as long as sufficient computer memory is available). When a model (or all models) is no longer needed, it may be erased. The following two commands erase a single ARMA model and all existing models, respectively:

CLEAR ARMA = Name
CLEAR ARMAS

If there are no MA terms (no **OUTPUT** statements with lags), the model weights are computed with a relatively fast deterministic algorithm. The user cannot interrupt this training. If there are any MA terms, an iterative algorithm must follow the deterministic initialization. If the user has previously issued a **PROGRESS ON** command, the error at the end of each iteration is printed on the screen. Iteration continues until the error has stabilized to a number of significant digits equal to the *sum* of the parameters specified by the following two commands:

ACCURACY = integer
REFINE = integer

The individual values of these two parameters are ignored. Only their sum is used. The default values (which may be found in the module DEFAULTS.CPP) are quite strict and may result in very long training times for large models. However, for small and moderate models, they should be perfect. In any event, the user may interrupt training by pressing ESCape. Training may be resumed where it left off by issuing another **TRAIN ARMA** command. Note that it is often the case that there is significant correlation among the weights. As a result, extended training may change the weights considerably while having no practical impact on the quality of the model.

The audit log file contains much information about the model. Before training starts, a line is written to this file stating the number of input weights, the number of MA weights, and the number of output signals. After training is complete, a line is written stating the pooled RMS error. This is the square root of the mean (across all outputs) of the squared errors. Then, a separate group of information appears for each output signal. The output is named, and its individual error variance and standard deviation are given. Under this line, every input signal and MA term appears with its lag and its computed weight.

If training is interrupted and then resumed, it is vital that the inputs and outputs be specified *exactly* the same way both times. The same is true if training is followed by prediction. It is legal to use different signal names. But the order of inputs and outputs and their lags must be identical. In fact, even differences that have no structural effect on the model may result in the user being issued error messages. The first two commands shown below are obviously equivalent to the third, but an error will result if one of these methods is used for initial training, and the other is used for subsequent retraining or prediction.

```
INPUT = x 1
INPUT = x 2

INPUT = x 1-2
```

In many cases, this will not be an issue. The user will simply issue the **INPUT** and **OUTPUT** commands once at the beginning of the session and not issue any more of them. The model will remain constant throughout the entire session. But there are some situations in which this issue may be important. If several different models are being compared, with all training done early and all testing being done later, the model will need to be redefined each time a different trained ARMA is invoked. If a trained ARMA model is saved to disk and then restored, the structure and signals will need to be specified before anything can be done with the restored model. Remember, except for signal names, the sequence and the parameters of all **INPUT** and **OUTPUT** commands must be absolutely consistent.

Predicting with the Model

Once an ARMA model has been trained, it can be used for prediction. The predictions start at the beginning of the output series, offset by the maximum *input* lag. MA lags do not influence the starting point. Points prior to the maximum input lag remain at their true values. For example, suppose the following model is used:

```
INPUT = x 1-2
OUTPUT = x 1
OUTPUT = x 10
```

The maximum input lag is 2. Therefore, prediction will start with the third point. The predicted values for the first two undefined points will be made equal to the original signal values. Box and Jenkins' method of *backcasting* could be used instead, but this method is not generally as good as might be hoped. Also, it is virtually never the case that these early predictions are needed, so there is no real reason to use questionable algorithms to compute them.

Predictions may be made beyond the ends of the known series. If the input series are different lengths, as much use as possible is made of each. If a series is used as both input and output, the known values are used for as long as they exist, then predicted values are used. If a series is used as input only, its final point is replicated for all subsequent predictions. (This should be avoided whenever possible, as it induces serious errors.) Shocks beyond the end of the known output series are assumed to be zero. ARMA model terms for prediction may be summarized with the following set of rules:

- If this is an input (AR) term...
    - If it is still within the known series, use the true value.
    - Else it is beyond the end of the known series...
        - If it is also a predicted output, use the prediction.
        - Else use the last known point in the series.
- Else this is an MA term...
    - If it is still within the known output series, use the shock.
    - Else use zero.

In order to use a trained ARMA model for prediction, three steps are necessary. First, the model's architecture and data must be specified by means of **INPUT** and **OUTPUT** statements. In nearly all situations, this step will not be explicitly performed because the information is already in place. The user will most likely have just trained the model, so the model and data will already be defined. But if different models have been used since the training of this model, or if this model is read from a disk file, the inputs and outputs must be specified again.

The second step is to name the series that will be generated as a result of the prediction. This is done with the following command:

NAME = Name1 , Name2 , ...

The third and final step is to issue the command to do the actual prediction. The length of the series to be generated and the name of the trained ARMA model are given on that command.

ARMA PREDICT = Length Name

There is a one-to-one correspondence between the model's output signals (in the order that **OUTPUT** statements appeared in the model specification) and the signals appearing on the **NAME** statement. The predictions made for the signal that was named first on an **OUTPUT** command will define the first signal listed on the **NAME** command, and so forth. Model outputs may be explicitly ignored by using a comma as a placeholder in the **NAME** command. If any of the named series already exist, they will be erased and replaced with this new series. This may be made more clear with an example.

```
INPUT = x 1
INPUT = x 10
INPUT = z 1-2
OUTPUT = x 1-2
OUTPUT = x 5
OUTPUT = y 1-2
OUTPUT = z

NAME = new_x , , new_z
```

The first signal named on an **OUTPUT** command is x, so the predictions for that signal will define the newly created signal *new_x*. The second signal named on an **OUTPUT** command is y, but the second signal on the **NAME** command is null, its place being taken by a comma. Thus, the predicted values of signal y will be discarded. Finally, the predicted signal z will define the new signal *new_z*.

It is legal for fewer names to appear on the **NAME** command than there are model outputs. Excess model output signals will be discarded. However, an error will be issued if the user supplies more names than there are model outputs.

Important note: Prediction includes a feature that is immensely useful for visually displaying results, but it may cause some confusion. Sometimes, the **NAME** command specifies signal names that are different from those used as data in the model. That way, the user can easily compare the true data with the predictions on a point-by-point basis. For example, in the example given a few paragraphs ago, comparing the plots of signals x and *new_x* may be very informative. However, sometimes the user wants to make use of a signal that consists of the true values for as far as they are known, with future predicted values appended to the end. This can be done by specifying on the **NAME** command a signal name that is identical to the corresponding model output. When this is done, the generated signal is equal to the original signal for the length of that original, followed by the predictions. Look at the following example that shows simple AR(1) prediction:

```
INPUT = x 1
OUTPUT = x
TRAIN ARMA = model_1
NAME = new_x
ARMA PREDICT = 110 model_1
NAME = x
ARMA PREDICT = 110 model_1
```

The created signal *new_x* will consist of 110 points. The first valid prediction is the second point. The first point remains equal to the first point in x. Suppose the original signal x contains 100 points. Computed points 2 through 101 will be based on known x values. Point 102 will be based on the previously predicted point 101, and so forth, with each subsequent prediction being recursively based on prior predictions.

The second prediction command, which replaces signal x, is similar to the first prediction. However, the resulting signal will be different in that its first 100 points (the length of the original signal x) will be set equal to the original x. The final 10 points will be equal to the final 10 points of *new_x*.

These two options have different uses. If we want to display a point-by-point comparison of predicted values versus true values across the known extent, we would generate a new signal. This can often vividly display areas of particularly good and bad performance. If, on the other hand, we want to display the known signal as far as it goes, with predictions appended, we would specify the same name for the predicted signal and the model output signal. This *must* be done if we will display confidence bands (discussed later).

When this name duplication is used, previous signal modifications (such as **CENTER**, **DETREND**, **DIFFERENCE**, and so forth) are preserved. This enables use of an **UNDO** command to reverse the prior operation through the predictions.

If more than one output is generated, and if a **NAME** output is deliberately specified the same as a model output, it is assumed that the user takes care that the name order makes sense. There is probably no use for the following command sequence:

```
OUTPUT = x
OUTPUT = y
NAME = y , x
```

This set of commands would produce a most confusing result. The future predictions for the first output, *x*, would be placed in the first **NAME** signal, *y*, but the original signal *y* would be preserved for its original extent. The future predictions for the second output, *y*, would be placed in the second **NAME** signal, *x*, but the original signal *x* would be preserved for its original extent. Most users would not want this to be done!

In conjunction with the prediction, some information will be written to the audit log file. The first line states the number of input terms, MA terms, and outputs in the model. Then, every input signal and MA output shock signal will be listed with its lag. Finally, all generated signals will be listed. The actual prediction is performed, and the pooled RMS error of the model is written. This is the square root of the mean (across all model outputs) of the squared errors. Predictions beyond the end of the known outputs are not included in this measure. Also, recall that the first few points (corresponding to input lags) cannot be predicted. These points are not included in the error measure. The pooled error includes all model outputs, regardless of whether they are named in a **NAME** statement. Last of all, this pooled error is broken down into individual output components. For each model output, its error variance and standard deviation are written.

Confidence Limits for ARMA Predictions

Robust confidence intervals, as discussed in Chapter 8, can be computed for multivariate ARMA predictions. Set up the input and output lists exactly as they will be used for prediction. The **NAME** command has no role in ARMA confidence computation. This is in contrast to neural network prediction confidence, in which a **NAME** command must precede confidence computation. After the input and output lists are in place, issue the following command:

ARMA CONFIDENCE = Npred ARMAname

The equal sign is followed by an integer that specifies the maximum number of future predictions whose confidence intervals are to be estimated. The confidence information is automatically attached to each of the signals in the output list. It is assumed that these same signals will be used for prediction and named in a **NAME** command.

In practice, what we usually do is insert the **ARMA CONFIDENCE** command before the **ARMA PREDICT** command and duplicate the output list signals on the **NAME** command. This way, the confidence information is associated with the signals that will be predicted.

Confidence information is associated *only* with predicted signals whose known extent is preserved. In other words, *any named output signal whose confidence is desired must also appear in the output list*.

By default, a 90 percent two-tailed confidence interval is computed. This means that there is a 5 percent chance that the true value exceeds the upper limit and a 5 percent chance that the true value is less than the lower limit. The user can specify the two-tailed probability with the following command:

CONFIDENCE PROBABILITY = Percent

The probability is a percent, and it must be in the range 50–100. Values larger than the default of 90 require a large number of test cases. Also, the computed interval may be so wide that it is useless.

After confidence computation is complete, two lines concerning the confidence in the confidence are printed in the audit log. Their exact meaning is discussed near the end of Chapter 8. The two lines are similar to the following:

There is a 5% chance each tail probability exceeds 11.62 %
There is a 99.60% chance the interval exceeds 80.00 %

These sample lines resulted from asking for a 90 percent confidence interval. The first line tells us that there is a 5 percent chance (fixed by the program) that each tail area is really as much as 11.62 percent of the error distribution, instead of the 5 percent requested by the user. The second line doubles the requested tail area, then states the probability that the computed confidence interval encompasses at least one minus that amount. In other words, the user asked for a 90 percent interval, implying that the tails together account for 10 percent of the error distribution. Doubling that causes the program to consider an 80 percent interval, and we are informed that there is a 99.6 percent chance that our interval encompasses at least 80 percent of the error distribution.

The user may receive an error message stating that too few cases exist. The following formulas may be useful in resolving this issue. Let maxlag be the maximum input lag, and let max_ma be the maximum

MA lag. Define offset as maxlag+2*max_ma. That many known values are skipped at the beginning of the series. Let npred be the number of future predictions specified in the command, and let shortest be the length of the shortest signal, including both inputs and outputs. Then, the number of cases that will go into the confidence estimates is computed by Equation (11.1).

$$ncases = shortest - offset - npred + 1 \qquad (11.1)$$

The number of cases must meet two criteria: First, it must be at least 20, an arbitrary but reasonable limit. Second, the number of tail cases, given by Equation (11.2), must be at least one. In that equation, prob is the two-tailed probability specified by the user or left at its default of 90.

$$tail = 0.5 * (1.0 - prob/100) * ncases \qquad (11.2)$$

It may be that one or more of the predicted outputs are in a modified domain. For example, the user may have taken the log of a series, then differenced it. If the ARMA model was trained on this modified data, the user is probably not very interested in confidence intervals in this domain. The predicted signal will probably be subjected to one or more **UNDO** commands to return it to the original domain. In order to compensate for these operations, the user must specify them prior to issuing the **ARMA CONFIDENCE** command. Legal compensations are the following:

CONFIDENCE CENTER = Signal [PER Other]
CONFIDENCE DETREND = Signal [PER Other]
CONFIDENCE OFFSET = Signal [PER Other]
CONFIDENCE SCALE = Signal [PER Other]
CONFIDENCE STANDARDIZE = Signal [PER Other]
CONFIDENCE DIFFERENCE = Signal [PER Other]
CONFIDENCE SEASONAL DIFFERENCE = Signal [PER Other]
CONFIDENCE LOG = Signal

For each of these commands, the user must specify the name of a signal in the output list (which will be named as a predicted output on a subsequent **NAME** command). This name is used to match operations to signals. It is *vital* that, for each different signal, these confidence compensation commands appear in exactly the same order as the operations were performed on the original signal. The subsequent **UNDO** commands will appear in the reverse order. By default, the differencing parameters for undoing are retrieved from the named signal. Optionally, a different signal may be named as the source of the undoing parameters by means of the **PER** keyword. (See page 385 for more on **PER**.)

If the first confidence compensation for a signal is **LOG**, multiplicative confidence intervals will be computed. Otherwise, additive intervals will be computed. See the discussion starting on page 323 for more details.

Sometimes the user will want to do several different predictions in the same program run. The confidence compensation list is cleared to make way for a new list by means of the following command:

CLEAR CONFIDENCE COMPENSATION

Example of ARMA Prediction Confidence

Here is a small example of how to compute confidence intervals for ARMA prediction. More examples illustrating a variety of techniques can be found elsewhere in this text. This example is meant to demonstrate the basic rules, not the full range of options.

We apply two modifications to the single series that will be recursively predicted. First, we take its log to convert multiplicative effects to additive effects, then we apply a seasonal difference. An AR(1) model is employed. We instruct it to compute confidences based on eventual undoing of the two modifications. The confidences are computed, and the predictions are made. Finally, the predicted signal is taken back to the original problem domain. A pointwise prediction is also made to facilitate study of the model's effectiveness.

```
LOG = x    ; Since LOG is first, multiplicative intervals computed
SEASONAL DIFFERENCE = 12 x
INPUT = x 1
OUTPUT = x
TRAIN ARMA = temp
CONFIDENCE LOG = x      ; Must be same order as applied above
CONFIDENCE SEASONAL DIFFERENCE = x
ARMA CONFIDENCE = 5 temp
```

```
NAME = pointwise
ARMA PREDICT = 200 temp
NAME = x
ARMA PREDICT = 200 temp

UNDO SEASONAL DIFFERENCE = x  ; Must be reverse order
EXP = x

DISPLAY CONFIDENCE = x  ; Intervals will be multiplicative: LOG first
```

Because the name *pointwise* does not appear in the output list, this predicted signal is a point-by-point prediction within the extent of the known x series. Predictions past the end of x are obtained by recursively using previous predictions (because x is an input and an output). The main use for this signal is to examine the prediction power of the model, a goal that is best achieved by staying in the prediction domain. Thus, there is no real need to undo the transformations. The predicted signal x does appear in the output list. Therefore, the original values of x are preserved within its known extent. Predictions past the end are appended, and the future values of x are identical to those in *pointwise*. It makes sense to undo the transformations on x.

Predicting with Known Shocks

This section describes one of the most powerful methods of hybridizing ARMA models with other prediction techniques: Train an ARMA model and compute the point-by-point prediction error within the known data. Then use another model, such as a neural network, to predict future values of the error. Finally, apply the ARMA model to the predicted shocks.

When ARMA prediction is done, the default operation uses shocks of zero beyond the end of the known output series. In order to specify that a particular signal is to be taken as the shocks for some output, the following command is used:

ARMA SHOCK = ShockName FOR OutName

The first signal named on this command is the signal that defines the shocks. This is typically a signal that has been predicted with another model. The second signal is the ARMA model's output that is to be predicted using those shocks. In virtually all cases, the user will want to have the shock signal be identical to the actual shocks for the extent of the known signal. This way, the predicted signal will also be identical to the known signal during that extent. If point-by-point predictions were used instead, the prediction would not be reliable. In neural network prediction, this equality is obtained by using the same name for the created output as for the recursively predicted input. See page 448 for more details on recursive prediction with neural networks.

Note that the known shocks can have *two* types of impact on the prediction. They are obviously added to each output as it is computed. If there are any AR terms, this implies a recursive impact. In addition, if there are any MA terms, the known shocks are also used for those components. This secondary effect can be important to the quality of the prediction.

ARMA SHOCK commands are recorded as part of the output list. The output named on this command must be a currently existing model output. Therefore, it is recommended that the user specify the inputs and outputs first, then specify shocks. These shock specifications are removed when a **CLEAR OUTPUT LIST** command is invoked.

The **ARMA SHOCK** command is legal for any number of predicted outputs. Known shocks may be used for a subset of the model's outputs. Those outputs that do not have known shocks will be treated in the traditional way, using shocks of zero beyond their end.

It is time for an example. For simplicity, a single series will be predicted. The complete text of this example can be found in the file ARMA_NN.CON in the EXAMPLES directory on the accompanying disk. Only the essential components are shown here. The known signal contains 175 points, and we want to predict (extravagantly and dangerously) another 25.

```
; Train ARMA and compute its shocks

INPUT = signal 1              ; One AR term
OUTPUT = signal 1             ; And one MA term
TRAIN ARMA = temp

NAME = within_pred            ; Pointwise prediction within known output
ARMA PREDICT = 175 temp
```

```
NAME = shock                ; Shock = actual minus predicted
SUBTRACT = signal AND within_pred

; Train a neural network to predict the shocks
; Exclude the first few cases, which are before MA stabilization

CLEAR INPUT LIST
CLEAR OUTPUT LIST
INPUT = shock 1 - 3         ; Arbitrarily choose to use 3 lags
OUTPUT = shock              ; to predict the current value
CUMULATE EXCLUDE = 5        ; Allow shocks to stabilize
CUMULATE TRAINING SET
TRAIN NETWORK = temp        ; OK to use same name for net and ARMA
NAME = shock                ; Recursive prediction is mandatory
NETWORK PREDICT = 200 temp  ; True values preserved within known
DISPLAY = shock             ; The predictions are at the end only

; Do two ARMA predictions.
; First is the traditional method, using zero shocks beyond the end.
; The second uses the predicted shocks.  Note the improvement.

CLEAR INPUT LIST
CLEAR OUTPUT LIST
INPUT = signal 1            ; The inputs and outputs are identical
OUTPUT = signal 1           ; to the originally trained model
NAME = predict_a            ; Traditional prediction with zero shocks
ARMA PREDICT = 200 temp
DISPLAY = predict_a
ARMA SHOCK = shock FOR signal  ; Use the predicted shocks
NAME = predict_b            ; Improved prediction with known shocks
ARMA PREDICT = 200 temp
DISPLAY = predict_b
```

The ARMA model uses one AR(1) term and one MA(1) term. It is trained, then its shocks are computed by subtracting the point-by-point predictions from the known values. A neural network is trained on the shock signal. The arbitrary decision was made to use three lags to predict the current value. In practice, the form of the input should be chosen carefully. As a general rule (somewhat violated here), the inputs for the neural network should be quite different from the ARMA model's inputs to avoid duplication of effort. The first five shock points are excluded from the training set. This gives the shock series time to stabilize. As a bare minimum, at least as many points as the longest ARMA lag should always be excluded.

The neural network prediction is recursive, which will always be the situation for hybrid models. Otherwise, the final known point in each input signal will be replicated as inputs for predictions beyond the end of the signal, an obviously worthless action. Furthermore, as pointed out on page 448, recursive prediction causes the predicted signal to equal the known signal within the extent of the known signal. Only points past the end are computed. The result is that the ARMA prediction based on these identical shocks is equal to the known output values within the known extent. This is good.

This version of the NPREDICT program does not support computation of confidence intervals for predictions made with predicted shocks. This is not an impossible task, but it is extremely complex. Perhaps this ability will be included in a future version of the program.

Saving and Restoring ARMA models

It is possible to preserve a trained ARMA model by writing it to a disk file. This model may be retrieved later. These operations are done with the following commands, which name the ARMA network and the file to which it will be saved and from which it will be restored:

SAVE ARMA = ArmaName TO FileName
RESTORE ARMA = ArmaName FROM FileName

When the file is read, much information about the model is written to the audit log file. The first line provides the pooled (across all outputs) RMS error. This error is then broken down according to individual outputs. For each output, the variance and standard deviation of the error is written. Then, the lag and weight for each input and MA term appear. The format of this text is almost identical to the audit log record created when the model was trained. However, there is one essential difference. To avoid possible confusion, neither the inputs nor the outputs are named. The training record contained those names. But since the same names may not make sense in the context under which the file is retrieved, names are not used. Instead, unique identification numbers are used.

After an ARMA model has been retrieved, it may be trained more, or it may be used for prediction. It is vital that the **INPUT** and **OUTPUT** statements that define the model and its data be structured *exactly* the same way as when the model was trained. Obviously, the signal names may be different. But the order of inputs and outputs, and their lags, must be absolutely consistent.

Neural Network Training and Test Sets

One of the principal uses for NPREDICT is to train and test various neural network prediction models. In order to train a model, one or more signals must be gathered together to form a *training set*. The trained model can be tested by means of a *test set*. The data in the test set is usually independent of the data in the training set. These two sets are generally identical in structure, and the same operations in NPREDICT are used to create both of them.

Specifying Inputs and Outputs

The first step in creating a training or test set is to designate the signals that comprise the inputs to the model. This is done by means of one or more **INPUT** commands for each signal. If lagged values of the signal are used as inputs, the lags are also specified. This command comes in three different forms, depending on whether zero, one, or several lags are used.

> **INPUT = SignalName**
> **INPUT = SignalName Lag**
> **INPUT = SignalName MinLag MaxLag**

The MinLag and MaxLag parameters may be separated by one or more spaces, or a dash may also be used to separate them. If several different lags are desired, but these lags are not defined by one contiguous range, then use multiple **INPUT** commands.

The second step in building a training or test set is to specify the output(s). This is slightly more complicated than the inputs, as two very different forms of output are possible. One form is straightforward prediction of current or future values of one or more signals. In this case, the **OUTPUT** command is analogous to the **INPUT** command. However, we are now dealing with lead times, rather than lags. The three forms of this command are

OUTPUT = SignalName
OUTPUT = SignalName Lead
OUTPUT = SignalName MinLead MaxLead

Class Names as Outputs

The other output form handles the case of classification. Instead of predicting numeric values, the user may wish to make discrete decisions, such as buy/sell/hold for stock transactions, or normal/abnormal for electrocardiograms. This is done by naming the class corresponding to the current inputs.

CLASS = ClassName

Some models allow the user to specify prior probabilities. To associate a prior probability with a class, use one of the following commands:

PRIOR = Number
PRIOR = N

The specified number must be positive and may optionally contain a decimal point. The prior probabilities need not sum to one, as they are treated only in a relative way. Internally, they will be added, and each will be divided by their sum. This automatic normalization frees the user from the burden of specifying normalized values.

As an alternative option, the **PRIOR=N** command indicates that the user wants the priors to be determined by the number of cases in each class. Thus, it is assumed that the makeup of the training set reflects the makeup of the general population.

If no prior probabilities are specified, then equal priors will be assumed. The effect of an unequal number of cases in each class will be removed.

It is the responsibility of the user to be consistent. If a **PRIOR** statement appears for at least one class during training set cumulation, then a **PRIOR** statement should appear for all. Similarly, mixing **PRIOR = number** and **PRIOR = N** statements should be avoided.

Not all models are able to incorporate prior probability information. At this time, only the PNN, SEPVAR, and SEPCLASS models accept this information. The MLFN model ignores priors.

Some models may be able to handle a *reject* class. For example, a military target recognition application might encompass several classes of threatening vehicles. But the program should not be restricted to choosing one of those classes as its decision. Letting the program have the option of choosing "none of the above" is mandatory in most cases. To designate members of the reject category, use the following command:

CLASS = REJECT

At this time, only the MLFN model can handle a reject class. The PNN, SEPVAR, and SEPCLASS models will generate an error message if any training cases appear in a reject class.

Cumulating the Training or Test Set

We come at last to the central goal of this section. The inputs and outputs have been defined. The named signals are used to construct a training or test set with one of the following commands:

CUMULATE TRAINING SET
CUMULATE TEST SET

If a training or test set already exists, it is preserved. The new cases are appended to any existing cases. Naturally, the new set must be compatible with the old set. The number of input variables must be the same. If the outputs are predicted values, the number of outputs must be the same. Also, if a **CLASS** command has been used at any time in the construction of this training or test set, it must be used every time.

In some applications, particularly hybrid ARMA/neural prediction, the first few points in the signal(s) defining the training set may be invalid. It is possible to exclude them with the following command:

CUMULATE EXCLUDE = integer

The specified number of points at the beginning of all signals will be ignored when the set is cumulated. Note that once this exclusion is invoked with the above command, it remains in effect for all subsequent **CUMULATE** commands. To turn it off, it is necessary to use a **CUMULATE EXCLUDE = 0** command.

When the user is building training and test sets together, it can be handy to specify that only a certain number of cases from the beginning are kept. This is done with the following command:

CUMULATE INCLUDE = integer

The specified number of cases is added to the training set (at most). Any excluded cases (from **CUMULATE EXCLUDE**) do not count. This command remains in effect for all subsequent **CUMULATE** commands. To return to the default of including all possible cases, use this command to set the count to a huge number.

The following example illustrates how these two commands may be used to collect the majority of points from the beginning of a series into a training set, while reserving some points at the end for a test set:

```
CLEAR TRAINING SET
CUMULATE INCLUDE = 828          ; Keep the first 828 observations
CUMULATE EXCLUDE = 0            ; Exclude nothing from the start
CUMULATE TRAINING SET           ; This defines the training set

CLEAR TEST SET
CUMULATE INCLUDE = 999999       ; Restore default of going to end
CUMULATE EXCLUDE = 828          ; Exclude the training set points
CUMULATE TEST SET               ; This defines the test set
```

Clearing Information

All of the commands described in this section are cumulative. Each **INPUT** and **OUTPUT** command adds a new signal to the list of inputs and outputs. Each **CUMULATE** command adds to the existing set. And each **CLASS** command adds to a list of classes. Many applications require different input, output, and class lists at different times. Several training and test sets may also be needed. There must be a way to get back to a clean slate. The following commands are used to do that:

CLEAR INPUT LIST
CLEAR OUTPUT LIST
CLEAR CLASSES
CLEAR TRAINING SET
CLEAR TEST SET

The user will make frequent use of the **CLEAR INPUT LIST** and **CLEAR OUTPUT LIST** commands. One or both of them will most likely be used after each **CUMULATE** command as preparation for defining a new group of inputs and outputs. (If the entire training or test set is built with just one **CUMULATE** command, clearing the input and output list is not needed.) The other commands will not be used so often. The **CLEAR TRAINING SET** and **CLEAR TEST SET** commands are needed only if the user wants to build a new set from scratch. They are also useful when memory is limited, as these two commands free the memory allocated to the sets.

The **CLEAR CLASSES** command should be used if a different set of classes is being used to train a new model. When a model is trained in classification mode, the number of outputs is equal to the number of classes that have been named. Suppose the user runs one application training session using classes of *normal* and *abnormal*. That model will have two outputs. Suppose another application is then trained, using classes of *tank*, *truck*, and *tree*. If no **CLEAR CLASSES** command is used, this new model will have five outputs, clearly not the user's intention!

Beware of an easy and disastrous error. The inputs and outputs for the model are ordered by the sequence in which the **INPUT** and **OUTPUT** commands appear. If more than one **CUMULATE** command is used, the order of inputs and outputs must be consistent! The following sequence of commands would not do what the user wants:

```
INPUT = HeightA
INPUT = CaloriesA
OUTPUT = WeightA
CUMULATE TRAINING SET
CLEAR INPUT LIST
CLEAR OUTPUT LIST
INPUT = CaloriesB    ; Order is reversed!
INPUT = HeightB
OUTPUT = WeightB
CUMULATE TEST SET
```

Examples

It's time for some examples of building a training (or test) set. Let's start out simply. We want to use the values of two variables to predict the value of a third. The following commands might be used:

```
INPUT = height
INPUT = calories
OUTPUT = weight
CUMULATE TRAINING SET
```

This example treats the signals as an unordered dataset. It is not really a time series application, but it is perfectly legal, nonetheless.

Now let's assume an order to the samples. We use lagged values of a variable to predict its future value. We want to consider the current value, the two previous values, and the value seven samples back, to predict the next value and the value seven samples ahead. The following command would set up an appropriate training set:

```
INPUT = price 0-2
INPUT = price 7
OUTPUT = price 1
OUTPUT = price 7
CUMULATE TRAINING SET
```

Finally, let's look at classification. We have two signals derived from a SONAR return. We want to use current and lagged values of these variables to discriminate between a submarine and a rock. Suppose we have one example of a submarine return and two examples of rock returns. The following commands might be appropriate:

```
CLASS = submarine
PRIOR = 0.1
INPUT = SubVarA 0-1
INPUT = SubVarB 0-3
CUMULATE TRAINING SET

CLASS = rock
PRIOR = 0.9
CLEAR INPUT LIST
```

```
INPUT = RockVarA1 0-1
INPUT = RockVarB1 0-3
CUMULATE TRAINING SET

CLEAR INPUT LIST
INPUT = RockVarA2 0-1
INPUT = RockVarB2 0-3
CUMULATE TRAINING SET
```

Neural Network Models

Several neural network models are available. These include members of the probabilistic neural network family and multiple-layer feed-forward networks. The model is selected with the following command:

NETWORK MODEL = Model

Different models have different strengths and weaknesses. Some train quickly but execute slowly. Others train slowly but execute quickly. Some can incorporate prior probabilities and easily generate confidence levels for decisions. Some can incorporate a reject classification category, while others cannot. Since there are such great differences among neural network models, the user should exercise care in choosing a model.

The Probabilistic Neural Network Family

The mathematics of the members of this family is discussed in detail in Masters (1995). Here, we will briefly review the salient features of PNN models. The specific family members will be shown, and an explanation of each will follow.

The principal claim-to-fame of the PNN family is that its members can be trained relatively quickly. Donald Specht's original PNN required little or no training at all. Recently developed versions of the PNN that are more sophisticated can require significant training time, but still not nearly as much as many competing models. For large, complex applications, this can be the deciding factor in choosing a model.

The other side of the coin is that these models have memory and time requirements that significantly exceed those of other models when they are put to use. Applications that must process voluminous data in real time often cannot make use of a PNN. These models lend themselves well to massive parallel processing accompanied by distributed memory, so hardware solutions to this dilemma are sometimes possible. But execution time can be a severe limitation of the PNN when specialized hardware is not available.

A major consideration in many applications is the need for confidence levels. In particular, neural networks used to solve military and medical problems are far more effective if they can provide confidence figures for their decisions. When the PNN is used as a classifier, and when the classes are mutually exclusive and exhaustive, confidences are easily obtained. There are ways to squeeze confidences from other models, but none of those methods has the solid mathematical foundation of the PNN. Since this model is directly based on Bayesian methods, confidences are part of the model.

The fact that the classes need to be mutually exclusive (no overlap of cases into multiple classes) and exhaustive (all possible classes are represented) makes use of the PNN a little trickier than other models. In particular, no explicit reject category is allowed. If the command **CLASS = REJECT** appears, training a PNN model will not be allowed. If the user believes that it is possible to define a specific reject category, and that by doing so the requirements of the PNN will be satisfied, then a different name for that class must be used. For example, the command **CLASS = RJCT** may be used prior to cumulating reject cases into the training or test set. Be aware that members of the PNN family cannot explicitly handle a reject category. This category, if used, will be treated as just another class.

The NPREDICT program contains three members of the PNN family. Each of these members is now named and briefly summarized.

NETWORK MODEL = PNN

This is the simplest probabilistic neural network. Exactly one sigma is used to cover all variables (and all classes in classification mode). Training is the fastest of all PNN models, but quality is usually lowest. It is particularly important that the scaling of all variables reflects their relative importance when this model is used.

NETWORK MODEL = SEPVAR

This middle-of-the-road PNN model is usually the best compromise between power and training speed. A separate sigma is used for each variable, but these are shared among all classes in classification mode. When in doubt as to which PNN model to use, this is a good choice. Remember that during the initial training phase, this model is, in effect, identical to the PNN (single-sigma) model. Thus, for a given training time, you have nothing to lose and possibly a lot to gain relative to the PNN model.

NETWORK MODEL = SEPCLASS

This is the most general model. Not only does it use a different sigma for each variable, but it also uses separate sigma vectors for each class. This implies that the SEPCLASS model can only be used in classification output mode. Training can be very slow for this model, and the increased possibility of accidental overfitting makes validation especially important when this model is used.

Kernel Functions

For the basic PNN probabilistic neural network model, the kernel function can be specified with the following command:

KERNEL = Kernel

The kernel functions that are available are now listed, along with a brief explanation for each. Note that the SEPVAR and SEPCLASS models are restricted to the Gaussian kernel. This command affects only the PNN model.

KERNEL = GAUSS

This is, by far, the most common kernel. The Gaussian function is widely accepted as being well behaved and reliable. Unless there is strong reason to use another, choose this one.

KERNEL = RECIPROCAL

This kernel function, described in Masters (1995), is an alternative that has much heavier tails than the GAUSS kernel. Distant cases in the training set exert relatively more influence in processing a case than they do for the GAUSS kernel.

Multiple-Layer Feedforward Networks

The NPREDICT program supports this ubiquitous family by providing the choice of zero, one, or two hidden layers. The hyperbolic tangent activation function is used for all hidden neurons. The user may elect to use that same function for output neurons, or linearity may be obtained by using no activation function at all for the outputs. This discussion will be limited to defining the architecture in NPREDICT, as further details are widely available in other texts, including Masters (1994), which covers complex-domain versions of the MLFN.

The number of hidden neurons is specified with the following two commands:

MLFN HID 1 = integer
MLFN HID 2 = integer

In order to define a network having no hidden layer, specify zero for both of these parameters. If the number of hidden neurons in the first layer is zero, the number in the second layer must also be zero.

The activation function for the output layer is specified by using one of the following commands:

OUTPUT ACTIVATION = LINEAR
OUTPUT ACTIVATION = NONLINEAR

If the former command is used, the activation function is the identity function. In other words, there is none. In many or most applications, this is the best choice. The absence of an output activation function implies that no limitations are imposed on the output values that may be attained. This relieves the user of the small but significant burden of choosing an effective output range. More importantly, in many applications the squeezing of the outputs near the extremes of the activation function's range results in distortion. If in doubt, use linear outputs.

If nonlinear outputs are specified, the hyperbolic tangent function is used. This has a theoretical range of from −1 to 1, with a practical range more on the order of −0.9 to 0.9. There is no point in attempting to train the network to attain outputs that are outside this range. Carelessness in this regard can lead to mysterious failures.

Four different versions of the MLFN allow the user to select the domain of each layer. The theory and applications of complex-domain neural networks are thoroughly discussed in Masters (1994). The domain model is specified with the following command:

MLFN DOMAIN = domain

The default domain, **REAL**, is the traditional version in which all inputs, outputs, and hidden neurons are real. A generally useless alternative, **COMPLEX INPUT**, assumes that all inputs are complex, but all hidden and output neurons are real. There appears to be no advantage to this version.

If the input data is inherently complex and its phase is meaningful, the **COMPLEX HIDDEN** domain version can be extremely effective. The inputs and the hidden neurons operate in the complex domain, while the outputs are real.

If the user is mapping to complex outputs, the **COMPLEX** domain version may be useful. In this version, all inputs, outputs, and hidden neurons operate in the complex domain.

All versions other than **REAL** forbid the use of a second hidden layer. Obviously, the **COMPLEX HIDDEN** version must have a hidden layer. Finally, the **COMPLEX** version may be used only in mapping mode, not classification.

General Training Considerations

Every trained neural network in NPREDICT has a user-supplied name. The ability to have multiple neural networks available at any given time facilitates all sorts of interesting and powerful techniques. Thus, the user must specify a name whenever a network is trained. This is accomplished with the following command:

TRAIN NETWORK = name

When this command is invoked, NPREDICT will first search for an existing network having that name. If one is found, its architecture will be compared with that currently specified by the user. They must be compatible, or an error will result. Two different models may not share the same name. If the named network exists, training will recommence, with the current weights serving as the starting point. If the named network does not exist, a new network will be created using the architecture specified by the user, and this network will be trained.

An existing neural network can be erased, freeing all memory allocated to that network. Also, it is possible to erase all neural networks in one fell swoop. The commands to effect these actions are the following:

CLEAR NETWORK = name
CLEAR NETWORKS

Sometimes the user does not wish to make unreasonable demands on the training algorithm. Satisfaction is obtained when the error drops to some sensible value. This is particularly true when NPREDICT is being used in batch mode. In this case, the user surely does not want the program to run indefinitely. A quitting point can be specified with the following command:

ALLOWABLE ERROR = number

If the training progresses to the point that the error becomes this small, training is halted. Usually, we would set this parameter equal to zero, then press ESCape to halt training when patience expires. But setting it to larger, realistic values can be useful when running multiple experiments in a command file.

There is a second way of permitting escape from a tedious training session when hope for improvement wanes. This is accomplished with the following two commands:

ACCURACY = integer
REFINE = integer

Both of these commands (roughly) specify the number of significant digits that need improvement. Smaller values result in shorter training times and lower accuracy in locating a good weight set.

Training continues until the error criterion fails to improve beyond the number of significant digits defined by the *sum* of the **ACCURACY** and **REFINE** values. For PNN family models, the individual values of these two parameters are irrelevant. Only their sum is used. For MLFN models, the separate values are important. This is discussed in Masters (1995). Typically, the **ACCURACY** would be set to something in the range 5–8, and the **REFINE** would be 2–3. Larger values are especially important for MLFN models, as their error functions frequently exhibit extensive flat plains with sharp cliffs on the horizon. PNN models are generally well behaved, so fewer significant digits are needed. The point of diminishing returns comes quickly for these models.

Cross Validation Training

Sometimes it is very difficult or expensive to collect known cases for training and testing. When relatively few known cases are available, it seems a shame to split them into tiny training and test sets. The technique of *cross validation training* can be used to combine training and validation into one operation. This is discussed in Masters (1995). To effect cross validation training, use the following command:

CV TRAIN NETWORK = name

This command causes the network to be immediately trained, exactly as it would be for ordinary training. Then, when training is complete, the training set will be cross validated. The first case will be held back, and the network will be retrained from scratch using the remaining cases. That network will be used to test the single case that was held back. This operation will be repeated for each case in the training set. The resulting mean squared error is reported as *CV=error* on the same line as the full training error. This report appears both on the computer screen and in the audit log file.

The user may press ESCape at any time. If ESCape is pressed before at least one cross validation test has been performed, the cross validation error will not be printed. But as long as both steps, initial training and one or more cross validations, are complete, the mean for whatever was done is printed. Beware that partial results are virtually guaranteed to be badly distorted and thus should be ignored. Take this number seriously only if the operation terminated on its own.

Cross validation training is probably valid from a strict mathematical point of view only for PNN family models that have been trained to reasonably high accuracy. The multiple minima inherent in MLFN models violates the assumptions of the method. However, if many retries are done, and high accuracy is requested, some users may wish to gamble with this method for MLFN models.

It should be obvious that the time requirements for cross validation training can be overwhelming. For PNN family models, the time is roughly proportional to the *cube* of the number of training cases. It's not quite so bad for MLFN models, being probably more like the square of the number of cases. The huge overhead of that model is the killer. But time is no serious impediment in many applications. Remember that this method is used only when there are few training cases available. If several hundred cases can be had, that collection should be split up into training and validation sets. It's only when just a dozen or two cases are in hand that we would consider cross validation, and its speed is usually reasonable then. For the gray area between several dozen and several hundred cases, it's a judgment call that depends on hardware and application requirements.

Specific PNN Training Parameters

There are several training parameters that apply only to members of the PNN family. This includes the PNN, SEPVAR, and SEPCLASS models. These parameters all relate to the initial univariate search for a good starting sigma. Since multiple minima are possible, with poor local minima being occasionally a serious problem, a reasonably thorough search across a wide range is important. If the range that is searched is not wide enough, with the result that the error function is decreasing at one endpoint, the training algorithm will automatically extend the search in that direction. But the user should not count on this insurance. A careless and unlucky user who bounds only an inferior local minimum will regret his or her stinginess in choosing a search range. Similarly, enough points in the search range should be tested so that the global minimum can be bounded with comfortably high probability.

Three commands are needed for specifying the initial search range and the number of trial points in that range. They are the following:

SIGMA LOW = number
SIGMA HIGH = number
SIGMA TRIES = integer

Most applications are satisfied by using a range from about 0.01 to 10.0, with at least ten trial points. Using even more points and/or a wider range is probably not a serious waste of time.

PNN Progress Reports

When the user invokes the **PROGRESS ON** rule, many intermediate results will be printed to the computer screen. First, the logarithm of each initial sigma will be printed, with its corresponding error value printed to its right. When the initial bounding search is complete, the optimal sigma and its error will be printed.

For the basic PNN model, output from the refinement phase is simple. It is just Brent's algorithm, as described on page 342. Each iteration of this algorithm causes six numbers to be printed across a line. These are the lower bound of the current interval, the location of the best value, the upper bound of the current interval, and the best, second best, and third best function values.

For the SEPVAR and SEPCLASS models, the conjugate gradient algorithm is used. Each iteration of this algorithm produces several different outputs. First, the global line search lists the parametric position, along with the error at that position for each function evaluation. This output is nearly identical to the initial bounding output, except that here the points are parameter values specifying line positions, instead of logs of sigma values. The second output from each conjugate gradient iteration is the progress of Brent's algorithm for refinement along the line. These six numbers are identical to those described in the previous paragraph. The last output generated from each iteration is the most interesting and informative. The parameter t is an excellent indicator of how well the second-derivative hints are doing at optimizing the global aspect of the linear search. Ideally, this should be near 1.0. The function value is the current error. Finally, the percent improvement relative to the previous iteration gives a good idea of how things are progressing.

Sometimes, the initial bounding search will advance to an extremely large sigma value, and then refinement will be skipped. This is an indication that the variables in the model, taken as a group, are generally worthless. This condition can be verified by using the **PRINT NETWORK** command and examining the sigma weights. If they are all huge, a new set of variables should be chosen. It may be that one or more of the individual variables is actually very good. It just happens that so many of its compatriots are so bad that the initialization algorithm gives up. This problem will probably be addressed in a future version of the program. The workaround is to try subsets of the variables and eliminate the bad predictors from the group.

Specific MLFN Training Parameters

There are several training algorithms available for MLFN models, and some of these have many parameters that can be specified by the user. This is a long section. Bear with it.

The first step is to specify the training algorithm. This is done with the following command:

MLFN LEARNING METHOD = method

The various learning methods are described in detail in Masters (1995). Here, we will only list the choices and briefly describe each.

AN1 — This is the primitive simulated annealing described on page 352. It is not recommended for general use, as it is quite slow. It may be useful as an initial search method in problems having a large number of inferior local minima.

AN2 — This is the generic simulated annealing described on page 354. The same considerations as those for **AN1** apply. This method seems to be slightly inferior to **AN1**.

SS — Simple stochastic smoothing is described on page 355. This method is almost never excellent and is included for experimentation only.

SSG — Stochastic smoothing with gradient hints, described on page 355, can be very good in problems that are fairly linearly separable and that are served by plenty of hidden neurons. In general, though, it is inferior to the next few methods.

REGRESS_CJ — Regression with conjugate gradient refinement is valid only when there is no hidden layer. In this situation either it or **REGRESS_LM** is definitely the method of choice.

REGRESS_LM — Regression with Levenberg-Marquardt refinement is valid only when there is no hidden layer. Its performance is very similar to **REGRESS_CJ**, except that it is probably the better choice for small models.

AN1_CJ — This method, described in detail in Masters (1995), is a standard workhorse. It can be counted on to give good performance in nearly all situations. The simulated annealing component makes location of the global minimum likely, while the conjugate gradient component ensures efficient descent to local minima, regardless of the number of weights.

AN2_CJ — This method is identical to **AN1_CJ**, except that the generic annealing algorithm is used in place of the primitive algorithm. It appears that **AN1** is slightly superior to **AN2**, but this is debatable.

AN1_LM, AN2_LM — These two methods are identical to the previous two methods, except that the Levenberg-Marquardt algorithm is used instead of conjugate gradients. The LM method is almost always superior to the CJ algorithm when the model is small (few weights). Its memory requirements and execution time are proportional to the square of the number of weights, so it is impractical for large problems. The exact line of demarcation is not clear. If in doubt, try both methods.

In addition to specifying the learning algorithm, the user must specify the error measure that is to be minimized. These are fully described in Masters (1994). The error is indicated with the following command:

MLFN ERROR TYPE = ErrorType

The choices for error type are now listed.

MEAN SQUARE — The mean (across all training cases and outputs) of the squared error is minimized. This is almost always the best choice, as its stability is excellent, and it has great intuitive and theoretical appeal. In fact, if another error measure is to be minimized, it is a good idea to start by minimizing the mean square error, then continuing training with the alternative error measure.

ABSOLUTE — The mean absolute value of each error is minimized. This may be of some value for industrial control problems in which minimizing the absolute error may lead to more stability in the controlled process.

KK — The Kalman-Kwasny error measure is minimized. This is of real value only for classification, and even then its utility is sporadic. Sometimes it helps avoid falling into poor local minima, while other times it exhibits so much instability that minimization algorithms have difficulty handling it.

CROSS ENTROPY — The cross entropy function is minimized. This is valid only for classification. The theory is that networks trained this way will have lower misclassification error, but many authorities have not yet been convinced. It can lead to great instability in optimization, and it definitely prefers nonlinear output activation functions.

The training algorithms that involve simulated annealing (**AN1, AN2, AN1_CJ, AN2_CJ, AN1_LM, AN2_LM**) need many special parameters to be specified. Details may be found in Masters (1995). They are given by these commands (plus an equal sign followed by the parameter):

ANNEALING INITIALIZATION TEMPERATURES — This many temperatures will be used.

ANNEALING INITIALIZATION ITERATIONS — This many iterations will be performed at each temperature.

ANNEALING INITIALIZATION SETBACK — The iteration counter is set back by this amount after each improvement.

ANNEALING INITIALIZATION START — This is the starting temperature (standard deviation for Gaussian perturbation).

ANNEALING INITIALIZATION STOP — This is the stopping temperature.

ANNEALING INITIALIZATION RANDOM — This specifies the probability distribution of the perturbations. It may be either **GAUSS** for a Gaussian distribution, or **CAUCHY** for a Cauchy distribution.

Those training methods that involve generic simulated annealing (**AN2, AN2_CJ, AN2_LM**) need several additional parameters. These are the following:

ANNEALING INITIALIZATION RATIO — This is the factor that controls the proportion of trials accepted at the start of annealing. Larger values result in higher acceptance, which gives a more global minimum. If little or no improvement is had as the temperature drops, decrease the ratio parameter.

ANNEALING INITIALIZATION REDUCTION — If this parameter is **EXPONENTIAL**, the usual exponential temperature reduction schedule is used. The value **FAST** causes Harold Szu's fast reduction to be used.

ANNEALING INITIALIZATION ACCEPT — If this parameter is **CLIMB**, the usual Metropolis acceptance criterion is used. If it is **NOCLIMB**, Szu's alternative criterion is used.

The four training methods that hybridize simulated annealing with a direct descent method, **AN1(2)_CJ(LM)**, use annealing in two different ways. First, they use it to find starting weights for direct descent. The parameters for that phase of training are specified with the commands that were just described. After descent to the nearest local minimum is complete, they use simulated annealing to attempt to escape from what may be a poor local minimum. The parameters for this phase of training are specified with these exact same commands, except that the word **ESCAPE** replaces the word **INITIALIZATION**.

These four hybrid training algorithms have two more special parameters that may be specified. Detailed explanations for these parameters may be found in Masters (1995). The first parameter is

MLFN RESTARTS = integer

This allows the user to limit the number of times the training algorithm restarts from scratch with a new random set of starting weights. This is useful when operating in batch mode, as it allows the program to give up and go on to something else. Instead of using the **ALLOWABLE ERROR** command to enable quitting when the error drops to a certain level, the user may wish to set the error limit to zero and use the **MLFN RESTARTS** command to provide the escape route. This is usually a good approach. Note that this parameter does not include the first try. It is in addition to the first try.

The other special parameter is the number of tries to **ACCURACY** significant digits that are performed before any refinement to an additional **REFINE** significant digits is attempted. This parameter is set with the following command:

MLFN PRETRIES = integer

These four hybrid training algorithms are described in detail in Masters (1995). A good general rule is to make **MLFN RESTARTS** a very large number, and specify **MLFN PRETRIES** in the range 4–7 or so. If **MLFN RESTARTS** is small (with the intention of completing all tries), maximum efficiency is obtained if **MLFN PRETRIES** is one greater than the number of restarts.

MLFN Progress Reports

When the command **PROGRESS ON** has been issued, extensive reporting of intermediate results will take place during MLFN training. Each training method has its own form of progress report, so they will be separately treated.

The AN1 algorithm, whether used alone or in a hybrid with a CJ or LM direct descent algorithm, is the simplest. Each line of the annealing algorithm lists the current temperature and the best error attained after processing at that temperature is complete. When all temperatures have been processed, the try number (1, 2, ...) is printed, along with the error attained in that try and the best error attained among all tries. This is repeated **MLFN RESTARTS** times.

The AN2 algorithm is only slightly more complicated. The sole difference is that *Initializing* is printed when the algorithm starts. When the initialization phase is complete, the error and its standard deviation are printed. The remainder of the reporting is identical to the AN1 algorithm.

When the conjugate gradient algorithm is hybridized with simulated annealing or regression, its output is substantial. Whenever glob_min, the global minimization routine, is called for rough line minimization, the parametric position along the line is reported each time glob_min evaluates the error criterion. In addition, the error at that position is reported. Then, when brentmin is called to refine the line minimization, it will print progress in sets of six numbers across a line. These numbers are the lower bound of the current interval, the location of the best point in the interval, the upper bound of the current interval, and the best, second-best, and third-best function values. When linear refinement is complete, a line summarizing the complete iteration is printed. This consists of the scale factor for jump optimization, the best error, the derivative vector length, and the percent improvement over the previous iteration. The last number is the most important, as it indicates the rate of progress. When several iterations in a row produce trivial improvements, convergence may be at hand.

When the Levenberg-Marquardt algorithm is used in a hybrid training algorithm, its progress output is relatively meager. At the end of every few iterations, the current error and lambda are printed. The frequency of printing is automatically adjusted according to the problem size.

The hybrid algorithms are relatively complex. The annealing phase prints its progress as just described. Then the try number and the best error so far are printed. The direct descent algorithm then prints its progress. When it is complete, its error is printed with the label *Gradient error*. Then annealing is used once more to attempt escape from the local minimum. If successful, the word *Escaped* is printed and direct descent is tried again. Otherwise, one of two routes is taken. Usually, the word *Restart* is printed, and a new try is begun with annealing for initialization. But under the right conditions, the word *Refining* is printed, and direct descent is used to refine the error to additional accuracy. This is always followed by a restart.

The stochastic smoothing algorithms, SS and SSG, have relatively simple progress reports. They start by printing the word *Initializing*. When initialization is complete, the average error and the best error are printed. Then, for each temperature, a line is printed listing that temperature, the mean error, and the best error obtained so far. The SSG algorithm additionally prints the gradient length.

Testing a Trained Network

The first step in testing a neural network is to cumulate a test set. This is discussed on page 423. Then the trained network is tested with the following command:

TEST NETWORK = NetName

Each case in the test set will be presented to the network, and the resulting outputs will be computed. Those outputs are compared to the target outputs, and the error of each case is computed. The mean squared error (across all cases and outputs) is reported as the principal test result. The audit log file will contain much supplementary information. If there is more than one output neuron, the error for each will be separately sumarized. Each neuron's mean squared error is followed by the standard error of the MSE in parentheses. As a rough rule of thumb, the reported MSE, plus and minus twice its standard error, has a high probability of encompassing the true MSE in the population represented by the test set. The mean absolute error (MAE) and its standard error are also reported.

The next line lists six percentiles of the error distribution: 5, 10, 25, 75, 90, and 95. As long as the user is confident that the test set represents the true population, these percentiles can be used to estimate confidence intervals for predicted values.

If classification mode is in effect, a confusion matrix will also be written to the audit log file. This matrix portrays the patterns of misclassification that are obtained from the test set. There is one row for each class, plus one more row if any implicit rejects are included in the test set. There are as many columns as there are classes, plus one additional column to accommodate the reject category. The element in row a and column b is the number of cases that are truly members of class a but that have been classified into class b. The number in the last column of row a is the number of cases in class a whose activation did not attain the minimum threshold and so were tossed into the reject category. Off-diagonal quantities represent error.

If desired, the user may set the classification threshold for rejection. The command to do this is the following:

THRESHOLD = number

The number is a percent (0–100) of full activation. When the threshold is set, it will remain at that value unless it is set with this command again. The default is zero, implying that no rejection will take place. This threshold controls the ease with which cases are banished to the reject category. Larger values will force more marginal cases to land in the reject heap. Using a threshold of zero will let all cases be classified, so that the reject column will be entirely zero.

The user can request even more information about the network error. This is done by using the following command:

EXTENDED TEST NETWORK = NetName

This command generates the same statistics as the **TEST NETWORK** command. In addition, it computes the 90th percentile of the absolute error. Then it uses both the jackknife and the bootstrap methods to compute the bias and standard error of this statistic. (These algorithms are discussed in detail in Masters (1995).) Be warned that these statistics can be very slow to compute, especially when the test set is large. Also keep in mind that the jackknife is generally unreliable compared with the bootstrap. It is included primarily for educational purposes. In practice, the bootstrap should be relied on for good estimates of the standard error. The jackknife should be ignored.

The 90th percentile of the absolute error is obviously valuable in itself. It allows the user to estimate confidence intervals for network predictions. But this is not the real reason for including the extended test option. This option is here to facilitate inclusion of custom error measures. Many readers will have some unusual error measure that they would just love to have evaluated for their neural network. These users need only modify the userstat subroutine in the module TESTNET.CPP. This subroutine currently sorts the data and returns the 90th percentile. By changing this routine to compute a custom error measure, the built-in jackknife and bootstrap routines will automatically evaluate the bias and standard error of the user's statistic—very convenient. (It would also be advisable to change the label of this statistic in PROCESS.CPP so that readers of the audit log file are not confused!)

The primary use for the **TEST NETWORK** (or **EXTENDED TEST NETWORK**) command is to test a validation set that is independent of the training set. In this way we can evaluate the performance of the network in an unbiased manner. Note that this testing is applicable to classification mode also. In this mode, the output targets are full activation for the neuron corresponding to the correct class and zero activation for all others.

Saving a Trained Network

We often want to preserve the weights of a trained network, particularly if the training required a long period of time. We may want to perform additional training, starting with a weight set that is already good. Or we may simply want to use the network to perform some useful task. The NPREDICT program saves the network in a compact internal format. This format is documented in the module WT_SAVE.CPP that can be found on the accompanying code disk. The weights are saved to and restored from a disk file with the following commands:

SAVE NETWORK = NetName TO FileName
RESTORE NETWORK = NetName FROM FileName

It is also possible to print the weights to an ASCII file for visual inspection. This file cannot be read by the program. It is for the user's convenience and edification only. The file is self-documenting, so no description of it is given here. The command to print a weight file is:

PRINT NETWORK = NetName TO FileName

Neural Network Prediction

A trained neural network can be used to generate signals based on other signals or recursively based on the same signals. If the network was trained for classification, the output signals represent approximate decision confidences for each class.

There are three separate steps necessary for using a neural network to predict outputs: The input signal(s) must be specified, the output signal(s) must be specified, and the command to predict with a named network must be given.

The input signals are specified exactly as they were for creating the training set. See page 423 for details. It is *mandatory* that the lags used for prediction inputs correspond *exactly* to those used in building the training set. For example, suppose that the training set inputs were defined as shown in the following two commands:

INPUT = TrainSig 1
INPUT = TrainSig 2

It is illegal to specify the prediction inputs as shown in the following command, even though the effect is equivalent:

INPUT = TestSig 1-2

The output signals are specified in the same way that signals are specified when reading them from a file. See page 382 for details of the **NAME** command. It is illegal to use more names than there are neural network outputs. If there are fewer names than outputs, the extra outputs will be ignored. As was the case for reading disk files, network outputs may be skipped by using extra commas as placeholders. For example, suppose the network has three outputs, but we do not care about the second. We can generate signals from the first and last with the following command:

NAME = FirstOut , , ThirdOut

Note that the current output list, as built by **OUTPUT** commands, is *almost* totally ignored for neural network prediction. It was used when the training set was constructed, but after the network has been trained, it is no longer needed for prediction. This is in contrast to ARMA models. However, it will be seen later that the output list may come into play in regard to preservation of known points in output signals.

After the input and output signals have been specified, the actual prediction is initiated by the following command:

NETWORK PREDICT = NetName

When the network was trained, the lead time (if any) for each output was recorded. (The lead time for classification models is always zero.) This same lead time will be used for prediction. The implication is that the time origin of the generated outputs corresponds exactly to the inputs. No time shift is induced. Another implication is that outputs that precede the lead time are not computed. They are fixed at zero. Note that if any of the output signals are named in the current output list, the leads recorded in that list are totally ignored. The leads that were in effect when the network was trained take precedence.

When the first few output signal points are computed, they will most likely be based on lagged input signal points that do not exist because they are prior to the start of the series. When this happens, the value of the first point in each input signal is used as the value of all prior points in that signal. Similarly, the user may generate an output signal that is so much longer than the inputs that some required inputs are beyond the end of the input series. If this happens, the final point in each input series is used as the value of all points past the end. Note that this is not the case when the same signal is recursively used for both input and output, as explained in the next paragraph. Also note that this practice is strongly discouraged. Replication of the final point nearly always causes terrible errors. As a general rule, *all* inputs should be recursively computed if the network will be used to predict the future beyond the lead time.

It is legal to use the same signal(s) for both input and output. This is useful for recursive prediction, in which the user wants to predict so far into the future that predicted outputs at some time slots are used as inputs for later time slots. When a signal is used for both input and output, values that exist in the input series are never overwritten by predictions. Updating of any such signals takes place only *after* the end of the input data. Newly predicted values are then available for subsequent predictions further into the future. Remember that errors can build up rapidly when this is done. Nevertheless, this is almost always vastly superior to letting some input series just expire!

When this recursive prediction is done, records of previous signal modifications (such as **CENTER**, **DETREND**, **DIFFERENCE**, and so forth) are preserved. This enables use of an **UNDO** command to reverse the prior operation all the way through the predictions.

Although most prediction is done in mapping output mode, it is perfectly legal to do prediction in classification mode. In this case, there are as many neural network outputs as there are classes. The signal generated by each output is roughly proportional to the confidence in each class membership. *The outputs appear in alphabetical order according to the name of each class.* (Capitalization is ignored.) This is worth repeating. The order in which classes appear during training is irrelevant. Alphabetical order is what determines the order of the outputs.

When the network model is a member of the PNN family, the outputs will range from zero to one and may be interpreted (under suitable conditions) as Bayesian probabilities of class membership. The outputs will sum to one. If the network is in the MLFN family, the outputs are not so easily interpretable. If all outputs have a low value, this is an indication that the input does not resemble any of the classes on which the network was trained. This powerful trait is not shared by the PNN, which is constrained to have all outputs sum to one.

The signals named on the **NAME** command are the generated outputs. If a named signal does not exist at the time the **NETWORK PREDICT** command is issued, the nature of this output signal is straightforward: It is a point-by-point prediction from beginning to end. The value in each time slot is the prediction based on the corresponding input data. However, if the named output signal already exists, there are two possibilities. It may be point-by-point, as happens when the signal does not already exist. However, it may also contain only future predictions. This happens if either of two conditions is true: If the named output signal is also in the current input list (via **INPUT** commands), this signal is being used recursively. In this case, all known values of the series are preserved. Predictions are appended to the end of the series. Also, if the named output signal is in the current output list (via **OUTPUT** commands), the known extent of the signal is preserved. Note that this is the *only* reason prediction looks at the output list. Except for making this one decision, the current output list is totally ignored.

Why does one want two prediction output options? They have entirely different uses. Point-by-point predictions are nice for displaying the network's capabilities across time. Even though the network may have been trained using the same signal(s) for both input and output (a common practice), it is interesting to create a new signal for the output. This way, the user can simultaneously examine the true and predicted signals throughout the entire time extent. Areas of particularly good or bad performance may become obvious.

The alternative of preserving the known extent is useful when the network is actually used to make and display predictions into the future. Preservation is also mandatory if we will **UNDO** previous modifications like differencing or transformations. The best way to display predictions is by displaying the known signal with the predicted value(s) appended. This makes for easy visual interpretation. For signals that are used recursively, this preservation is automatically obtained by using the same name in both the input list and the named output. But sometimes we may be predicting an auxiliary output that is not used for input. To preserve the known extent of this signal, make sure it appears in the output list. The lead specified in the output list, if any, is ignored.

An Example of Neural Network Prediction

The commands needed for neural network prediction can be clarified with a small example. The complete text of this example is in the file NNPRED.CON in the EXAMPLES directory of the accompanying disk.

Suppose we want to use the current and previous values of a series to predict the next value. So that we have something concrete to work with, let us specify a neural network model and also generate a short test series.

```
NETWORK MODEL = PNN
NAME = sinewave
GENERATE = 32 SINE ( 1.0 , 16 , 0.0 )
```

For this example, we will use the intuitive approach of predicting at a lead of one based on lags of zero and one. We could just as well predict the current value of the series based on lags of one and two points. The effect would be the same. The training set is cumulated with the following commands:

```
INPUT = sinewave 0-1
OUTPUT = sinewave 1
CUMULATE TRAINING SET
```

The neural network can now be trained, and its capabilities can be visually ascertained with the following commands:

```
TRAIN NETWORK = demo
NAME = predicted
NETWORK PREDICT = 35 demo
DISPLAY = predicted
```

The computed signal, *predicted*, uses the trained neural network to do a case-by-case prediction of the input series. We hope that the predicted signal closely resembles the training signal. The mean

Neural Network Prediction

Figure 11.1 Simple neural network prediction.

squared error reported at the conclusion of training tells us numerically how good the prediction is, and the video display lets us identify specific problems.

Figure 11.1 illustrates how the predicted signal is computed. Since training was done with an output lead of one, the first point in the predicted signal is not computed. It is fixed at zero. The second point (at lead one from the current point at slot zero) is based on the current input and the input at a lag of one. That latter point is prior to the start of the input series, so the first point is duplicated. The implication is that the output at slot one will probably be in error since it is not entirely based on correct data.

Starting with output slot two, all predictions will be based on completely valid input data. This continues through slot 32. If the user requests that more predictions be made, these will probably be less than excellent. Output slot 33 is based on one fudged input value, and slots 34 and all that follow are based on two fudged values.

Most users would not be satisfied with the ability to make only the one fully legitimate prediction available in slot 32. There is obviously no totally correct means of using this model to make more distant predictions. But it is possible to improve on the simplistic approach of duplicating endpoints. Instead of using those fixed values as network inputs, use the output predicted at one stage as the input for the next stage. This is done by using the same signal for both input and output, as implemented by the following rules:

```
NAME = sinewave
NETWORK NPREDICT = 201 demo
```

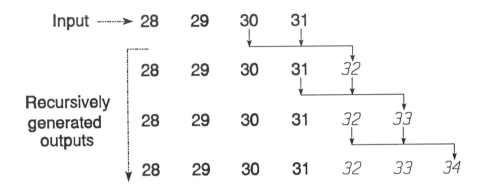

Figure 11.2 Recursive prediction.

Figure 11.2 illustrates how this recursive prediction works. The original points in the signal remain untouched. Prediction starts with the slot just past the end of the input series. After that slot, each successive prediction is based on prior predictions. It should be obvious that errors can build up rapidly. On the other hand, the ability of neural networks to handle recursive prediction can sometimes be almost frightening. Interested readers should display the recursively predicted signal computed with the previous commands.

Confidence Intervals for Predictions

Robust confidence intervals, as discussed in Chapter 8, can be computed for neural network predictions. Set up the input list and named outputs (**NAME** command) exactly as they will be used for prediction. Remember that all named outputs must be signals that already exist, as these will provide the known values needed by the confidence algorithm. Then issue the following command:

NETWORK CONFIDENCE = Npred NetName

The equal sign is followed by an integer that specifies the maximum number of future predictions whose confidence intervals are to be estimated. The confidence information is automatically attached to the named output signals. It is assumed that these same signals will be used for prediction. In practice, what we usually do is insert the **NETWORK CONFIDENCE** command just after the **NAME** command

and before the **NETWORK PREDICT** command. This way, the confidence information is associated with the signals that will be predicted.

Confidence information is associated *only* with predicted signals whose known extent is preserved during prediction. In other words, any named output signal whose confidence is desired must also appear in either the input list (for recursive use) or in the output list (to manually force preservation). Even though all named signals exist at the time confidence is computed, *only those that appear in the input or output list will take part in the computation*. The discussion starting on page 446 covers preservation of output signals.

By default, a 90 percent two-tailed confidence interval is computed. This means that there is (on average) a 5 percent chance that the true signal value exceeds the upper limit, and a 5 percent chance that the true value is less than the lower limit. The user can specify the two-tailed probability with the following command:

CONFIDENCE PROBABILITY = Percent

The probability is a percent, and it must be in the range 50–100. Values larger than the default of 90 require a large number of test cases. Also, the computed interval may be so wide that it is useless.

After confidence computation is complete, two lines concerning the confidence in the confidence are printed in the audit log. Their exact meaning is discussed at the end of Chapter 8. The two lines are similar to the following:

There is a 5% chance each tail probability exceeds 11.62%
There is a 99.60% chance the interval exceeds 80.00%

These sample lines resulted from asking for a 90 percent confidence interval. The first line tells us that there is a 5 percent chance (fixed by the program) that each tail area is really as much as 11.62 percent of the error distribution, instead of the 5 percent requested by the user. The second line doubles the requested tail area, then states the probability that the computed confidence interval encompasses at least one minus that amount. In other words, the user asked for a 90 percent interval, implying that the tails together account for 10 percent of the error distribution. Doubling that causes the program to consider an 80 percent interval, and we are informed that there is a 99.6 percent chance that our interval encompasses at least 80 percent of the error distribution.

The user may receive an error message stating that too few cases exist. The following formulas may be useful in resolving this issue. Let maxlag be the maximum input lag, and let maxlead be the maximum output lead. Define offset as maxlag+maxlead. That many known values are skipped at the beginning of the series. Let npred be the number of future predictions specified in the command, and let shortest be the length of the shortest signal, including both inputs and outputs. Then the number of cases that will go into the confidence estimates is computed by Equation (11.3).

$$\text{ncases} = \text{shortest} - \text{offset} - \text{npred} + 1 \qquad (11.3)$$

The number of cases must meet two criteria. First, it must be at least 20, an arbitrary but reasonable limit. Second, the number of tail cases, given by Equation (11.4), must be at least one. In that equation, prob is the two-tailed probability specified by the user or left at its default of 90.

$$\text{tail} = 0.5 * (1.0 - \text{prob}/100) * \text{ncases} \qquad (11.4)$$

It may be that one or more of the predicted outputs are in a modified domain. For example, the user may have taken the log of a series, then differenced it. If the network was trained on this modified data, the user is probably not very interested in confidence intervals in this domain. The predicted signal will probably be subjected to one or more **UNDO** commands to return it to the original domain. This is the domain in which the user will undoubtedly want confidence bounds. In order to compensate for these preprocessing operations, the user must specify them prior to issuing the **NETWORK CONFIDENCE** command. Legal compensations are:

> **CONFIDENCE CENTER = Signal [PER Other]**
> **CONFIDENCE DETREND = Signal [PER Other]**
> **CONFIDENCE OFFSET = Signal [PER Other]**
> **CONFIDENCE SCALE = Signal [PER Other]**
> **CONFIDENCE STANDARDIZE = Signal [PER Other]**
> **CONFIDENCE DIFFERENCE = Signal [PER Other]**
> **CONFIDENCE SEASONAL DIFFERENCE = Signal [PER Other]**
> **CONFIDENCE LOG = Signal**

For each of these commands, the user must specify the name of a signal that will be designated as a predicted output on a subsequent **NAME** command. This name is used to match operations to signals. It is *vital* that, for each different signal, these confidence compensation commands appear in exactly the same order as the operations were performed on the original signal. The subsequent **UNDO** commands will appear in the reverse order. By default, the parameters for undoing are retrieved from the named signal. Optionally, a different signal may be named as the source of the undoing parameters by means of the **PER** keyword. (See page 385 for more on **PER**.)

If the first confidence compensation for a signal is **LOG**, multiplicative confidence intervals will be computed. Otherwise, additive intervals will be computed. See the discussion starting on page 323 for more details.

Sometimes the user will want to do several different types of predictions in the same program run. The confidence compensation list is cleared to make way for a new list by means of the following command:

CLEAR CONFIDENCE COMPENSATION

Example of Network Prediction Confidence

Here is a small example of how to compute confidence intervals for neural network prediction. In deference to economists, we apply two modifications to the single series that will be recursively predicted. First, we take its log to convert multiplicative effects to additive effects. Then we difference the series to render it stationary. The two most recent values will be used to predict the current value. The training set is cumulated, and the network is trained. We instruct it to compute confidences based on eventual undoing of the two modifications. Since the first confidence compensation is **LOG**, multiplicative intervals are computed. Finally, the predicted signal is taken back to the original problem domain. We also do a pointwise prediction for display.

```
LOG = x
DIFFERENCE = 1 x
INPUT = x 1-2
OUTPUT = x
```

CUMULATE TRAINING SET
TRAIN NETWORK = temp

CONFIDENCE LOG = x
CONFIDENCE DIFFERENCE = x

NAME = pointwise
NETWORK PREDICT = 200 temp

NAME = x
NETWORK CONFIDENCE = 5 temp
NETWORK PREDICT = 200 temp

UNDO DIFFERENCE = x
EXP = x

Alphabetical Glossary of Commands

ACCURACY = integer — This is the approximate number of error significant digits optimized during the direct descent part of the main loop in MLFN hybrid training algorithms (AN1(2)_CJ(LM)). When training members of the probabilistic neural network family (PNN, SEPVAR, SEPCLASS), the sum of ACCURACY and REFINE is the number of significant digits optimized. The default is 6.

ADD = Signal1 AND Signal2 — The two signals (which must be the same length) are added point-by-point. The result defines the signal(s) appearing on a previous **NAME** command.

ALLOWABLE ERROR = number — When the error falls this low during learning, stop trying to improve it. The default is zero.

ANNEALING ESCAPE ACCEPT = CLIMB / NOCLIMB — If CLIMB is specified, error improvements are always accepted, as is traditional. For NOCLIMB, even improvements may be rejected. This is valid for the AN2 method only, and this rule applies to simulated annealing local minimum escape for hybrid algorithms (AN2_CJ(LM)). The default is CLIMB.

ANNEALING ESCAPE ITERATIONS = integer — This is the number of trials that will be performed at each temperature for simulated annealing local minimum escape. The default is 20.

ANNEALING ESCAPE RANDOM = GAUSSIAN / CAUCHY — This specifies the probability distribution function for the random perturbations during simulated annealing local minimum escape. The default is GAUSSIAN.

ANNEALING ESCAPE RATIO = number — This is the ratio of the temperature scale factor (in the acceptance probability equation) to the scale factor for the random perturbations. It should be set so that improvements happen slowly but steadily. If no improvement occurs, reduce this value. This is valid for the AN2 method only, and this rule applies to simulated annealing local minimum escape for hybrid algorithms (AN2_CJ(LM)). The default is 0.1.

ANNEALING ESCAPE REDUCTION = EXPONENTIAL / FAST — The schedule for reducing the temperature in simulated annealing local minimum escape is determined by this command. In general, FAST should be used only in conjunction with CAUCHY random perturbations. This is valid for the AN2 method only, and this rule applies to simulated annealing local minimum escape for hybrid algorithms (AN2_CJ(LM)). The default is EXPONENTIAL.

ANNEALING ESCAPE SETBACK = integer — Each time improvement is had at a temperature, set back the iteration counter this amount to keep trying to do better. This is for simulated annealing local minimum escape. The default is 20. Typically, this would be set to about half of the number of iterations.

ANNEALING ESCAPE START = number — The temperature is the standard deviation of the weight perturbation. This is the starting temperature for simulated annealing local minimum escape. The default is 4.0.

ANNEALING ESCAPE STOP = number — This specifies the final temperature for simulated annealing local minimum escape. The default is 0.1.

ANNEALING ESCAPE TEMPERATURES = integer — This is the number of temperatures for simulated annealing local minimum escape. The default is 4.

ANNEALING INITIALIZATION ACCEPT = CLIMB / NOCLIMB — If CLIMB is specified, error improvements are always accepted, as is traditional. For NOCLIMB, even improvements may be rejected. This is valid for the AN2 method only, and this rule applies to simulated annealing weight initialization. The default is CLIMB.

ANNEALING INITIALIZATION ITERATIONS = integer — This is the number of trials that will be performed at each temperature for simulated annealing weight initialization. The default is 50.

ANNEALING INITIALIZATION RANDOM = GAUSSIAN / CAUCHY — This specifies the probability distribution function for the random perturbations during simulated annealing weight initialization. The default is GAUSSIAN.

ANNEALING INITIALIZATION RATIO = number — This is the ratio of the temperature scale factor (in the acceptance probability equation) to the scale factor for the random perturbations. It should be set so that improvement occurs slowly but steadily. Excessively large values prevent improvement. This is valid for the AN2 method only, and this rule applies to simulated annealing weight initialization for hybrid algorithms (AN2_CJ(LM)). The default is 0.1.

ANNEALING INITIALIZATION REDUCTION = EXPONENTIAL / FAST — The schedule for reducing the temperature in simulated annealing weight initialization is determined by this command. In general, FAST should be used only in conjunction with CAUCHY random perturbations. This is valid for the AN2 method only, and this rule applies to simulated annealing weight initialization for hybrid algorithms (AN2_CJ(LM)). The default is EXPONENTIAL.

Alphabetical Glossary of Commands

ANNEALING INITIALIZATION SETBACK = integer — Each time improvement is had at a temperature, set back the iteration counter this amount to keep trying to do better. This is for simulated annealing weight initialization. The default is 50.

ANNEALING INITIALIZATION START = number — The temperature is the standard deviation of the weight perturbation. This is the starting temperature for simulated annealing weight initialization. The default is 3.0.

ANNEALING INITIALIZATION STOP = number — This specifies the final temperature for simulated annealing weight initialization. The default is 0.2.

ANNEALING INITIALIZATION TEMPERATURES = integer — This is the number of temperatures for simulated annealing weight initialization. The default is 5.

APPLY ORTHOGONALIZATION = Name — The previously defined orthogonalization whose name is given is used in conjunction with the signals in the current input list to create a new set of signals as specified on the most recent **NAME** command.

ARMA CONFIDENCE = Npredictions ARMAname — Confidence intervals for all signals in the output list are computed for the specified number of future predictions. A **NAME** command is not needed for this operation.

ARMA FIXED = NO — This, the default, decrees that the ARMA training algorithm is to estimate an optimal constant offset for each output series.

ARMA FIXED = YES — The outputs in an ARMA model are assumed to have an offset fixed at zero, so no offset will be estimated.

ARMA PREDICT = Points ArmaName — One or more signals (specified on a previous **NAME** command) are generated using the named ARMA model. The length of the signal is the specified number of points. The model's inputs are determined by the current input list, and the MA terms and outputs are determined by the current output list. Lags in these lists must correspond exactly with those in training the model. For any predicted signal (specified on the **NAME** command), if the signal is also in the output list, the known extent of the signal will be preserved, with predictions being appended to the end. Otherwise, the predicted signal will be a point-by-point prediction.

ARMA SHOCK = ShockName FOR OutName — This allows the user to specify that a signal be used for the shock values in an ARMA prediction, as opposed to the default of using shocks of zero beyond the end of the known output.

AUDIT LOG = FileName — The default name of the audit log file is AUDIT.LOG. This command can be used to change the filename, or to turn off the log by entering just AUDIT LOG with no equal sign or filename. The AUDIT.LOG file in the current directory is always erased when the program starts. However, if the filename is changed with this command and the newly named file already exists, it is preserved, and new lines are appended to its end.

AUTOCORRELATION = Lags SignalName — The autocorrelations of the named signal from a lag of one through the specified limit are computed. A previous **NAME** statement supplies the name of the created signal.

BANDPASS = Frequency Width Signal — A bandpass filter having the specified frequency (0.0–0.5) and width (typically 0.01–0.2) is applied to the signal. The filtered output is sent to the most recent signal on a **NAME** command.

BYE — Quit the NPREDICT program.

CENTER = Signal — The signal is centered by having its mean subtracted.

CLASS = name — This sets the class name for subsequent CUMULATE TRAINING (TEST) SET commands.

CLEAR ALL — Every aspect of the NPREDICT program is reset as if the program were just starting.

Alphabetical Glossary of Commands

CLEAR ARMA = name — Erase the named ARMA model.

CLEAR ARMAS — Erase all ARMA models.

CLEAR CLASSES — The list of named classes is cleared. If a training or test set has been built based on one set of classes, this should be done before building a training or test set based on a different collection of classes.

CLEAR CONFIDENCE COMPENSATION — The list of active confidence compensations (such as **CONFIDENCE DIFFERENCE**) is erased.

CLEAR INPUT LIST — This resets the signal list defining the inputs for a training or test set.

CLEAR NETWORK = name — Erase the named network.

CLEAR NETWORKS — Erase all neural networks.

CLEAR ORTHOGONALIZATION = Name — The named orthogonalization is erased.

CLEAR ORTHOGONALIZATIONS — All orthogonalizations are erased.

CLEAR OUTPUT LIST — This resets the signal list defining the outputs for a training or test set.

CLEAR TEST SET — Erase the entire test set. This frees memory, so it should be done when the test set is no longer needed.

CLEAR TRAINING SET — Erase the entire training set. This frees memory, so it should be done when the training set is no longer needed.

COMMAND FILE = filename — This processes an ASCII text file of commands.

CONFIDENCE CENTER = Signal — Specifies that confidence intervals computed for the named signal by subsequent **ARMA**

CONFIDENCE or **NETWORK CONFIDENCE** commands are to be based on the assumption that the predicted signal will be operated on by an **UNDO CENTER** command.

CONFIDENCE DETREND = Signal — Specifies that confidence intervals computed for the named signal by subsequent **ARMA CONFIDENCE** or **NETWORK CONFIDENCE** commands are to be based on the assumption that the predicted signal will be operated on by an **UNDO DETREND** command.

CONFIDENCE DIFFERENCE = Signal — Specifies that confidence intervals computed for the named signal by subsequent **ARMA CONFIDENCE** or **NETWORK CONFIDENCE** commands are to be based on the assumption that the predicted signal will be operated on by an **UNDO DIFFERENCE** command.

CONFIDENCE LOG = Signal — Specifies that confidence intervals computed for the named signal by subsequent **ARMA CONFIDENCE** or **NETWORK CONFIDENCE** commands are to be based on the assumption that the predicted signal will be operated on by an **EXP** command.

CONFIDENCE OFFSET = Signal — Specifies that confidence intervals computed for the named signal by subsequent **ARMA CONFIDENCE** or **NETWORK CONFIDENCE** commands are to be based on the assumption that the predicted signal will be operated on by an **UNDO OFFSET** command.

CONFIDENCE SCALE = Signal — Specifies that confidence intervals computed for the named signal by subsequent **ARMA CONFIDENCE** or **NETWORK CONFIDENCE** commands are to be based on the assumption that the predicted signal will be operated on by an **UNDO SCALE** command.

CONFIDENCE SEASONAL DIFFERENCE = Signal — Specifies that confidence intervals computed for the named signal by subsequent **ARMA CONFIDENCE** or **NETWORK CONFIDENCE** commands are to be based on the assumption that the predicted signal will be operated on by an **UNDO SEASONAL DIFFERENCE** command.

CONFIDENCE STANDARDIZE = Signal — Specifies that confidence intervals computed for the named signal by subsequent **ARMA CONFIDENCE** or **NETWORK CONFIDENCE** commands are to be based on the assumption that the predicted signal will be operated on by an **UNDO STANDARDIZE** command.

COPY = Count Signal — This signal is copied to one or more previously named signals (on a **NAME** command). If the count is omitted, an exact copy is made. Otherwise, the length of the created signal is the absolute value of the count. If the count is positive, the copy starts at the beginning. If it is negative, the end of the source signal is the end of the destination signal. If the count exceeds the length of the source, zeros are appended (count positive) or prepended (count negative).

CROSSCORRELATION = Lags MainSignal AND LaggedSignal — The crosscorrelations of the first named signal with the second named signal from a lag of one through the specified limit are computed. This is one-sided in that only positive lags of the second signal with respect to the first are computed. A previous **NAME** statement supplies the name of the created signal.

CUMULATE EXCLUDE = integer — For all subsequent **CUMULATE TRAINING SET** and **CUMULATE TEST SET** commands, exclude this many valid cases from the beginning of every signal. The default is zero.

CUMULATE INCLUDE = integer — For all subsequent **CUMULATE TRAINING SET** and **CUMULATE TEST SET** commands, allow this many cases. Any excluded by means of **CUMULATE EXCLUDE** are not included in this count. The default is to keep all possible cases. To return to the default, set this to a gigantic number.

CUMULATE TEST SET — This is the command used to build a test set. It relies on previous INPUT and OUTPUT or CLASS commands to define the composition of the test set. If a test set already exists, it will be preserved and the new data will be appended.

CUMULATE TRAINING SET — This is the command used to build a training set. It relies on previous INPUT and OUTPUT or CLASS commands to define the composition of the training set. If a training set already exists, the new data will be appended.

CV TRAIN NETWORK = name — The named neural network is trained. If it does not already exist, it is created. Then, after training is complete, cross validation is used to estimate the population error based on the training set. This algorithm is extremely slow when there are more than a few training cases, but it makes efficient use of scarce data.

DEFINE ORTHOGONALIZATION = Name — The current training set is used to define an orthogonalization. A name for the orthogonalization object, by which it will be referenced later, must be specified.

DETREND = SignalName — The signal is detrended by subtracting the least-squares trend line. The mean is not changed.

DIFFERENCE = degree SignalName — The signal is differenced the specified number of times (one to three).

DISPLAY = SignalName — The named signal is graphically displayed.

DISPLAY CONFIDENCE = SignalName — If confidence intervals have been computed for a neural or ARMA prediction, using this command to display the predicted signal causes the confidence intervals to appear as well.

DISPLAY ORIGIN = Number — This is the time origin of the first point in the series. It is used for labeling the display. This command is ignored when displaying a spectrum or lagged correlation.

DISPLAY RANGE = OPTIMAL — The vertical axis is scaled to provide optimal display of the signal in regard to spacing.

Alphabetical Glossary of Commands

DISPLAY RANGE = SYMMETRIC — The vertical axis is scaled for optimal use of space, except that the center of the display corresponds to a signal level of zero.

DISPLAY RANGE = Min Max — The user explicitly specifies minimum and maximum display values for the vertical axis. Signal values outside these limits are plotted as if they were at the limit.

DISPLAY RATE = SampleRate — The user may specify the number of points sampled per unit of time. This affects the labeling of the horizontal axis for signal and spectrum displays. The display rate is ignored for lagged correlations.

DIVIDE = Signal1 AND Signal2 — The two signals (which must be the same length) are divided point-by-point (first divided by second). The result defines the signal(s) appearing on a previous **NAME** command.

EXP = Signal — Every point in the signal is replaced by its exponential.

EXTENDED TEST NETWORK = NetName — This is identical to the **TEST NETWORK** command except that some additional (slow!) testing is performed. The 90th percentile of the absolute error is computed. Then, the jackknife and the bootstrap algorithms are used to estimate the bias and standard error of this statistic. The jackknife values are reported primarily for education. They are not nearly as reliable as the bootstrap estimates.

GENERATE = Points ARMA AR1 AR2 AR10 MA1 MA2 MA10 — An ARMA signal (specified in a previous **NAME** statement) having the specified number of points is generated. The user specifies six weights: three AR and three MA, at lags of 1, 2, and 10.

GENERATE = Points NORMAL Mean StdDev — A signal (specified in a previous **NAME** statement) having the specified number of points is generated using a normal (Gaussian) random number generator. The mean and the standard deviation must be provided as shown.

GENERATE = Points RAMP Amplitude Period — A signal having the specified number of points is generated. It is a sawtooth wave that starts at zero and has the given amplitude and period.

GENERATE = Points SINE Amplitude Period Phase — A signal (specified in a previous **NAME** statement) having the specified number of points is generated. It is a sine wave whose amplitude, period (signal points), and phase (degrees) are as specified.

GENERATE = Points UNIFORM Low High — A signal (specified in a previous **NAME** statement) having the specified number of points is generated using a uniform random number generator. The lower and upper range limits must be provided as shown.

HIGHPASS = Frequency Width Signal — A highpass filter having the specified frequency (0.0–0.5) and width (typically 0.01–0.2) is applied to the signal. The filtered output is sent to the most recent signal on a **NAME** command.

INPUT = SignalName MinLag MaxLag — The named signal is added to the list of signals defining the inputs for subsequent training or test set cumulation. A single lag may optionally be specified, or a range of lags specified.

KERNEL = GAUSS — This, the default, specifies that the Gaussian function is to be the kernel for the PNN model. Unless there is strong reason to use another, choose this one. This kernel is mandatory for the **SEPVAR** and **SEPCLASS** models.

KERNEL = RECIPROCAL — The kernel function described in Masters (1995) is an alternative that has much heavier tails than the GAUSS kernel. Distant cases in the training set exert relatively more influence in processing a case than they do for the GAUSS kernel. Expect the unexpected if this kernel is used.

LOG = Signal — Every point in the signal is replaced by its base *e* log.

LOWPASS = Frequency Width Signal — A lowpass filter having the specified frequency (0.0–0.5) and width (typically 0.01–0.2) is applied to the signal. The filtered output is sent to the most recent signal on a **NAME** command.

MAXENT = Resolution Order Signal — The maximum entropy power spectrum of the signal is computed. The spectrum is placed in the signal named on the most recent **NAME** command. The resolution should usually be several times the length of the source signal, and the order should be as small as possible, typically 20 to 100.

MEDIAN CENTER = Signal — The signal is centered by having its median subtracted.

MLFN DOMAIN = COMPLEX — All aspects of the MLFN operate strictly in the complex domain. This includes inputs, outputs, and all hidden neurons. This version may not be used for classification.

MLFN DOMAIN = COMPLEX INPUT — The input of the MLFN is complex, but all other parts of the model operate strictly in the real domain. This version is not generally recommended.

MLFN DOMAIN = COMPLEX HIDDEN — This is an extremely powerful model. The inputs and the hidden neurons are fully complex, but the outputs are real.

MLFN DOMAIN = REAL — This, the default, specifies that all aspects of the MLFN operate strictly in the real domain. This is the traditional model.

MLFN ERROR TYPE = ABSOLUTE — The mean absolute error is minimized during MLFN training. It is occasionally useful.

MLFN ERROR TYPE = CROSS ENTROPY — The cross entropy is minimized during learning. It is legal only for classification, and it is a frequently recommended (though a bit unstable) alternative to mean squared error.

MLFN ERROR TYPE = KK — The Kalman-Kwasny error is minimized during learning. This is appropriate only for classification, and it is apparently useful only when there are a large number of classes.

MLFN ERROR TYPE = MEAN SQUARE — The traditional mean squared error is minimized during learning. This is the default and is a reliable all-purpose method.

MLFN HID 1 = integer — This defines the number of neurons in the first hidden layer. If there is to be no hidden layer, set this to zero. The default is two.

MLFN HID 2 = integer — This defines the number of neurons in the second hidden layer. If there is to be no hidden layer, set this to zero. If the first hidden layer has zero neurons, this must be zero also. The default is zero.

MLFN LEARNING ALGORITHM = AN1 — Use pure simulated annealing of type AN1 to train the network. This is a slow technique that is included for educational purposes only.

MLFN LEARNING ALGORITHM = AN1_CJ — Use a hybrid of simulated annealing type AN1 with conjugate gradients to train the network. This, the default, is the recommended method if there are any hidden layers and there are a lot of weights.

MLFN LEARNING ALGORITHM = AN1_LM — Use a hybrid of simulated annealing type AN1 with the Levenberg-Marquardt algorithm to train the network. This is the recommended method if there are any hidden layers and there are few weights.

MLFN LEARNING ALGORITHM = AN2 — Use pure simulated annealing of type AN2 to train the network. This is a slow technique that is included for educational purposes only.

MLFN LEARNING ALGORITHM = AN2_CJ — Use a hybrid of simulated annealing type AN2 with conjugate gradients to train the network.

MLFN LEARNING ALGORITHM = AN2_LM — Use a hybrid of simulated annealing type AN2 with the Levenberg-Marquardt algorithm to train the network.

MLFN LEARNING ALGORITHM = REGRESS_CJ — Use linear regression to initialize the weights, then continue with conjugate gradients to train the network. This can be used only if there is no hidden layer. In that case, it is excellent.

MLFN LEARNING ALGORITHM = REGRESS_LM — Use linear regression to initialize the weights, then continue with the LM algorithm to train the network. This can be used only if there is no hidden layer and relatively few inputs. In that case, it is excellent.

MLFN OUTPUT ACTIVATION = LINEAR — The output neurons are to have an identity activation function. This is the default.

MLFN OUTPUT ACTIVATION = NONLINEAR — The output neurons are to have the hyperbolic tangent activation function.

MLFN PRETRIES — For MLFN hybrid training algorithms (AN1(2)_CJ(LM)), this is the number of trials performed just to ACCURACY significant digits before refinement. This is ignored when training members of the probabilistic neural network family (PNN, SEPVAR, SEPCLASS). The default is 5.

MLFN RESTARTS = integer — For MLFN hybrid training algorithms (AN1(2)_CJ(LM)), this is the number of tries after the first that are performed. This is ignored when training members of the probabilistic neural network family (PNN, SEPVAR, SEPCLASS). The default is 32,767.

MORLET = Frequency Width Signal — A quadrature-mirror filter of the Morlet wavelet type is computed for the named signal with the specified center frequency (0.0–0.5) and width (typically 0.01–0.2). A previous **NAME** command should have specified up to four output signals: in-phase, in-quadrature, amplitude, and phase.

MOVING AVERAGE = Period Signal — A moving average filter with the specified period is computed for the named signal. A previous **NAME** command should have specified the output signal.

MULTIPLY = Signal1 AND Signal2 — The two signals (which must be the same length) are multiplied point-by-point. The result defines the signal(s) appearing on a previous **NAME** command.

NAME = NameList — This command lists one or more signal names. These signals will be created with a subsequent **READ SIGNAL FILE** command. This command is also used to define the signals created by **ARMA PREDICT** and **NETWORK PREDICT** commands, as well as many other commands that create new signals.

NETWORK CONFIDENCE = Npred NetName — Confidence intervals for signals specified on a previous **NAME** command *and* also appearing in the current input or output list are computed. These intervals are internally associated with their corresponding named signals. The user specifies the number of future predictions to be evaluated and the name of the neural network.

NETWORK MODEL = PNN — The original basic probabilistic neural network model is used. This employs one sigma for all variables in all classes.

NETWORK MODEL = SEPCLASS — This probabilistic neural network has a separate sigma for each variable and class. This is valid for classification mode only.

NETWORK MODEL = SEPVAR — The probabilistic neural network used has a separate sigma for each variable. In classification mode, all classes share the same sigma vector.

NETWORK PREDICT = Points NetName — One or more signals (specified on a previous **NAME** command) are generated using the named neural network. The length of the signal is the specified number of points. The network inputs are determined by the current input list. Lags in this input list must exactly correspond to those in the training set used to train the network. Leads in that training set output induce the same leads in the generated signal. For any predicted signals that already exist, the known extent will be preserved if and only if the signal also appears in the current input or output list.

OFFSET = constant Signal — The constant is added to every point in the signal.

ORTHOGONALIZATION FACTORS = number — Subsequent **DEFINE ORTHOGONALIZATION** commands will compute at most this many factors. The default is to keep all factors.

ORTHOGONALIZATION LIMIT = number — Subsequent **DEFINE ORTHOGONALIZATION** commands will compute factors whose quantity is as few as possible while still capturing at least the specified percentage (0–100) of the total variance. The default is to keep all factors.

ORTHOGONALIZATION STANDARDIZE = NO — Subsequent **DEFINE ORTHOGONALIZATION** commands will compute factors based on raw variable scaling. This is ignored for the **DISCRIMINANT** type, and it is usually recommended for the **CENTROIDS** type.

ORTHOGONALIZATION STANDARDIZE = YES — Subsequent **DEFINE ORTHOGONALIZATION** commands will compute factors based on standardized variable scaling. This is ignored for the **DISCRIMINANT** type, and it is usually recommended for the **PRINCIPAL COMPONENTS** type. This is the default.

ORTHOGONALIZATION TYPE = CENTROIDS — Subsequent **DEFINE ORTHOGONALIZATION** commands will compute the factors based on the class centroids. This is only rarely useful.

ORTHOGONALIZATION TYPE = DISCRIMINANT — Subsequent **DEFINE ORTHOGONALIZATION** commands will compute the factors based on the optimal linear discriminant function separating the classes.

ORTHOGONALIZATION TYPE = PRINCIPAL COMPONENTS — Subsequent **DEFINE ORTHOGONALIZATION** commands will compute the factors based on the principal components of the complete training set.

OUTPUT = SignalName MinLead MaxLead — The named signal is added to the list of signals defining the outputs for subsequent training or test set cumulation. A single lead may optionally be specified, or a range of leads specified.

PADDING = DETREND — This, the default, is relevant to lowpass, highpass, bandpass, and QMF (including Morlet) filters. It specifies that the input series is centered and detrended so that the first and last points are zero. This is appropriate for series that wander about due to deterministic trends or non-stationarity.

PADDING = MEAN — This is the complement to the command above. It specifies that no detrending is done and that the input series is padded with its mean. This is appropriate for series that are centered around a stable mean value.

PARTIAL AUTOCORRELATION = Lags SignalName — The partial autocorrelations of the named signal from a lag of one through the specified limit are computed. A previous **NAME** statement supplies the name of the created signal.

PARTIAL CROSSCORRELATION = Lags MainSignal AND LaggedSignal — The partial crosscorrelations of the first named signal with the second named signal from a lag of one through the specified limit are computed. This is one-sided in that only positive lags of the second signal with respect to the first are computed. A previous **NAME** statement supplies the name of the created signal. *This operation can be extremely slow if more than a few lags are computed.*

PRINT NETWORK = NetName TO FileName — A neural network weight file is written to an ASCII file for visual examination by the user. If a file of this name already exists, it is erased.

PRIOR = N — This indicates that the user wants to let the number of training cases in each class determine the prior probabilities. This is relevant only for members of the probabilistic neural network family (PNN, SEPVAR, SEPCLASS).

Alphabetical Glossary of Commands

PRIOR = number — This specifies the prior probability that will be assumed for all subsequent classes (until another PRIOR statement occurs). This is relevant only for members of the probabilistic neural network family (PNN, SEPVAR, SEPCLASS).

PROGRESS OFF — This, the default, limits screen output to the minimum necessary to keep the user informed. This is ignored in the Windows version of NPREDICT, which always displays progress.

PROGRESS ON — For slow operations like neural network training, extensive intermediate output is printed to the screen to keep the user from worrying about progress. This is ignored in the Windows version of NPREDICT, which always displays progress.

QMF = Frequency Width Signal — A quadrature-mirror filter is computed for the named signal with the specified center frequency (0.0–0.5) and width (typically 0.01–0.2). A previous **NAME** command should have specified up to four output signals: in-phase, in-quadrature, amplitude, and phase.

READ SIGNAL FILE = FileName — An ASCII data file is read to create one or more previously named signals.

REFINE = integer — This is the additional approximate number of error significant digits computed after **PRETRIES** tries are done and after any subsequent improvement. That explanation applies only to MLFN training with hybrid algorithms (AN1(2)_CJ(LM)). When training members of the probabilistic neural network family (PNN, SEPVAR, SEPCLASS), the sum of ACCURACY and REFINE is the number of significant digits optimized. The default is 2.

RESTORE ARMA = ArmaName FROM FileName — An ARMA model file is read and used to define a model. If a model of this name already exists, it is erased.

RESTORE NETWORK = NetName FROM FileName — A neural network weight file is read and used to define a network. If a network of this name already exists, it is erased.

RESTORE ORTHOGONALIZATION = OrthName FROM FileName — An orthogonalization weight file is read and used to define an orthogonalization. If one of this name already exists, it is erased.

SAVE ARMA = ArmaName TO FileName — An ARMA model file is written to a disk file in order to preserve the named model. If a file of this name already exists, it is erased.

SAVE NETWORK = NetName TO FileName — A neural network weight file is written to preserve the named network. If a file of this name already exists, it is erased.

SAVE ORTHOGONALIZATION = OrthName TO FileName — An orthogonalization weight file is written to preserve the named orthogonalization. If a file of this name already exists, it is erased.

SAVE SIGNAL = SignalName TO FileName — An ASCII file is written containing the named signal. There is one line per sample. If a file of this name already exists, it is erased.

SAVGOL = HalfLength Degree Signal — A Savitzky-Golay filter is applied to the named signal (which is not changed). The filtered signal is placed in the signal named on the most recent **NAME** command. The half-length should be large for good smoothing, but it should not significantly exceed the width of the narrowest feature to be resolved. The degree should be even and in the range two to six.

SCALE = constant Signal — Every point in the signal is multiplied by the constant.

SEASONAL DIFFERENCE = period SignalName — The signal is seasonally differenced using the specified period.

SIGMA HIGH = number — This suggests a maximum value of sigma for the PNN, SEPVAR, and SEPCLASS models. Initial global scanning for a rough starting point will not try values above this number unless the error criterion is at a minimum here. In that case, the search will continue to as high a value as is needed. The default is 10.0.

SIGMA LOW = number — This is similar to SIGMA HIGH except that it specifies the minimum value of sigma for the initial search. The default is 0.01.

SIGMA TRIES = number — This is the number of sigma values ranging from SIGMA LOW to SIGMA HIGH that will be tried in the initial search. The minimum is three, but values of ten or more are usually reasonable to ensure that the global minimum is located. The default is ten.

SPECTRUM = Signal — The Fourier transform of the signal is computed. A previous **NAME** command should specify up to five signals. These receive the real part of the transform, the imaginary part, the power, the phase, and the cumulative spectrum deviation from equality, respectively.

SPECTRUM WINDOW = NONE — No data window is applied for subsequent **SPECTRUM** commands. This is not recommended.

SPECTRUM WINDOW = WELCH — This, the default, specifies that a Welch data window is applied for subsequent **SPECTRUM** commands.

STANDARDIZE = Signal — The signal is standardized to have zero mean and unit variance.

SUBTRACT = Signal1 AND Signal2 — The two signals (which must be the same length) are subtracted point-by-point (first minus second). The result defines the signal(s) appearing on a previous **NAME** command.

TEST NETWORK = name — The current test set is sent to the named neural network. Mean squared (and absolute) error is computed, as well as some percentiles. If classification is being done, a confusion matrix is computed. This information is output to the audit log file.

THRESHOLD = number — This is a percent (0–100) of full activation. When a confusion matrix is computed via the **TEST NETWORK** command, cases will be tossed into the reject bin if the winning neuron's activation is not at least this quantity. The default of zero causes all cases to be classified, resulting in no rejections.

TRAIN ARMA = name — The named ARMA model is trained. If it already exists, the current architecture must be compatible with the existing model. If it does not already exist, it is created.

TRAIN NETWORK = name — The named neural network is trained. If it already exists, the current architecture must be compatible with the existing model. If it does not already exist, it is created.

UNDO CENTER = SignalName [PER Other] — The previously computed mean or median is added to the signal.

UNDO DETREND = SignalName [PER Other] — The previously computed trend line is added to the signal.

UNDO DIFFERENCE = SignalName [PER Other] — The signal is summed as many times as the original differencing was done. The former first point is restored before each summing.

UNDO OFFSET = Signal [PER Other] — The previously added offset is subtracted from the signal.

UNDO SCALE = Signal [PER Other] — Every point in the signal is divided by the constant that previously multiplied them.

UNDO SEASONAL DIFFERENCE = SignalName [PER Other] — The signal is seasonally summed. The points in the first period of the original signal are restored before summing begins.

UNDO STANDARDIZE = Signal [PER Other] — The effects of a previous standardization are removed by a linear transformation.

Validation Suite

A set of command control files is included to provide a fairly complete (though certainly not exhaustive) validation suite. These files are especially useful to users who recompile the source code, as this suite exercises the majority of the algorithms in the NPREDICT program. A secondary benefit of these files is to provide the reader with relatively complex examples of operation. They are not excellent for this use, because their primary goal is testing, not education. The examples distributed throughout the text are far more suitable for education. Nevertheless, there is much to be gained by carefully studying these files. A brief explanation of each validation file is now given.

VALID1.CON — This is a very long and complex (but repetitive) test of every neural network model and every learning algorithm, using both classification and mapping output modes.

VALID2.CON — This tests saving and restoring all network models. Each model is trained and saved, then restored and retested.

VALID3.CON — All members of the probabilistic neural network family are tested for their use of prior probabilities.

VALID4.CON — All domains of the MLFN model are tested, including both training and save/restore.

VALID5.CON — This tests neural network prediction. All major varieties are tested, including pointwise, recursive, and even prediction in classification mode.

VALID6.CON — Resampling techniques are covered here. PNN and MLFN models are used in both classification and mapping mode to test cross validation training and the extended test algorithm.

VALID7.CON — A representative variety of ARMA models is tested. These range from primitive one-parameter models to extremely complex multivariate models. Each model is saved to disk and restored to test that aspect of operation. This file executes very slowly because of the size of the multivariate model.

VALID8.CON — This long and complex file requires three subfiles: VALID8A/B/C.CON. These comprise a set of tests for hybrid ARMA/neural prediction.

VALID9.CON — All types of orthogonalization models are tested, including saving them to disk and restoring them.

VALID10.CON — This is a thorough test of auto and partial correlation and crosscorrelation. A few of the tests (those involving partial crosscorrelation) can be fairly slow.

VALID11.CON — This tests computation of robust confidence intervals for ARMA prediction.

VALID12.CON — This is a wild and crazy test of robust confidence intervals for neural network prediction. The PNN family is used throughout. Hardcore students of PNN confidence computation will gain a lot from carefully studying the file. Many bizarre and unexpected (by amateurs) results are revealed here.

VALID13.CON — This exercises the various methods of computing and displaying a power spectrum. Both the ordinary DFT (with and without a Welch window) and the maximum entropy method are tested. The Savtizky-Golay filter is also invoked.

VALID14.CON — All filters, including lowpass, highpass, bandpass, QMF, and Morlet wavelet, are tested.

Appendix

This appendix contains information concerning the files supplied on the accompanying CD-ROM. The CD-ROM contains the complete source code as well as DOS and Windows NT executables of a time-series prediction program that illustrates many of the techniques discussed in the text. This CD-ROM also contains a variety of real-life time series for testing purposes.

Disclaimer

The algorithms, programs, and datasets listed in this text and/or supplied on the accompanying CD-ROM are for educational purposes only. They are not commercial products, and they are exclusively intended for personal use. Although this software has been tested extensively and is believed to be correct, no warranties of its correct operation are made. In fact, given the size and complexity of much of the code, *hidden bugs are likely*. It is even possible that some of the underlying algorithms are fundamentally flawed. Therefore, all readers must heed the following warning:

THIS SOFTWARE MUST NOT BE USED IN ANY APPLICATION IN WHICH THERE IS RISK OF INJURY OR SERIOUS LOSS.

The user of this software bears all responsibility for the consequences of its use. Neither the publisher nor the author accepts any liability for losses incurred as a result of the use of, or inability to use, the software or data.

CD-ROM Contents

The enclosed CD-ROM contains an installation program that loads fully functioning DOS and Windows NT versions of the NPREDICT program, along with all source files required to compile them. The C++ source files should be able to be compiled using any ANSI C++ compiler. The author has tested them with Borland C++ 4.5 and Symantec C++ 6.1. The contents of the CD-ROM are now listed. Instructions for compiling and linking the program follow this file list.

\NPREDICT\COMMON

CLASSES.H	FLRAND.CPP	PROCESS.CPP
CONST.H	GENERATE.CPP	QMF_SIG.CPP
FUNCDEFS.H	GLOB_MIN.CPP	QSORT.CPP
ACTIVITY.CPP	GRADIENT.CPP	RANDOM .CPP
ACT_FUNC.CPP	GRAPHLAB.CPP	READSIG.CPP
ANNEAL1.CPP	IN_OUT.CPP	REGRESS.CPP
ANNEAL2.CPP	LIMIT.CPP	REGRS_DD.CPP
ANX.CPP	LEV_MARQ.CPP	SAVGOL.CPP
ANX_DD.CPP	LM_CORE.CPP	SEPCLASS.CPP
ARMA.CPP	MAXENT.CPP	SEPVAR.CPP
ARMACONF.CPP	MEM.CPP	SHAKE.CPP
ARMAPRED.CPP	MLFN.CPP	SIGNAL.CPP
ARMASAVE.CPP	MORLET.CPP	SIG_SAVE.CPP
AUTOCORR.CPP	MOV_AVG.CPP	SPECTRUM.CPP
BRENTMIN.CPP	MRFFT.CPP	SSG.CPP
BURG.CPP	MRFFT_K.ASM	SSG_CORE.CPP
COMBINE.CPP	MRFFT_K.C	STRINGS.CPP
CONJGRAD.CPP	MRFFT_P.CPP	SVDCMP.CPP
CONTROL.CPP	NET_CONF.CPP	TESTNET.CPP
COPY.CPP	NET_PRED.CPP	TRAIN.CPP
CVTRAIN.CPP	NETWORK.CPP	WT_SAVE.CPP
DEFAULTS.CPP	NP_CONF.CPP	VECLEN.CPP
DERMIN.CPP	ORTHOG.CPP	ASM.BAT
DOTPROD.CPP	ORTHSAVE.CPP	COMPALLB.BAT
DOTPRODC.CPP	PARSDUBL.CPP	COMPALLS.BAT
EIGEN.CPP	PNNBASIC.CPP	COMMON.RSP
FILTER.CPP	PNNET.CPP	
FILT_SIG.CPP	POWELL.CPP	

\NPREDICT\DOS

DISPLAY.CPP	PROG_WIN.CPP	NPREDICT.LNK
GRAPHICS.CPP	COMPALLS.BAT	NPREDICT.EXE
NPREDICT.CPP	PL.BAT	

\NPREDICT\WINDOWS

ARITH.H	PROG_WIN.H	ORTHOG.CPP
ARMA.H	READSAVE.H	MODIFY.CPP
CLEAR.H	SPECTRUM.H	PRDCMDI1.CPP
COM_FILE.H	TEST_NET.H	PRDCMDIC.CPP
CONFCOMP.H	TEXTLIST.H	PRDCTAPP.CPP
COPY.H	ARITH.CPP	PROG_WIN.CPP
CUMULATE.H	ARMA.CPP	READSAVE.CPP
DISP.H	CLEAR.CPP	SPECTRUM.CPP
EXTERN.H	COM_FILE.CPP	TEST_NET.CPP
FILTER.H	CONFCOMP.CPP	TEXTLIST.CPP
GENERATE.H	COPY.CPP	APPLMDI.ICO
IN_OUT.H	CUMULATE.CPP	COMMAND.ICO
LAG_CORR.H	DISP.CPP	MDICHILD.ICO
MODIFY.H	DISPLAY.CPP	PROGRESS.ICO
NETMODEL.H	FILTER.CPP	PRDCTAPP.RC
NET_PRED.H	GENERATE.CPP	PRDCTAPP.RH
NET_TRN.H	IN_OUT.CPP	PRDCTAPP.DEF
ORTHOG.H	LAG_CORR.CPP	NPREDICT.IDE
PRDCMDI1.H	NETMODEL.CPP	NPREDICT.EXE
PRDCMDIC.H	NET_PRED.CPP	
PRDCTAPP.H	NET_TRN.CPP	

\NPREDICT\VALIDATE

VALID1.CON	VALID8.CON	VALID12.CON
VALID2.CON	VALID8A.CON	VALID13.CON
VALID3.CON	VALID8B.CON	VALID14.CON
VALID4.CON	VALID8C.CON	CLASS1.DAT
VALID5.CON	VALID9.CON	CLASS2.DAT
VALID6.CON	VALID10.CON	
VALID7.CON	VALID11.CON	

\EXAMPLES

ARMACONF.CON	NNPRED.CON	QMF_DIFF.CON
ARMA_NN.CON	ORTHOG.CON	
NET_CONF.CON	PREDPREC.CON	

\FINANCE

READ.ME	LTGOVTBD.DAT	M3NS.DAT
BKACCEPT.DAT	M1NS.DAT	TB3MA.DAT
EURO.DAT	M2NS.DAT	TB3MO.DAT

\WEATHER

READ.ME	TEMP.NY	PREC.MI
TEMP.CA	TEMP.TN	PREC.MN
TEMP.GA	TEMP.TX	PREC.NM
TEMP.IN	TEMP.WA	PREC.NY
TEMP.KS	PREC.CA	PREC.TN
TEMP.ME	PREC.GA	PREC.TX
TEMP.MI	PREC.IN	PREC.WA
TEMP.MN	PREC.KS	
TEMP.NM	PREC.ME	

\SUNSPOT

READ.ME
SUNSPOT.DAT

Hardware and Software Requirements

The NPREDICT programs included on CD-ROM must be run on a standard IBM-compatible computer having an 80386 or higher processor and a math coprocessor. The DOS version requires a VESA-compatible video display. The Windows NT version requires NT version 3.1 or later, or Windows 3.1 with the Win32s extensions. It is likely that this program will run under Windows 95, but this has not been tested as of the publication date.

Compiling and Linking the NPREDICT Programs

The NPREDICT program source code is divided into three subdirectories. The \COMMON subdirectory contains core routines that are used by both the DOS and Windows NT versions of the program. Most modules are in this subdirectory. The \DOS subdirectory contains the DOS main program and the few subroutines limited to that version. The \WINDOWS subdirectory contains a relatively large number of files. These include not only the main program and a few special subroutines for this version, but also the many files associated with any Windows program. Everything needed to do a complete build under Borland C++ 4.5 is in this subdirectory, including the resource files, icon files, and project file.

Readers who wish to compile and link a DOS version of the NPREDICT program are faced with one special problem. The high resolution graphics used to display signals require special video routines. The compiled version of the program supplied with this text is linked with the FlashTek *Flash Graphics* video library. Therefore, readers who wish to replicate the program precisely must purchase this library. However, there are several alternatives. For the reader's convenience, all calls to video display routines are handled by a small set of routines in the module GRAPHICS.CPP. These routines, in turn, call the *Flash Graphics* routines. Therefore, readers who already own a different graphics library need only make slight modifications to those intermediaries, changing the *Flash Graphics* calls to the equivalent calls in their own library. Another alternative is to modify the intermediary routines to avoid all video display calls. This will produce a program that cannot display signals. However, the program will function correctly in every other respect.

The DOS version supplied with this text is also linked with the FlashTek *X-32VM* virtual memory DOS extender. This allows the program to access free space on the hard CD-ROM for use as additional memory. When NPREDICT starts, the DOS environment is searched for a variable called TMP or TEMP. If such a variable is found, the swap file is placed in the specified directory. Otherwise, the swap file is placed in the drive having the largest amount of free space. Note that this DOS extender is not required for successfully linking the program.

The DOS Version

Readers who do not own the FlashTek *Flash Graphics* toolbox must modify the GRAPHICS.CPP subroutine to accommodate the video tools that they wish to use. The following steps are necessary to use the Symantec C++ development system (version 6) to compile and link the DOS version of the NPREDICT program:

1. Compile all subroutines in the \COMMON subdirectory. This is easily accomplished with the batch file COMPALLS.BAT.

2. Compile all subroutines in the \DOS subdirectory. This is easily accomplished with the batch file COMPALLS.BAT. Note that the command for compiling the *Flash Graphics* initialization routine FG_CSTM.C is slightly different from the others because this routine does not follow ANSI standards.

3. Link the program. The batch file PL.BAT invokes the Symantec linker, which, in turn, uses the response file NPREDICT.LNK.

The response file NPREDICT.LNK includes two references to the FlashTek *X-32VM* virtual memory system. Readers who do not own this product need to remove these references. No other actions are needed. The response file also contains references to FlashTek *Flash Graphics* video display routines. Readers who use a different library or who wish to disable all video display must make the appropriate changes to NPREDICT.LNK.

The Windows NT Version

The Windows NT version of the NPREDICT program is best created by using the Borland IDE. The program supplied with this text was prepared with version 4.5 of the Borland C++ development system. Unlike the DOS version, no special libraries are needed to compile and link the program for use under Windows NT. The following steps are necessary to compile and link the program:

1. Compile all subroutines in the \COMMON subdirectory, and place each one in the library BOR_WIND.LIB, which must reside in the \COMMON subdirectory. This filename and directory are explicitly referenced in the project file, so this exact name must be used for the library. This task is easily accomplished by invoking the batch file COMPALLB.BAT.

2. Use the Borland IDE to open and build the project NPREDICT.

Data Files

Many sample datasets are supplied on this CD-ROM. Copyright restrictions, if any, accompany the description of each dataset. Please take special note of the following:

These datasets are derived from databases that are believed to be correct. Great care has been taken to preserve the integrity of the data. However, errors in a collection this large are quite possible. Therefore, *no warranty is made that the data is correct.* Neither the publisher nor the author accept any responsibility for injury or damages due to errors in any of these datasets. Sole responsibility for their use lies with the reader.

Financial Data

This set of files includes historical data originating with the Federal Reserve System and the United States Government. A brief description of each dataset may be found in the READ.ME file. All of these datasets are in the public domain and may be copied freely. Updated files are available from the FRED BBS. The telephone number of the FRED BBS is currently (314) 621-1824. Set your communications program to 2400 baud, no parity, 8 data, 1 stop, full duplex. For more information, write to the following address:

Federal Reserve Bank of St. Louis.
P.O. box 442
St. Louis, Missouri 63166

Climate Data

This set of files includes monthly temperature and precipitation data for a few randomly selected states. The years range from 1895 through 1989. This data is supplied by:

> NOAA/National Climatic Data Center
> 151 Patton Avenue Room 120
> Asheville, NC 28801-5001

All of these temperature and precipitation datasets are in the public domain and may be freely copied. NOAA/NCDC publishes an extensive variety of databases on magnetic tape, floppy disk, CD-ROM, and via BBS. If you request their catalog, you will be overwhelmed with the vast wealth of material available at nominal cost.

These files share a common format. Each monthly observation is on a new line. The first field in each line is the year and the month, expressed as a single floating-point number. The remaining fields are for the different climatic regions of the state. Maps of the exact boundaries of these regions are available from NOAA. These regions (in the order that they appear on each line in the files) are the following:

California
> North Coast Drainage; Sacramento Drainage; Northeast Interior Basins; Central Coast Drainage; San Joachin Drainage; South Coast Drainage; Southeast Desert Basins

Georgia
> Northwest; North Central; Northeast; West Central; Central; East Central; Southwest; South Central; Southeast

Indiana
> Northwest; North Central; Northeast; West Central; Central; East Central; Southwest; South Central; Southeast

Kansas
> Northwest; North Central; Northeast; West Central; Central; East Central; Southwest; South Central; Southeast

Maine
: Northern; Southern Interior; Coastal

Michigan
: West Upper; East Upper; Northwest Lower; Northeast Lower; West Central Lower; Central Lower; East Central Lower; Southwest Lower; South Central Lower; Southeast Lower

Minnesota
: Northwest; North Central; Northeast; West Central; Central; East Central; Southwest; South Central; Southeast

New Mexico
: Northwestern Plateau; Northern Mountains; Northeastern Plains; Southwestern Mountains; Central Valley; Central Highlands; Southeastern Plains; Southern Desert

New York
: Western Plateau; Eastern Plateau; Northern Plateau; Coastal; Hudson Valley; Mohawk Valley; Champlain Valley; St. Lawrence Valley; Great Lakes; Central Lakes

Tennessee
: Eastern; Cumberland Plateau; Middle; Western

Texas
: High Plains; Low Rolling Plains; North Central; East Texas; Trans Pecos; Edwards Plateau; South Central; Upper Coast; Southern; Lower Valley

Washington
: West Olympic Coastal; NE Olympic San Juan; Puget Sound Lowlands; E Olympic Cascade Foothills; Cascade Mountains West; East Slope Cascades; Okanogan Big Bend; Central Basin; Northeastern; Palouse Blue Mountains

Sunspot Data

One file of sunspot activity is included. This file contains final monthly means of American relative sunspot numbers since 1945. The data in the file SUNSPOT.DAT is derived from the following source:

AAVSO - Solar Division
4523 Thurston Lane - Apt. 5
Madison, WI 53711-4738

This dataset is *not* in the public domain. Permission to copy this file must be obtained from AAVSO. However, they were extremely cooperative, and it appears likely that they will continue to share their data willingly. In particular, Peter O. Taylor of AAVSO was very kind in his fast and courteous response to my request for sunspot data.

Each line of this file contains two numbers. The first number is the date, with the year and month expressed as a single floating point number. The second number is the sunspot measurement.

Extra Data

The CD-ROM includes a directory named \EXTRA. This directory contains many datasets that became available just before press time. Please see the READ.ME files in this directory for additional information.

Bibliography

Aarts, E., and van Laarhoven, P. (1987). *Simulated Annealing: Theory and Practice*. John Wiley and Sons, New York.

Abe, S., Kayama, M., Takenaga, H., and Kitamura, T. (1992). "Neural Networks as a Tool to Generate Pattern Classification Algorithms." *International Joint Conference on Neural Networks*, (Baltimore, MD).

Acton, Forman S. (1959). *Analysis of Straight-Line Data*. Dover Publications, New York.

Acton, Forman S. (1970). *Numerical Methods That Work*. Harper & Row, New York.

Anderson, James, and Rosenfeld, Edward, eds. (1988). *Neurocomputing: Foundations of Research*. MIT Press, Cambridge, MA.

Anderson, T. W. (1958). *An Introduction to Multivariate Statistical Analysis*. John Wiley and Sons, New York.

Austin, Scott (1990). "Genetic Solutions to XOR Problems." *AI Expert* (December), 52–57.

Avitzur, Ron (1992). "Your Own Handprinting Recognition Engine." *Dr. Dobb's Journal* (April), 32–37.

Azencott, R., ed. (1992). *Simulated Annealing: Parallelization Techniques*. John Wiley and Sons, New York.

Baba, Norio (1989). "A New Approach for Finding the Global Minimum of Error Function of Neural Networks." *Neural Networks*, **2**:5, 367–373.

Baba, N., and Kozaki, M. (1992). "An Intelligent Forecasting System of Stock Price Using Neural Networks." *International Joint Conference on Neural Networks*, (Baltimore, MD).

Barmann, Frank, and Biegler-Konig, Friedrich (1992). "On a Class of Efficient Learning Algorithms for Neural Networks." *Neural Networks*, **5**: 139–144.

Barnard, Etienne, and Casasent, David (1990). "Shift Invariance and the Neocognitron." *Neural Networks*, **3**: 403–410.

Barr, Avron, Cohen, Paul R., and Feigenbaum, Edward A., eds. (vol. I, 1981; vol. II, 1982; vol. III, 1982; vol. IV, 1989). *The Handbook of Artificial Intelligence*. Addison-Wesley, Reading, MA.

Bartlett, E. B. (1991). "Chaotic Time-series Prediction Using Artificial Neural Networks." *Abstracts from 2nd Government Neural Network Applications Workshop* (September), Session III.

Barton, D. E., and Dennis, K. E. (1952). "The Conditions Under Which Gram-Charlier and Edgeworth Curves are Positive Definite and Unimodal." *Biometrika*, **39**: 425.

Battiti, R., and Colla, M. (1994). "Democracy in Neural Nets: Voting Schemes for Classification." *Neural Networks*, **7**: 691–707.

Birx, D., and Pipenberg, S. (1992). "Chaotic Oscillators and Complex-Mapping Feedforward Networks for Signal Detection In Noisy Environments." *International Joint Conference on Neural Networks* (Baltimore, MD).

Birx, D., and Pipenberg, S. (1993). "A Complex Mapping Network for Phase-Sensitive Classification." *IEEE Transactions on Neural Networks*, 4:1, 127-135.

Blum, A. L., and Rivest, R. L. (1992). "Training a 3-Node Neural Network Is NP-Complete." *Neural Networks*, 5:1, 117–127.

Blum, Edward, and Li, Leong (1991). "Approximation Theory and Feedforward Networks." *Neural Networks*, **4**: 511–515.

Booker, L. B., Goldberg, D. E., and Holland, J. H. (1989). "Classifier Systems and Genetic Algorithms." *Artificial Intelligence*, **40**: 235–282.

Box, George, and Jenkins, Gwilym (1976). *Time-series Analysis, Forecasting and Control*. Prentice Hall, Englewood Cliffs, NJ.

Bracewell, Ronald N. (1986). *The Fourier Transform and Its Applications*. McGraw-Hill, New York.

Brent, Richard (1973). *Algorithms for Minimization without Derivatives*. Prentice-Hall, Englewood Cliffs, NJ.

Brillinger, David R. (1975). *Time Series, Data Analysis and Theory*. Holt, Rinehart and Winston, New York.

Burgin, George (1992). "Using Cerebellar Arithmetic Computers." *AI Expert* (June), 32–41.

Cacoullos, T. (1966). "Estimation of a Multivariate Density." *Annals of the Institute of Statistical Mathematics* (Tokyo), **18**:2, 179–189.

Cardaliaguet, Pierre, and Euvrard, Guillaume (1992). "Approximation of a Function and its Derivative with a Neural Network." *Neural Networks*, **5**:2, 207–220.

Carpenter, Gail A., and Grossberg, Stephen (1987). "A Massively Parallel Architecture for a Self-Organizing Neural Pattern Recognition Machine." Academic Press *(Computer Vision, Graphics, and Image Processing)*, **37**: 54–115.

Carpenter, Gail A., Grossberg, Stephen, and Reynolds, John H. (1991). "ARTMAP: Supervised Real-Time Learning and Classification of Nonstationary Data by a Self-Organizing Neural Network." *Neural Networks*, **4**: 565–588.

Caruana, R. A., and Schaffer, J. D. (1988). "Representation and Hidden Bias: Gray vs. Binary Coding for Genetic Algorithms" in Laird, J. (ed.) *Proceedings of the Fifth International Congress on Machine Learning*. Morgan Kaufmann, San Mateo, CA.

Caudill, Maureen (1988). "Neural Networks Primer, Part IV—The Kohonen Model." *AI Expert* (August).

Caudill, Maureen (1990). "Using Neural Nets: Fuzzy Decisions." *AI Expert* (April), 59–64.

Chambers, J. M. (1967). "An Extension of the Edgeworth Expansion to the Multivariate Case." *Biometrika*, **54**: 367–383.

Childers, D. G., ed. (1978). *Modern Spectrum Analysis*. IEEE Press, New York.

Chin, Daniel (1994). "A More Efficient Global Optimization Algorithm Based on Styblinski and Tang." *Neural Networks*, **7**: 573.

Chui, C. (1992). *An Introduction to Wavelets*. Academic Press, New York.

Cooley, William, and Lohnes, Paul (1971). *Multivariate Data Analysis*. John Wiley and Sons, New York.

Cotter, Neil E., and Guillerm, Thierry J. (1992). "The CMAC and a Theorem of Kolmogorov." *Neural Networks*, **5**: 221–228.

Cottrell, G., Munro, P., and Zipser, D. (1987). "Image Compression by Backpropagation: An Example of Extensional Programming." *ICS Report 8702*, University of California at San Diego.

Cox, Earl (1992). "Solving Problems with Fuzzy Logic." *AI Expert* (March), 28–37.

Cox, Earl (1992). "Integrating Fuzzy Logic into Neural Nets." *AI Expert* (June), 43–47.

Cramer, Harald (1926). "On Some Classes of Series Used in Mathematical Statistics." *Skandinaviske Mathematikercongres*, Copenhagen.

Cramer, Harald (1946). *Mathematical Methods of Statistics*. Princeton University Press, Princeton, NJ.

Crooks, Ted (1992). "Care and Feeding of Neural Networks." *AI Expert* (July), 36–41.

Daubechies, Ingrid (1990). "The Wavelet Transform, Time-Frequency Localization, and Signal Analysis." *IEEE Transactions on Information Theory*, **36**:5, 961–1005.

Davis, D. T., and Hwang, J. N. (1992). "Attentional Focus Training by Boundary Region Data Selection." *International Joint Conference on Neural Networks* (Baltimore, MD).

Davis, Lawrence (1991). *Handbook of Genetic Algorithms*. Van Nostrand Reinhold, New York.

Devroye, L. (1986). *Non-Uniform Random Number Generation*. Springer-Verlag, New York.

Dracopoulos, D., and Jones, A. (1993). "Modeling Dynamic Systems." *World Congress on Neural Networks*, (Portland, OR).

Draper, N. R., and Smith, H. (1966). *Applied Regression Analysis*. John Wiley and Sons, New York.

Duffin, R. J., and Schaeffer, A. C. (1952). "A Class of Nonharmonic Fourier Series." *Transactions of the American Mathematical Society*, **72**: 341-366.

Eberhart, Russell C., and Dobbins, Roy W., eds. (1990). *Neural Network PC Tools, A Practical Guide*. Academic Press, San Diego, CA.

Efron, Bradley (1982). *The Jackknife, the Bootstrap, and Other Resampling Plans*. Society for Industrial and Applied Mathematics, Philadelphia, PA.

Elliot, D. F. (1987). *Handbook of Digital Signal Processing*. Academic Press, San Diego, CA.

Fahlmann, Scott E. (1988). "An Empirical Study of Learning Speed in Backpropagation Networks." *CMU Technical Report CMU-CS–88–162* (June 1988).

Fakhr, W., Kamel, M., and Elmasry, M. I. (1992). "Probability of Error, Maximum Mutual Information, and Size Minimization of Neural Networks." *International Joint Conference on Neural Networks* (Baltimore, MD).

Finkbeiner, Daniel T., II (1972). *Elements of Linear Algebra.* W. H. Freeman, San Francisco, CA.

Foley, James D., van Dam, Andries, Feiner, Steven K., and Hughes, John F. (1990). *Computer Graphics: Principles and Practice (Second Edition).* Addison-Wesley, Reading, MA.

Forsythe, George E., Malcolm, Michael A., and Moler, Cleve B. (1977). *Computer Methods for Mathematical Computations.* Prentice-Hall, Englewood Cliffs, NJ.

Freeman, James A., and Skapura, David M. (1992). *Neural Networks: Algorithms, Applications, and Programming Techniques.* Addison-Wesley, Reading, MA.

Fu, K. S., ed. (1971). *Pattern Recognition and Machine Learning.* Plenum Press, New York.

Fukunaga, Keinosuke (1972). *Introduction to Statistical Pattern Recognition.* Academic Press, Orlando, FL.

Fukunaga, Keinosuke (1987). "Bayes Error Estimation Using Parzen and k-NN Procedures." *IEEE Transactions on Pattern Analysis and Machine Intelligence,* **9**: 634–643.

Fukushima, Kunihiko (1987). "Neural Network Model for Selective Attention in Visual Pattern Recognition and Associative Recall." *Applied Optics* (December), **26**: 23.

Fukushima, Kunihiko (1989). "Analysis of the Process of Visual Pattern Recognition by the Neocognitron." *Neural Networks*, **2**: 413–420.

Gallant, Ronald, and White, Halbert (1992). "On Learning the Derivatives of an Unknown Mapping with Multilayer Feedforward Networks." *Neural Networks*, **2**: 129–138.

Gallinari, P., Thiria, S., Badran, F., and Fogelman-Soulie, F. (1991). "On the Relations between Discriminant Analysis and Multilayer Perceptrons." *Neural Networks*, **4**:3, 349–360.

Garson, David G. (1991). "Interpreting Neural-Network Connection Weights." *AI Expert* (April), 47–51.

Georgiou, G. (1993). "The Multivalued and Continuous Perceptrons." *World Congress on Neural Networks* (Portland, OR).

Gill, Philip E., Murray, Walter, and Wright, Margaret H. (1981). *Practical Optimization*. Academic Press, San Diego, CA.

Glassner, Andrew S., ed. (1990). *Graphics Gems*. Academic Press, San Diego, CA.

Goldberg, David E. (1989). *Genetic Algorithms in Search, Optimization and Machine Learning*. Addison-Wesley, Reading, MA.

Gori, M., and Tesi, A. (1990). "Some Examples of Local Minima during Learning with Back-Propagation." *Third Italian Workshop on Parallel Architectures and Neural Networks (E. R. Caianiello, ed.)*. World Scientific Publishing Co.

Gorlen, Keith E., Orlow, Sanford M., and Plexico, Perry S. (1990). *Data Abstraction and Object-Oriented Programming in C++*. John Wiley & Sons, Chichester, England.

Grossberg, Stephen (1988). *Neural Networks and Natural Intelligence*. MIT Press, Cambridge, MA.

Guiver, John P., and Klimasauskas, Casimir, C. (1991). "Applying Neural Networks, Part IV: Improving Performance." *PC AI* (July/August).

Hald, A. (1952). *Statistical Theory with Engineering Applications*. John Wiley and Sons, New York.

Hansen, L., and Salamon, P. (1990). "Neural Network Ensembles." *IEEE Transactions on Pattern Analysis and Machine Intelligence*. **12**: 993–1000.

Haralick, R. M. (1979). "Statistical and Structural Approaches to Texture." *Proceedings of the IEEE*, **67**: 786-804.

Harrington, Steven (1987). *Computer Graphics, A Programming Approach* (Second Edition). McGraw-Hill, New York.

Hashem, M. (1992). "Sensitivity Analysis for Feedforward Neural Networks with Differentiable Activation Functions." *International Joint Conference on Neural Networks* (Baltimore, MD).

Hastings, Cecil, Jr. (1955). *Approximations for Digital Computers*. Princeton University Press, Princeton, NJ.

Hecht-Nielsen, Robert (1987). "Nearest Matched Filter Classification of Spatiotemporal Patterns." *Applied Optics* (May 15), **26**: 10.

Hecht-Nielsen, Robert (1991). *Neurocomputing*. Addison-Wesley, Reading, MA.

Hecht-Nielsen, Robert (1992). "Theory of the Backpropagation Network." *Neural Networks for Perception, vol. 2 (Harry Wechsler, ed.)* Academic Press, New York.

Hirose, A. (1992). "Proposal of Fully Complex-Valued Neural Networks." *International Joint Conference on Neural Networks* (Baltimore, MD).

Hirose, A. (1993). "Simultaneous Learning of Multiple Oscillations of Recurrent Complex-Valued Neural Networks." *World Congress on Neural Networks* (Portland, OR).

Hirose, Yoshio, Yamashita, Koichi, and Hijiya, Shimpei (1991). "Back-Propagation Algorithm Which Varies the Number of Hidden Units." *Neural Networks*, **4**:1, 61-66.

Ho, T. K., Hull, J. J., and Srihari, S. N. (1994). "Decision Combination in Multiple Classifier Systems." *IEEE Transactions on Pattern Analysis and Machine Intelligence*, **16**: 66–75.

Hornik, Kurt (1991). "Approximation Capabilities of Multilayer Feedforward Networks." *Neural Networks*, **4**:2, 251–257.

Hornik, Kurt, Stinchcombe, Maxwell, and White, Halbert (1989). "Multilayer Feedforward Networks are Universal Approximators." *Neural Networks*, **2**:5, 359–366.

Howell, Jim (1990). "Inside a Neural Network." *AI Expert* (November), 29–33.

Hu, M. K. (1962). "Visual Pattern Recognition By Moment Invariants." *IRE Transactions on Information Theory*, **8**:2, 179–187.

IEEE Digital Signal Processing Committee, eds. (1979). *Programs for Digital Signal Processing*. IEEE Press, New York.

Ito, Y. (1991a). "Representation of Functions by Superpositions of a Step or Sigmoid Function and Their Applications to Neural Network Theory." *Neural Networks*, **4**:3, 385–394.

Ito, Y. (1991b). "Approximation of Functions on a Compact Set by Finite Sums of a Sigmoid Function without Scaling." *Neural Networks*, **4**:6, 817–826.

Ito, Y. (1992). "Approximation of Continuous Functions on $\mathbf{R}^d$ by Linear Combinations of Shifted Rotations of a Sigmoid Function with and without Scaling." *Neural Networks*, **5**:1, 105–115.

Jain, A. K., Dubes, R. C., and Chen, C. C. (1987). "Bootstrap Techniques for Error Estimation." *IEEE Transactions on Pattern Analysis and Machine Intelligence*, **9**: 628–633.

Kalman, B. L., and Kwasny, S. C. (1991). "A Superior Error Function For Training Neural Networks." *International Joint Conference on Neural Networks* (Seattle, WA).

Kalman, B. L., and Kwasny, S. C. (1992). "Why Tanh? Choosing a Sigmoidal Function." *International Joint Conference on Neural Networks* (Baltimore, MD).

Karr, Chuck (1991a). "Genetic Algorithms for Fuzzy Controllers." *AI Expert* (February), 26–33.

Karr, Chuck (1991b). "Applying Genetics to Fuzzy Logic." *AI Expert* (March), 39–43.

Kendall, M., and Stuart, A. (vol. I, 1969; vol. II, 1973; vol. III, 1976). *The Advanced Theory of Statistics*. Hafner, New York.

Kenue, S. K. (1991). "Efficient Activation Functions for the Back-Propagation Neural Network." *SPIE, Proceedings from Intelligent Robots and Computer Vision X: Neural, Biological, and 3-D Methods* (November).

Kim, M. W., and Arozullah, M. (1992a). "Generalized Probabilistic Neural Network-Based Classifiers." *International Joint Conference on Neural Networks* (Baltimore, MD).

Kim, M. W., and Arozullah, M. (1992b). "Neural Network Based Optimum Radar Target Detection in Non-Gaussian Noise." *International Joint Conference on Neural Networks* (Baltimore, MD).

Kim, M. W. (1993). "Handwritten Digit Recognition Using Gram-Charlier and Generalized Probabilistic Neural Networks." *World Conference on Neural Networks* (Portland, OR).

Klimasauskas, Casimir C. (1987). *The 1987 Annotated Neuro-Computing Bibliography*. NeuroConnection, Sewickley, PA.

Klimasauskas, Casimir C. (1992a). "Making Fuzzy Logic 'Clear'." *Advanced Technology for Developers*, 1 (May), 8–12.

Klimasauskas, Casimir C. (1992b). "Hybrid Technologies: More Power for the Future." *Advanced Technology for Developers*, 1 (August), 17–20.

Klir, George J., and Folger, Tina A. (1988). *Fuzzy Sets, Uncertainty, and Information*. Prentice Hall, Englewood Cliffs, NJ.

Knuth, Donald (1981). *Seminumerical Algorithms*. Addison-Wesley, Reading, MA.

Kohonen, Teuvo (1982). "Self-Organized Formation of Topologically Correct Feature Maps." *Biological Cybernetics*, **43**: 59–69.

Kohonen, Teuvo (1989). *Self-organization and Associative Memory*. Springer-Verlag, New York.

Kosko, Bart (1987). "Fuzziness vs. Probability." *Air Force Office of Scientific Research (AFOSR F49620-86-C-0070) and Advanced Research Projects Agency (ARPA Order No. 5794)*, (July).

Kosko, Bart (1988a). "Bidirectional Associative Memories." *IEEE Transactions on Systems, Man, and Cybernetics* (Jan./Feb.), **18**:1.

Kosko, Bart (1988b). "Hidden Patterns in Combined and Adaptive Knowledge Networks." *International Journal of Approximate Reasoning*, vol. 1.

Kosko, Bart (1992). *Neural Networks and Fuzzy Systems*. Prentice Hall, Englewood Cliffs, NJ.

Kotz, Samuel, and Johnson, Norman, eds. (1982). *Encyclopedia of Statistical Sciences*. John Wiley and Sons, New York.

Kreinovich, Vladik Ya (1991). "Arbitrary Nonlinearity Is Sufficient to Represent All Functions by Neural Networks: A Theorem." *Neural Networks*, 4:3, 381–383.

Kuhl, Frank, Reeves, Anthony, and Taylor, Russell (1986). "Shape Identification With Moments and Fourier Descriptors." *Proceedings of the 1986 ACSM-ASPRS Convention* (March), 159–168.

Kurkova, Vera (1992). "Kolmogorov's Theorem and Multilayer Neural Networks." *Neural Networks*, **5**:3, 501–506.

Lawton, George (1992). "Genetic Algorithms for Schedule Optimization." *AI Expert* (May), 23–27.

Levin, A. (1993). "Predicting With Feedforward Networks." *World Congress on Neural Networks* (Portland, OR).

Lim, Jae S. (1990). *Two-Dimensional Signal and Image Processing*. Prentice Hall, Englewood Cliffs, NJ.

Lo, Zhen-Ping, Yu, Yaoqi, and Bavarian, Behnam (1993). "Analysis of the Convergence Properties of Topology-Preserving Neural Networks." *IEEE Transactions on Neural Networks*, **4**:2, 207–220.

Lu, C. N., Wu, H. T., and Vemuri, S. (1992). "Neural Network Based Short-Term Load Forecasting." *IEEE/PES 1992 Winter Meeting, New York* (92 WM 125-5 PWRS).

Maren, Alianna, Harston, Craig, and Pap, Robert (1990). *Handbook of Neural Computing Applications*. Academic Press, New York.

Masters, Timothy (1993). *Practical Neural Network Recipes in C++*. Academic Press, New York.

Masters, Timothy (1994). *Signal and Image Processing with Neural Networks*. John Wiley & Sons, New York.

Masters, Timothy (1995). *Advanced Algorithms for Neural Networks*. John Wiley & Sons, New York.

Matsuba, I., Masui, H., and Hebishima, S. (1992). "Optimizing Multilayer Neural Networks Using Fractal Dimensions of Time-Series Data." *International Joint Conference on Neural Networks* (Baltimore, MD).

McClelland, James, and Rumelhart, David (1988). *Explorations in Parallel Distributed Processing*. MIT Press, Cambridge, MA.

Meisel, W. (1972). *Computer-Oriented Approaches to Pattern Recognition*. Academic Press, New York.

Miller, J., Goodman, R., and Smyth, P. (1991). "Objective Functions for Probability Estimation." *International Joint Conference on Neural Networks* (Seattle, WA).

Minsky, Marvin, and Papert, Seymour (1969). *Perceptrons*. MIT Press, Cambridge, MA.

Mougeot, M., Azencott, R., and Angeniol, B. (1991). "Image Compression with Back Propagation: Improvement of the Visual Restoration Using Different Cost Functions." *Neural Networks*, **4**:4, 467–476.

Mucciardi, A., and Gose, E. (1970). "An Algorithm for Automatic Clustering in N-Dimensional Spaces Using Hyperellipsoidal Cells." *IEEE Sys. Sci. Cybernetics Conference* (Pittsburgh, PA).

Musavi, M., Kalantri, K., and Ahmed, W. (1992). "Improving the Performance of Probabilistic Neural Networks." *International Joint Conference on Neural Networks* (Baltimore, MD).

Musavi, M., Kalantri, K., Ahmed, W., and Chan, K. (1993). "A Minimum Error Neural Network (MNN)." *Neural Networks*, **6**: 397–407.

Negoita, Constantin V., and Ralescu, Dan (1987). *Simulation, Knowledge-Based Computing, and Fuzzy Statistics*. Van Nostrand Reinhold, New York.

Nitta, T. (1993a). "Three-Dimensional Backpropagation." *World Congress on Neural Networks* (Portland, OR).

Nitta, T. (1993b). "A Complex-Numbered Version of the Backpropagation Algorithm." *World Congress on Neural Networks* (Portland, OR).

Pao, Yoh-Han (1989). *Adaptive Pattern Recognition and Neural Networks*. Addison-Wesley, Reading, MA.

Parzen, E. (1962). "On Estimation of a Probability Density Function and Mode." *Annals of Mathematical Statistics*, **33**: 1065–1076.

Pethel, S. D., Bowden, C. M., and Sung, C. C. (1991). "Applications of Neural Net Algorithms to Nonlinear Time Series." *Abstracts from 2nd Government Neural Network Applications Workshop* (September), Session III.

Polak, E. (1971). *Computational Methods in Optimization.* Academic Press, New York.

Polzleitner, Wolfgang, and Wechsler, Harry (1990). "Selective and Focused Invariant Recognition Using Distributed Associative Memories (DAM)." *IEEE Transactions on Pattern Analysis and Machine Intelligence* (August), **12**:8.

Pratt, William K. (1991). *Digital Image Processing.* John Wiley and Sons, New York.

Press, William H., Flannery, B., Teukolsky, S., and Vetterling, W. (1992). *Numerical Recipes in C.* Cambridge University Press, New York.

Raudys, Sarunas J., and Jain, Anil K. (1991). "Small Sample Size Effects in Statistical Pattern Recognition: Recommendations for Practitioners." *IEEE Transactions on Pattern Analysis and Machine Intelligence* (March), **13**:3.

Reed, R., Oh, S., and Marks, R. J. (1992). "Regularization Using Jittered Training Data." *International Joint Conference on Neural Networks* (Baltimore, MD).

Reeves, A., Prokop, R., Andrews, S., and Kuhl, F. (1988). "Three-Dimensional Shape Analysis Using Moments and Fourier Descriptors." *IEEE Transactions on Pattern Analysis and Machine Intelligence* (November), **10**: 937–943.

Rich, Elaine (1983). *Artificial Intelligence.* McGraw-Hill, New York.

Rosenblatt, Frank (1958). "The Perceptron: A Probabilistic Model for Information Storage and Organization in the Brain." *Psychological Review*, **65**: 386–408.

Rosenfeld, A., and Kak, A. (1982). *Digital Picture Processing*. Academic Press, New York.

Rumelhart, David, McClelland, James, and the PDP Research Group (1986). *Parallel Distributed Processing*. MIT Press, Cambridge, MA.

Sabourin, M., and Mitiche, A. (1992). "Optical Character Recognition by a Neural Network." *Neural Networks*, **5**: 843–852.

Samad, Tariq (1988). "Backpropagation Is Significantly Faster if the Expected Value of the Source Unit Is Used for Update." *1988 Conference of the International Neural Network Society*.

Samad, Tariq (1991). "Back Propagation with Expected Source Values." *Neural Networks*, **4**:5, 615–618.

Schioler, H., and Hartmann, U. (1992). "Mapping Neural Network Derived from the Parzen Window Estimator." *Neural Networks*, **5**:6, 903–909.

Schwartz, Tom J. (1991). "Fuzzy Tools for Expert Systems." *AI Expert* (February), 34–41.

Sedgewick, Robert (1988). *Algorithms*. Addison-Wesley, Reading, MA.

Shapiro, Stuart C., ed. (1990). *Encyclopedia of Artificial Intelligence*. John Wiley & Sons, New York.

Siegel, Sidney (1956). *Nonparametric Statistics for the Behavioral Sciences*. McGraw-Hill, New York.

Singleton, R. C. (1969). "An Algorithm for Computing the Mixed-Radix Fast Fourier Transform." *IEEE Trans. Audio and Electroacoust.*, **AU-17**:2, 93–100.

Soulie, Francoise Fogelman, Robert, Yves, and Tchuente, Maurice, eds. (1987). *Automata Networks in Computer Science*. Princeton University Press, Princeton, NJ.

Specht, Donald (1967). "Generation of Polynomial Discriminant Functions for Pattern Recognition." *IEEE Transactions on Electronic Computers*, **3**: 308–319.

Specht, Donald (1988). "Probabilistic Neural Networks for Classification, Mapping, or Associative Memory." *IEEE International Conference on Neural Networks*, San Diego, CA.

Specht, Donald (1990a). "Probabilistic Neural Networks." *Neural Networks*, **3**: 109–118.

Specht, Donald (1990b). "Probabilistic Neural Networks and the Polynomial Adeline as Complementary Techniques for Classification." *IEEE Transactions on Neural Networks*, **1**:1 111–121.

Specht, Donald (1991). "A General Regression Neural Network." *IEEE Transactions on Neural Networks*, **2**:6 568–576.

Specht, Donald (1992). "Enhancements to Probabilistic Neural Networks." *International Joint Conference on Neural Networks* (Baltimore, MD).

Specht, Donald F., and Shapiro, Philip D. (1991). "Generalization Accuracy of Probabilistic Neural Networks Compared with Back-Propagation Networks." *Lockheed Missiles & Space Co., Inc. Independent Research Project RDD 360*, I-887-I-892.

Spillman, Richard (1990). "Managing Uncertainty with Belief Functions." *AI Expert* (May), 44–49.

Stork, David G. (1989). "Self-Organization, Pattern Recognition, and Adaptive Resonance Networks." *Journal of Neural Network Computing* (Summer).

Strand, E. M., and Jones, W. T. (1992). "An Adaptive Pattern Set Strategy for Enhancing Generalization While Improving Backpropagation Training Efficiency." *International Joint Conference on Neural Networks* (Baltimore, MD).

Styblinski, M. A., and Tang, T.-S. (1990). "Experiments in Nonconvex Optimization: Stochastic Approximation with Function Smoothing and Simulated Annealing." *Neural Networks*, **3**: 467–483.

Sudharsanan, Subramania I., and Sundareshan, Malur K. (1991). "Exponential Stability and a Systematic Synthesis of a Neural Network for Quadratic Minimization." *Neural Networks*, **4**: 599–613.

Sultan, A. F., Swift, G. W., and Fedirchuk, D. J. (1992). "Detection of High Impedance Arcing Faults Using a Multi-Layer Perceptron." *IEEE/PES 1992 Winter Meeting, New York* (92 WM 207-1 PWRD).

Sussmann, Hector J. (1992). "Uniqueness of the Weights for Minimal Feedforward Nets with a Given Input-Output Map." *Neural Networks*, **5**:4, 589–593.

Szu, Harold (1986). "Fast Simulated Annealing." *AIP Conference Proceedings 151: Neural Networks for Computing* (Snowbird, UT).

Szu, Harold (1987). "Nonconvex Optimization by Fast Simulated Annealing." *Proceedings of the IEEE*, **75**:11, 1538–1540.

Tanimoto, Steven L. (1987). *The Elements of Artificial Intelligence*. Computer Science Press, Rockville, MD.

Taylor, Russell, Reeves, Anthony, and Kuhl, Frank (1992). "Methods For Identifying Object Class, Type, and Orientation, in the Presence of Uncertainty." *Remote Sensing Reviews*, **6**:1, 183–206.

Ulmer, Richard, Jr., and Gorman, John (1989). "Partial Shape Recognition Using Simulated Annealing." *IEEE Proceedings, 1989 Southeastcon.*

Unnikrishnan, K. P., and Venugopal, K. P. (1992). "Learning in Connectionist Networks Using the Alopex Algorithm." *International Joint Conference on Neural Networks* (Baltimore, MD).

van Ooyen, A., and Nienhuis, B. (1992). "Improving the Convergence of the Back-Propagation Algorithm." *Neural Networks*, **5**:3, 465–471.

von Mises, Richard (1964). *Mathematical Theory of Probability and Statistics.* Academic Press, New York.

Wallace, Timothy P., and Wintz, Paul A. (1980). "An Efficient Three-Dimensional Aircraft Recognition Algorithm Using Normalized Fourier Descriptors." *Computer Graphics and Image Processing*, **13**: 99-126.

Wang, Kaitsong, Gorman, John, and Kuhl, Frank (1992). "Spherical Harmonics and Moments for Recognition of Three-Dimensional Objects." *Remote Sensing Reviews*, **6**:1, 229–250.

Wayner, Peter (1991). "Genetic Algorithms." *BYTE* (January), 361–368.

Webb, Andrew R., and Lowe, David (1990). "The Optimized Internal Representation of Multilayer Classifier Networks Performs Nonlinear Discriminant Analysis." *Neural Networks*, **3**:4, 367–375.

Wenskay, Donald (1990). "Intellectual Property Protection for Neural Networks." *Neural Networks*, **3**:2, 229–236.

Weymaere, Nico, and Martens, Jean-Pierre (1991). "A Fast and Robust Learning Algorithm for Feedforward Neural Networks." *Neural Networks*, **4**:3, 361–369.

White, Halbert (1989). "Neural-Network Learning and Statistics." *AI Expert* (December), 48–52.

Wiggins, Ralphe (1992). "Docking a Truck: A Genetic Fuzzy Approach." *AI Expert* (May), 29–35.

Wirth, Niklaus (1976). *Algorithms + Data Structures = Programs*. Prentice-Hall, Englewood Cliffs, NJ.

Wolpert, David H. (1992). "Stacked Generalization." *Neural Networks*, **5**: 241–259.

Yau, Hung-Chun, and Manry, Michael T. (1991). "Iterative Improvement of a Nearest Neighbor Classifier." *Neural Networks*, **4**: 517–524.

Zadeh, Lotfi A. (1992). "The Calculus of Fuzzy If/Then Rules." *AI Expert* (March), 23–27.

Zeidenberg, Matthew (1990). *Neural Network Models in Artificial Intelligence*. Ellis Horwood, New York.

Zhang, Y., Chen, G. P., Malik, O. P., and Hope, G. S. (1992). "An Artificial Neural Network-Based Adaptive Power System Stabilizer." *IEEE/PES 1992 Winter Meeting, New York* (92 WM 018-2 EC).

Zhou, Yi-Tong, and Chellappa (1992). *Artificial Neural Networks for Computer Vision*. Springer-Verlag, New York.

Zornetzer, Steven, Davis, Joel, and Lau, Clifford, eds. (1990). *An Introduction to Neural and Electronic Networks*. Academic Press, New York.

Index

A

absolute value, 70, 129
ACCURACY, 410, 434, 442, 456
ADD, 456
adjacent-point differencing, 252
aliasing, 74
ALLOWABLE ERROR, 434, 456
ANNEALING, 440, 441
ANNEALING ESCAPE, 456-458
ANNEALING INITIALIZATION, 458
APPLY ORTHOGONALIZATION, 398, 459
AR model, 182
ARIMA model, 261
ARMA CONFIDENCE, 415, 459
ARMA FIXED, 409, 459
ARMA model, 182, 185, 406
    collecting errors, 284
    multivariate, 200, 211, 232
    prediction, 191
    saving, 422
ARMA PREDICT, 412, 459
ARMA SHOCK, 419, 460
AUDIT LOG, 379, 460
audit log file, 379
autocorrelation, 223, 234, 256, 405, 460
    partial, 223, 235
Autoregressive (AR) model, 182

B

backcasting, 189
BANDPASS, 403, 460
bandpass filter, 30, 33, 38, 112, 115, 403
binom (code), 319
binomial distribution, 308
Box-Jenkins model, 406
Burg's algorithm, 364
BYE, 379, 460

C

Cauchy random number, 336
CENTER, 386, 460
center frequency, 33, 112
centering, 9
centroid, 23
CENTROIDS, 396
CLASS, 395, 424, 425, 430, 460
class membership, 8, 22
classification, 22
CLEAR ALL, 460
CLEAR ARMA, 409, 461
CLEAR ARMAS, 409, 461
CLEAR CLASSES, 395, 427, 461
CLEAR CONFIDENCE COMPENSATION,
    418, 455, 461
CLEAR INPUT LIST, 394, 409, 427, 461
CLEAR NETWORK, 434, 461
CLEAR NETWORKS, 434, 461
CLEAR ORTHOGONALIZATION, 398, 461
CLEAR ORTHOGONALIZATIONS, 398, 461
CLEAR OUTPUT LIST, 409, 420, 427, 461
CLEAR TEST SET, 427, 461
CLEAR TRAINING SET, 395, 427, 461
cluster, 5
command control file, 381
COMMAND FILE, 381, 461
comment, 381
compressing transformation, 5
confidence:
    ARMA, 415
    interval, 325
    neural, 452
CONFIDENCE CENTER, 417, 461
CONFIDENCE DETREND, 417, 462
CONFIDENCE DIFFERENCE, 417, 461-463
CONFIDENCE LOG, 417, 462
CONFIDENCE OFFSET, 417, 462
CONFIDENCE PROBABILITY, 416, 453

CONFIDENCE SCALE, 417, 462
CONFIDENCE SEASONAL DIFFERENCE, 417, 462
CONFIDENCE STANDARDIZE, 417, 463
conjugate gradient algorithm, 344
continuous series, 66
convolution, 103
COPY, 388, 463
copying a signal, 388
correlation, 390
cross validation training, 435
crosscorrelation, 232, 235, 405, 463
    partial, 232, 237
cube root transformation, 5
CUMULATE EXCLUDE, 425, 463
CUMULATE INCLUDE, 426, 463
CUMULATE TEST SET, 425, 463
CUMULATE TRAINING SET, 395, 425, 464
cumulative spectrum, 83, 390
    deviation, 85
CV TRAIN NETWORK, 435, 464

D

data reduction, 16, 362, 393
data window, 75
    Welch, 77, 391
Daubechies wavelet, 167
DEFINE ORTHOGONALIZATION, 397, 464
deterministic optimization, 339
DETREND, 386, 464
detrending, 9, 12
DFT, 66
DIFFERENCE, 387, 464
differencing, 40
    adjacent-point, 252
    seasonal, 47, 252, 261, 265
discrete Fourier transform, 66
discrete series, 66
DISCRIMINANT, 396
discriminant function, 24, 25
DISPLAY, 388, 464

DISPLAY CONFIDENCE, 464
DISPLAY DOMAIN, 389
DISPLAY ORIGIN, 388, 464
DISPLAY RANGE, 389, 464, 465
DISPLAY RATE, 389, 390, 465
displaying correlations, 390
displaying signals, 388
displaying spectra, 390
DIVIDE, 465
domain:
    frequency, 63
    time, 63
duality, 184

E

eigenvalue, 357
endpoint flattening, 108
energy, 71
event flag, 8
EXP, 387, 465
EXTENDED TEST NETWORK, 445, 465

F

factor, 16, 18
fast Fourier transform, 69, 358
FFT, 69, 358
    mixed-radix, 70
filter, 402
    bandpass, 30, 33, 38, 112, 115, 403
    center frequency, 112
    half-length, 82
    highpass, 30, 33, 114, 403
    in-phase, 121
    in-quadrature, 121
    lowpass, 30, 33, 114, 403
    parameters, 33
    QM, 129
    quadrature-mirror, 129, 404
    Savitzky-Golay, 81, 92, 392

Index

shape, 110
width, 112, 145, 403
Filter (code), 117
Filter::lowpass (code), 119
Filter::morlet (code), 176
Filter::qmf (code), 132
filtering, 29
flattening, 108
Fourier transform, 66
frequency, 72
frequency domain, 63

G

Gaussian random number, 336
GENERATE, 384, 465, 466
generating a signal, 384

H

half-length, 82
harmonic, 44
Heisenberg uncertainty principle, 173
HIGHPASS, 403, 466
highpass filter, 30, 33, 114, 403
histogram, 4
histogram equalization, 4
homogeneity, 185, 253

I

imaginary part, 67, 129
in-phase filter, 121
in-quadrature filter, 121
INPUT, 394, 407, 423, 447, 449, 466
INTEGRATE, 387
integration, 261
invertability, 184, 200

K

KERNEL, 431, 466
Kolmogorov-Smirnov distribution, 85
Kolmogorov-Smirnov test, 390, 391

L

lagged correlation, 390
leakage, 77
Levenberg-Marquardt algorithm, 347
linear discriminant function, 24
linear regression, 211
LOG, 387, 466
log transformation, 5
LOWPASS, 403, 466
lowpass filter, 30, 33, 114, 403

M

MA model, 183
magnitude, 70
MAXENT, 393, 467
maximum entropy spectrum, 48, 90, 93, 393
MEDIAN CENTER, 386, 467
mixed-radix FFT, 70
MLFN DOMAIN, 433, 467
MLFN ERROR TYPE, 439, 467, 468
MLFN HID 1, 432, 468
MLFN HID 2, 432, 468
MLFN LEARNING ALGORITHM, 468
MLFN LEARNING METHOD, 438
MLFN OUTPUT ACTIVATION, 469
MLFN PRETRIES, 442, 469
MLFN RESTARTS, 441, 469
modifying signals, 385
MORLET, 404, 469
Morlet wavelet, 167, 172, 175, 404
mother wavelet, 171
moving average, 45, 103

MA model, 183
multiple-layer network, 373, 432
    training, 438
multiplicative response, 5
MULTIPLY, 470
multivariate ARMA model, 200, 211, 232

N

NAME, 382, 384, 388, 391-393, 398, 403-405, 412, 447, 470
NETWORK CONFIDENCE, 452, 470
NETWORK MODEL, 429, 430, 470
NETWORK PREDICT, 447, 470
neural network, 429
    MLFN, 432, 438
    multiple-layer, 373, 432, 438
    PNN, 430
    prediction, 446
    probabilistic, 374, 429, 436
    saving, 446
    SEPCLASS, 431
    SEPVAR, 431
    test set, 423
    training set, 423
nonstationarity, 259
NORMAL, 384
normal random number, 336
Nyquist frequency, 74, 120
Nyquist sampling theorem, 75

O

octave, 146
OFFSET, 386, 471
optimization
    conjugate gradient, 344
    deterministic, 339
    Levenberg-Marquardt, 347
    Powell's algorithm, 350
    simulated annealing, 352, 354
    stochastic, 351
    stochastic smoothing, 355
order of ME spectrum, 91
order relationship, 8
order statistic, 303, 307
orthogonal, 18
orthogonalization, 16, 362, 393
ORTHOGONALIZATION FACTORS, 396, 471
ORTHOGONALIZATION LIMIT, 396, 471
ORTHOGONALIZATION STANDARDIZE, 397, 471
ORTHOGONALIZATION TYPE, 395, 471
OUTPUT, 407, 424, 447, 449, 472
OUTPUT ACTIVATION, 432

P

p_limit (code), 320
padding, 106, 108, 116, 152, 404, 472
partial autocorrelation, 223, 235, 405, 472
partial crosscorrelation, 232, 237, 405, 472
period, 72
periodicity, 185
periodogram, 71
phase, 70, 121
Powell's algorithm, 216, 350
power spectrum, 48, 70, 92, 99, 152, 229, 390
    cumulative, 83
    maximum entropy, 48, 90, 93, 393
    smoothing, 392
prediction:
    ARMA, 191, 406, 411
    neural network, 446
    predicted shocks, 193
    recursive, 191
    sampled errors, 276
presence of a feature, 137
principal components, 6, 17, 395
PRINT NETWORK, 446, 472
PRIOR, 424, 472

probabilistic neural network, 6, 374, 429
    kernel, 431
    PNN, 430
    SEPCLASS, 431
    SEPVAR, 431
    training, 436
PROGRESS OFF, 380, 473
PROGRESS ON, 380, 410, 437, 442, 473

Q

QM filter, 129, 404
QMF, 404, 473
quadrature-mirror filter, 129, 404

R

RAMP, 384
random number, 334
    Cauchy, 336
    Gaussian, 336
    normal, 336
    uniform, 334
READ SIGNAL FILE, 383, 473
reading a signal, 382
real part, 67, 129
recursive prediction, 191
REFINE, 410, 434, 442, 473
regression, 211
reject category, 425, 430
resolution, 74
RESTORE ARMA, 422, 473
RESTORE NETWORK, 446, 473
RESTORE ORTHOGONALIZATION, 397, 474
restoring
    ARMA models, 422
    neural networks, 446
    orthogonalizations, 397

S

sample distribution function, 303
sample rate, 73
sampling prediction errors, 276
    ARMA models, 284
    neural network, 277
SAVE ARMA, 422, 474
SAVE NETWORK, 446, 474
SAVE ORTHOGONALIZATION, 397, 474
SAVE SIGNAL, 383, 474
SAVGOL, 392, 474
saving a signal, 383
saving and restoring
    ARMA models, 422
    neural network, 446
    orthogonalizations, 397
Savitzky-Golay filter, 81, 92, 392
SCALE, 386, 474
scaling, 6
seasonal component, 44, 48
SEASONAL DIFFERENCE, 387, 474
seasonal differencing, 47, 252, 261, 265
shape (of filter), 110
shock, 182
sigma, 6
SIGMA HIGH, 437, 474
SIGMA LOW, 437, 475
SIGMA TRIES, 437, 475
simulated annealing
    primitive, 352
    traditional, 354
SINE, 384
singular value decomposition, 338
smoothing the power spectrum, 392
spectrum, 48, 70, 92, 99, 152, 229, 390, 391, 475
    cumulative, 83
    maximum entropy, 48, 90, 93, 393
SPECTRUM WINDOW, 392, 475
square root transformation, 5
STANDARDIZE, 386, 396, 475
state of a feature, 137

stationarity, 185, 253, 259
status flag, 8
stochastic optimization, 351
stochastic smoothing, 355
subsetting a signal, 388
SUBTRACT, 475
summing, 261

T

t_limit (code), 323
TEST NETWORK, 444, 475
test set, 368, 423
THRESHOLD, 444, 476
time domain, 63
tolerance interval, 317, 323
TRAIN ARMA, 409, 476
TRAIN NETWORK, 433, 476
training set, 368, 423
transformation, 3
    compressing, 5
    cube root, 5
    log, 5
    square root, 5
trend, 12

U

uncertainty principle, 173
UNDO, 385, 418, 448
    CENTER, 386
    DETREND, 386
    DIFFERENCE, 387
    OFFSET, 386
    SCALE, 386
    SEASONAL DIFFERENCE, 387
    STANDARDIZE, 386
UNDO CENTER, 476
UNDO DETREND, 476
UNDO DIFFERENCE, 476
UNDO OFFSET, 476
UNDO SCALE, 476
UNDO SEASONAL DIFFERENCE, 476
UNDO STANDARDIZE, 476
UNIFORM, 384
uniform random number, 334

W

wavelet, 166, 171, 177
    Daubechies, 167
    Morlet, 167, 172, 175
    mother, 171
Welch data window, 77, 391
width, 33, 112, 145, 403
wraparound, 106

Y

Yule-Walker equations, 190, 211

CUSTOMER NOTE: IF THIS BOOK IS ACCOMPANIED BY SOFTWARE, PLEASE READ THE FOLLOWING BEFORE OPENING THE PACKAGE.

This software contains files to help you utilize the models described in the accompanying book. By opening the package, you agree to be bound by the following agreement:

This software is protected by copyright and all rights are reserved by the author and John Wiley & Sons, Inc. You are licensed to use this software on a single computer. Copying the software to another medium or format for use on a single computer does not violate the U.S. Copyright Law. Copying the software for any other purpose is a violation of the U.S. Copyright Law.

This software product is sold as is without warranty of any kind, either expressed or implied, including but not limited to the implied warranty of merchantability and fitness for a particular purpose. Neither the author nor Wiley nor its dealers or distributors assumes any liability of any alleged or actual damages arising from the use of or the inability to use this software. (Some states do not allow the exclusion of implied warranties, so the exclusion may not apply to you.)